Handbook of Rare Earth Elements

Handbook of Rare Earth Elements

Editor: Blossom Jenkins

www.callistoreference.com

Callisto Reference,
118-35 Queens Blvd., Suite 400,
Forest Hills, NY 11375, USA

Visit us on the World Wide Web at:
www.callistoreference.com

ISBN: 978-1-64116-803-8 (Hardback)

Cataloging-in-Publication Data

Handbook of rare earth elements / edited by Blossom Jenkins.
p. cm.
Includes bibliographical references and index.
ISBN 978-1-64116-803-8
1. Rare earths. 2. Rare earth metals. 3. Nonferrous metals. I. Jenkins, Blossom.
QE390.2.R37 H36 2023
338.274 94--dc23

Table of Contents

Preface

The rare earth elements are a group of nearly indistinguishable and lustrous silvery-white soft heavy metals. They react slowly with cold water to form hydroxides and ignite at high temperatures. These elements are used in the production of highly efficient magnets, glasses, alloys and electronics. They are also used in the manufacture of fuel cells, mobile phones, computers, optical fiber cables, hybrid cells, and LEDs. In the agricultural sector, rare earth elements are used to increase the growth of the plants, increase crop yield, and improve stress resistance in plants. Some of the rare earth elements are scandium, yttrium, cerium, promethium, europium, dysprosium and ytterbium. This book provides significant information to help develop a good understanding of the rare earth elements. A number of latest researches have been included to keep the readers up-to-date with the global concepts in this area of study. The book aims to serve as a resource guide for students and experts alike and contribute to the growth of research on these elements.

The information shared in this book is based on empirical researches made by veterans in this field of study. The elaborative information provided in this book will help the readers further their scope of knowledge leading to advancements in this field.

Finally, I would like to thank my fellow researchers who gave constructive feedback and my family members who supported me at every step of my research.

Editor

1

Rare Earth Element Distributions in Continental Shelf Sediment, Northern South China Sea

Qian Ge [1,2,*], Z. George Xue [3,4,5] and Fengyou Chu [1,2]

[1] Key Laboratory of Submarine Geosciences, Ministry of Natural Resources, Hangzhou 310012, China; chu@sio.org.cn
[2] Second Institute of Oceanography, Ministry of Natural Resources, Hangzhou 310012, China
[3] Department of Oceanography and Coastal Sciences, Louisiana State University, Baton Rouge, LA 70803, USA; zxue@lsu.edu
[4] Center for Computation and Technology, Louisiana State University, Baton Rouge, LA 70803, USA
[5] Coastal Studies Institute, Louisiana State University, Baton Rouge, LA 70803, USA
* Correspondence: qge@sio.org.cn

Abstract: A total of 388 surface sediment samples taken from the northern South China Sea (SCS) continental shelf were analyzed to characterize the signature of their rare earth elements (REEs). The average REEs concentration was 192.94 μg/g, with a maximum of 349.07 μg/g, and a minimum of 32.97 μg/g. The chondrite-normalized REEs pattern exhibits a remarkably light REEs accumulation, a relatively flat heavy REEs pattern, and a negative Eu anomaly. We subdivided the study area into three zones using the characteristics of REEs and statistical characteristics. Zone I: continental shelf off western Guangdong Province. Here, the sediment provenance is mainly river-derived from the Pearl River, Taiwanese rivers, and those in the adjacent area. Zone II: Qiongzhou Strait and Leizhou Peninsula. Here, the sediment provenance consists of the Qiongzhou Strait and the Hainan Island. Zone III: Hainan Island and SCS slope sediments are dominated. The REEs compositions are mainly controlled by source rock properties, hydrodynamic conditions, and an intensity of chemical weathering. We reconstructed the sediment dispersal and transport route using the REEs compositions, grain size, and other geochemical characteristics throughout the study area.

Keywords: rare earth elements; Pearl River; provenance; transportation and depositional processes

---

1. Introduction

The source of sediment is one of the key topics in modern marine sedimentary geology [1]. It is important to establish the provenance of sediment to understand the geological and paleoclimatic history presented therein. Terrigenous materials retain the information of the source rock properties, the sedimentation affects, and changes in the composition of the sediments. The unique stability of rare earth elements (REEs) makes them strong inheritance to the parent rock, and thus, they could be used to indicate the evolution of sediment provenance [2]. Once the REEs enter the marine environment, the REEs compositions and distribution patterns for the sediments basically do not change significantly [3]. Therefore, the study of the REEs in marine sediments can provide help for provenance determination, environmental evolution, and stratigraphic correlation [4–8].

Previous studies systematically analyzed the characteristics of REEs in seafloor sediments of the Northwestern Pacific (Bohai Sea, Yellow Sea, and East China Sea [6,9–11]). The South China Sea (SCS) contains significant information about the materials and energy exchange between the Pacific Ocean and the marginal sea in Chinese coastal waters. In recent years, a breakthrough has been made in understanding the evolution of the paleoclimate and the history of the East Asian monsoon in this region [12–15]. As the area with strong weathering and erosion, the origination, transport, and deposition of the SCS sediments are highly localized and complicated. Comprehensive concepts of the transport and accumulation processes of the weathering and erosion products still remain poorly understood [16]. The northern SCS, which is adjacent to the Chinese mainland and Taiwan, receives significant amounts of terrigenous materials [17]. The northern SCS sediments are good proxies for analyzing the sedimentary processes from source to sink. This study selects 388 surface sediment samples (Figure 1) for REEs analysis to (1) delineate the REEs distribution and concentrations in surface sediments of the northern SCS continental shelf; (2) evaluate controlling factors of the REEs distribution patterns; and (3) investigate modern sediment provenance and their transportation processes.

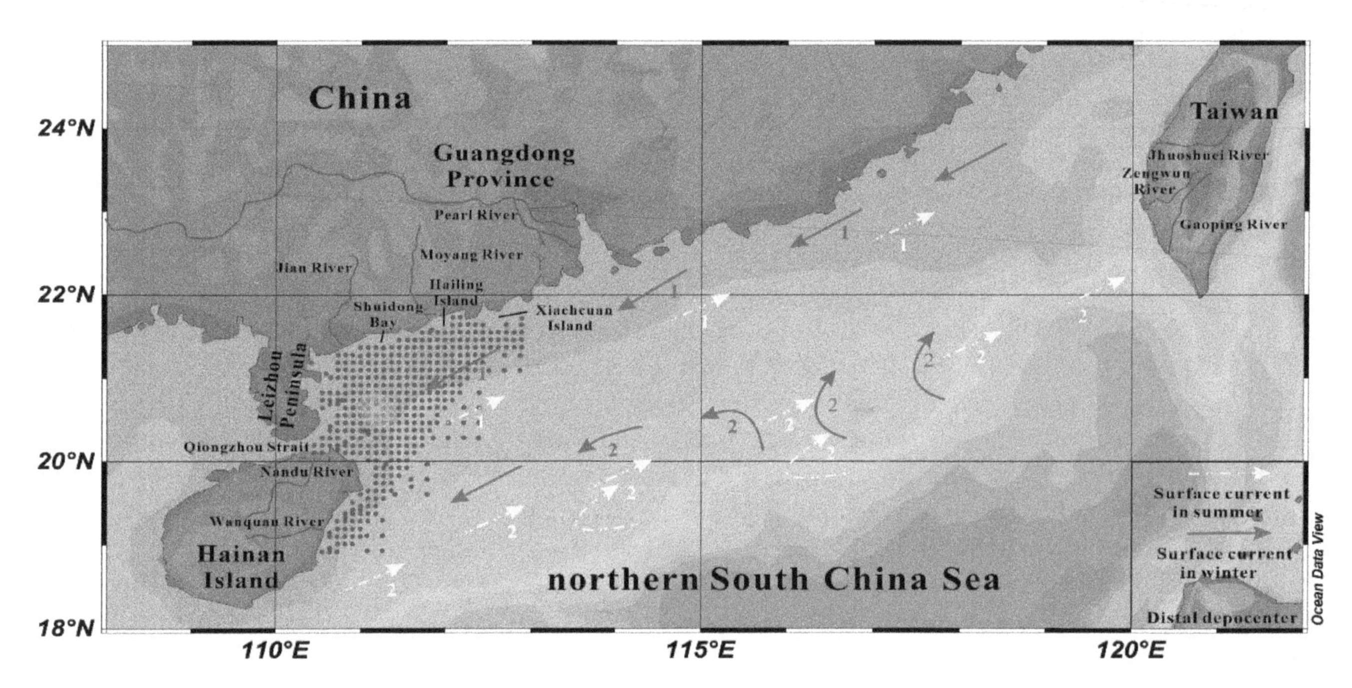

Figure 1. Location of surface sediment samples on the northern South China Sea continental shelf. The location of distal depocenter is cited from Ge et al. [17]. Seasonal variations of surface circulation patterns on the northern South China Sea continental shelf are revised from Liu et al. [16]. 1. Chinese Coastal Current; 2. South China Sea Warm Current.

2. Materials and Methods

During a joint cruise between the Second Institute of Oceanography, Ministry of Natural Resources, and South China Sea Bureau of Ministry of Natural Resources in 2008, we collected 388 surface sediment samples using a box corer at water depth around 4–135 m (for the samples locations, see Figure 1). The samples were stored at 4 °C immediately after collection. The grain size was measured using a Mastersizer-2000 laser particle size analyzer in Nanjing University and South China Sea Bureau of Ministry of Natural Resources. The procedure of grain size analysis is reported in Ge et al. [The identification of REEs constituents, La, Ce, Pr, Nd, Sm, Eu, Gd, Tb, Dy, Ho, Er, Tm, Yb, Lu, and Y were carried out at the Institute of Geophysical and Geochemical Exploration, Chinese Academy of Geological Sciences, China, using an Inductively Coupled Plasma Mass Spectrometer (ICP-MS) (iCAP Qc, Thermo Fisher Scientific, Waltham, MA, USA). The sample was dried at 105 °C for 12 h and was crushed by agate mortar (200 mesh, ≈0.074 mm pore size). Then, the powdered samples (0.05 g) were digested using high-purity HNO_3 (1.5 mL) and HF (hydrofluoric acid, 1.5 mL) in a tightly closed Teflon bomb for 48 h at 190 °C. To remove residual HF, the dry samples were retreated with HNO_3 (1 mL). Then, the samples were digested with 50% HNO_3 (3 mL) in a tightly closed Teflon bomb for another 12 h (190 °C). After cooling, the final solution was transferred to a 100 mL polyethylene bottle, to which we added 1 mL (Rh + Re) of mixed standard solution and then diluted it by the addition of Milli-Q water for analysis. The GSD-9 and GSD-10 sediment samples were analyzed for quality assurance and control, which showed a less than 5% REEs error.

3. Results

3.1. Grain Size Parameters

The results of grain size are reported in Ge et al. [18], and they show that the average values of sand, silt, and clay fractions are 25%, 62%, and 13%, respectively. The contents of clay and silt fractions exhibit a similar pattern, having high values on the shelf off western Guangdong Province, which are opposite to those of sand fractions. The mean grain size (Mz) ranges from 0.05 φ to 7.28 φ, with a mean value of 5.60 φ.

3.2. Concentrations and Spatial Characteristics of REEs

Table 1 shows the total concentrations of REEs ($\sum$REE). The mean $\sum$REE of surface sediments is 192.94 μg/g, with a minimum and maximum value of 32.97 and 349.07 μg/g, respectively. The light REEs (LREE, from La to Eu) exhibit a mean value of 154.36 μg/g, dominating the REEs content; the mean value of heavy REEs (HREE, from Gd to Lu) is 16.11 μg/g. The LREE/HREE values range from 5.91 to 10.93 with an averaged value of 9.50. The $\sum$REE exhibit significant spatial variabilities in the study area (Figure 2). The high $\sum$REE values are found in the western Guangdong coastal waters. Two accumulation centers are identified around the Xiachuan and Hailing Islands (Figures 1 and 2), where the $\sum$REE values can reach more than 250 μg/g. Another high $\sum$REE value area was found and correlated to a distal depocenter (Figures 1 and 2), which is mentioned in Ge et al. [17]. The values decrease as waters are moving off the Guangdong coast. The lowest value is located in the continental shelf off the eastern Leizhou Peninsula and Qiongzhou Strait, where the $\sum$REE values are mostly lower than 100 μg/g. The LREE and HREE exhibit similar spatial characteristics with $\sum$REE, yet with a slightly smaller spatial gradient (Figure 2).

3.3. The REEs Fractionation Characteristics

To further investigate the fractionation characteristics of REEs, the chondrite-normalized REEs of sediments were analyzed [19]. The trends reflect a remarkable accumulation of LREE. The even distribution of HREE is relatively flat (Figure 3), with the $(La/Yb)_N$ ranging from 5.87 to 15.21 (Table 1). The fractionation level of REEs was determined following, where cerium and europium anomalies (δCe and δEu) are derived by the comparison between the concentrations of Ce and Eu and those of their neighboring elements [20]:

$$\delta Ce = Ce_N/(La_N \times Pr_N)^{1/2} \quad (1)$$

$$\delta Eu = Eu_N/(Sm_N \times Gd_N)^{1/2} \quad (2)$$

where N is the normalization to chondrite [19]. The δCe ranges from 0.79 to 1.08 with a mean value of 0.97. No notable Ce anomaly was identified. The δEu ranges from 0.28 to 1.05 with a mean value of 0.65 (Table 1), exhibiting a negative anomaly.

Table 1. Statistical data of rare earth elements (REEs) values (μg/g) in surface sediments taken from the northern South China Sea continental shelf, together with some important fractionation parameters.

Statistics	La	Ce	Pr	Nd	Sm	Eu	Gd	Tb	Dy	Ho	Er	Tm	Yb	Lu	Y	∑REE	LREE	HREE	δEu	δCe	$(La/Yb)_N$
Min	6.80	10.00	1.40	4.70	0.97	0.21	0.95	0.16	0.91	0.18	0.51	0.08	0.51	0.07	5.40	32.97	24.20	3.41	0.28	0.79	5.87
Max	73.40	125.40	14.90	49.40	9.39	1.83	8.79	1.42	8.37	1.68	5.15	0.77	4.58	0.71	44.30	349.07	273.30	31.47	1.05	1.08	15.21
Average	38.50	71.50	8.32	29.30	5.64	1.11	4.85	0.78	4.31	0.83	2.32	0.38	2.28	0.37	22.46	192.94	154.36	16.11	0.65	0.97	12.12
Standard deviation	12.40	23.60	2.68	9.50	1.84	0.37	1.56	0.26	1.40	0.27	0.76	0.13	0.75	0.13	7.21	62.14	50.03	5.22	0.06	0.06	1.36

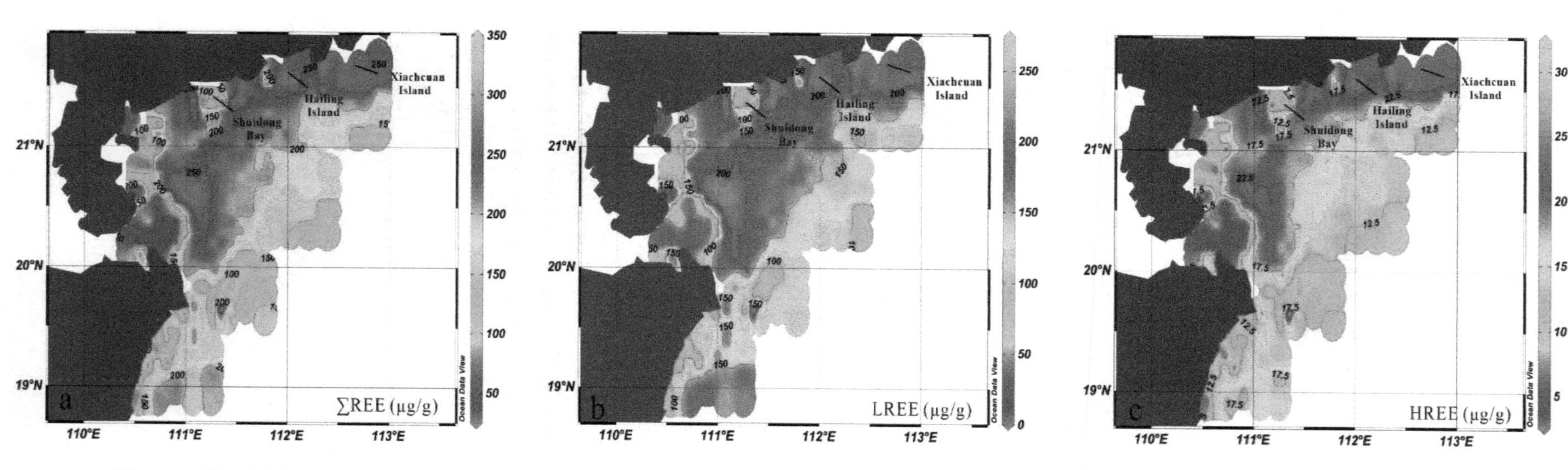

Figure 2. Spatial characteristics of REEs in the surface sediments of the northern South China Sea continental shelf ((**a**) ∑REE; (**b**) LREE; (**c**) HREE).

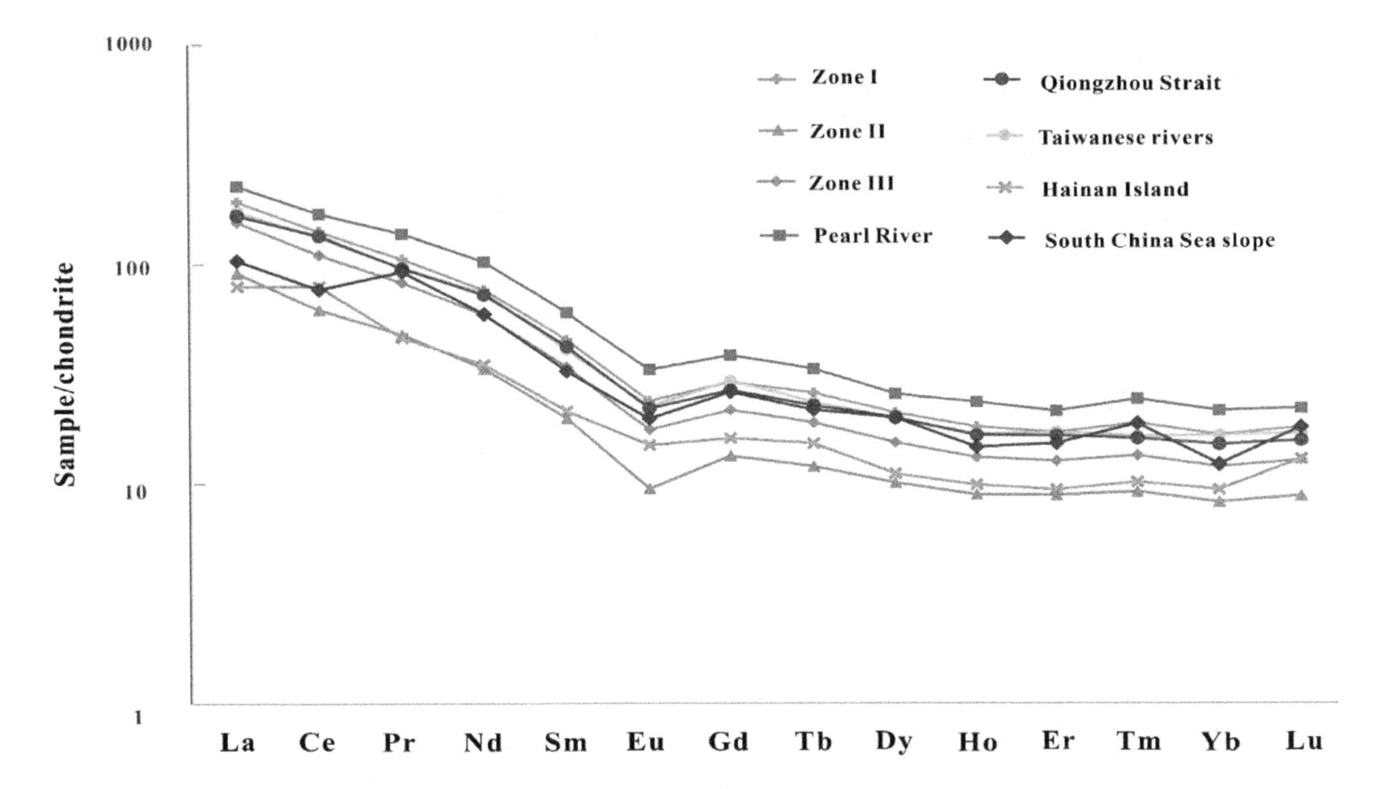

Figure 3. Chondrite-normalized REEs distribution patterns of the samples in the study area (three different zones) and the samples from the Pearl River [21], Qiongzhou Strait [22], Taiwanese rivers [23], Hainan Island [24], and South China Sea slope [25].

3.4. Spatial Zones of REEs

The Q-cluster analysis has been widely used for the sediment distribution analyses [18,26,27]. We performed the Q-cluster analysis to investigate the spatial relationships among all REEs. These analyses are carried out on the original data using Statistical Product and Service Solutions software. The Q-cluster analysis indicated that the study area can be grouped into three zones (Figure 4), which is similar to our previous study based on major elements [18]. Zone I designates the continental shelf off the western Guangdong Province and distal depocenter extending to the southwest. The mean $\sum$REE value of sediments in this zone is 233.96 μg/g. Zone II comprises the coastal area off Leizhou Peninsula and the eastern exit of Qiongzhou Strait. The average value of $\sum$REE is 107.00 μg/g. Zone III is located on the outer shelf (water depth >50 m) to the east of Hainan Island. The mean $\sum$REE value of sediments in this zone is 180.81 μg/g (Table 2). In addition, the chondrite-normalized REEs parameters in different zones, such as δCe, δEu, $(La/Sm)_N$, and $(Gd/Lu)_N$, are shown in Table 2.

Table 2. REEs characteristic parameters and mean contents of heavy minerals in three zones.

Zones	δCe	δEu	$\sum$REE (μg/g)	LREE/HREE	$(La/Sm)_N$	$(Gd/Lu)_N$	Contents of Heavy Minerals [a] (%)
I	0.99	0.66	223.96	9.32	4.24	1.74	0.95
II	0.92	0.63	107.00	8.20	4.46	1.62	2.76
III	0.97	0.65	180.81	9.94	4.59	1.81	0.89

[a] cited from [28].

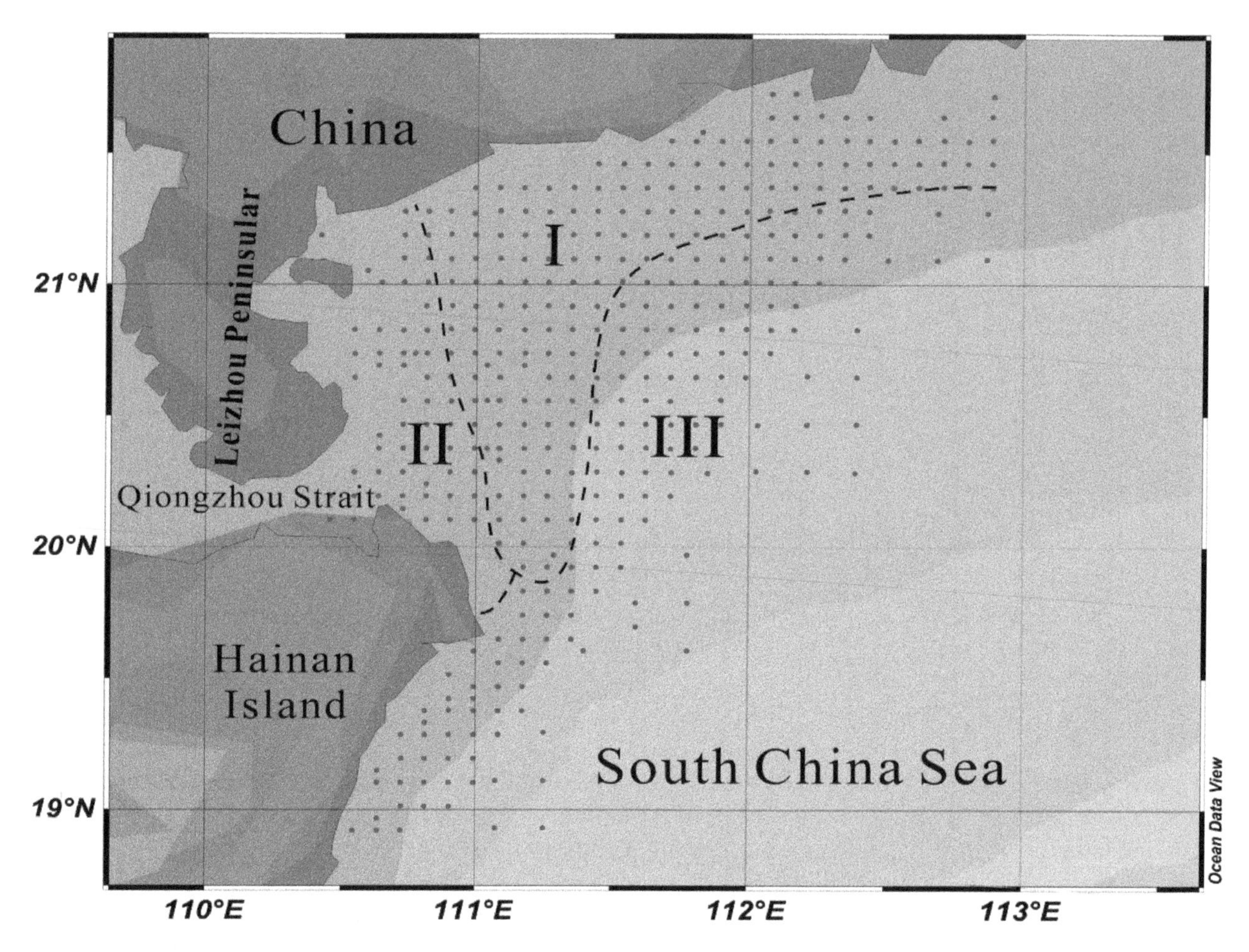

Figure 4. Three REEs geochemical zones on the northern South China Sea continental shelf (dashed lines represent the estimated boundary of REEs zones).

4. Discussion

4.1. Controlling Factors of REEs Compositions

Many factors are responsible for REEs compositions in sediments, including source rock properties, intensity of chemical weathering, mineralogy, grain size, and diagenesis [6,29]. Among these factors, source rock properties (sediment provenance) are regarded as the most important controlling factor [2,6], which will be discussed in the next section.

We found strong correlations between $\sum$REE and granulometric data (contents of sand, silt, and clay, Mz, Figure 5) [18]. In general, low $\sum$REE values are found in coarse-grained sediments, while high $\sum$REE values are found in fine-grained sediments. There is a strong correlation between the $\sum$REE of the fine-grained fraction and coarse-grained fraction. Such a correlation is consistent with the "element granularity control rate" mechanism, which indicates that $\sum$REE values are rich in the fine fractions and deficit in the coarse sediments [30]. Hydrodynamics controls the transportation, re-suspension, and deposition of sediments, and thus, they can explain the spatial characteristics of $\sum$REE and the dominant effect of grain size on the $\sum$REE [27]. The surface sediments are mainly transported from the surrounding rivers. The riverine sediments are deposited on the continental shelf around the estuaries in summer. In the following winter, the coastal current transports the re-suspended sediments southwestwardly [17,18]. The $\sum$REE values in this region are high. The low $\sum$REE values in Shuidong Bay (Figures 1 and 2) can be ascribed to the residual sand deposits [31]. The $\sum$REE values are low in the coastal area off Leizhou Peninsula and the eastern exit of Qiongzhou Strait, where the seismic sub-bottom profiles are dominated by sand waves [17]. These sandy deposits resulted from sediment eroded from the old stratum in the sea floor [32], implying a high-energy depositional environment. The low $\sum$REE value area off the eastern Hainan Island is associated with the coarse-grained terrigenous sediments from the Wanquan River.

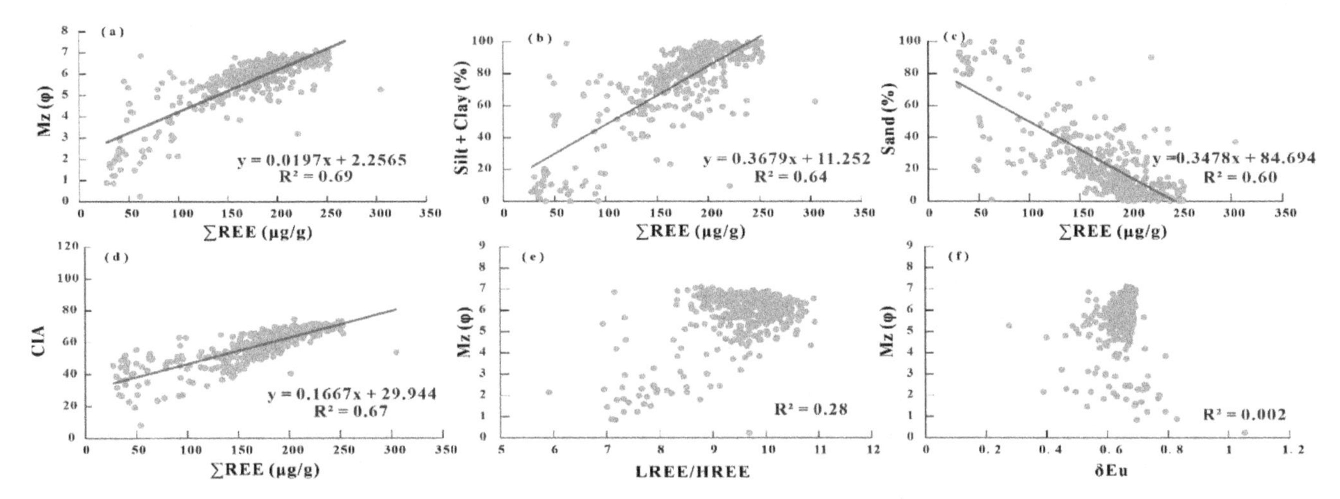

Figure 5. Correlations between total concentrations of REEs ($\sum$REE) with (**a**) Mz, (**b**) silt + clay, (**c**) sand and (**d**) chemical index of alteration (CIA), and Mz with (**e**) light REEs (LREE)/heavy REEs (HREE), and (**f**) δEu. The granulometric data and CIA are cited from Ge et al. [18].

As unstable minerals have been removed by acid digestion, the REEs reported in this study are dominated by the weathering processes of siliciclastic minerals [29]. Chemical weathering affects the residual fractions of the REEs compositions, and such an effect is most salient in LREE, which has high contents of essential weathering minerals [33]. The chemical index of alteration (CIA) can be utilized to quantify the weathering states of riverine sediments [34]. The plot of $\sum$REE and CIA [18] indicates a strong positive correlation, with a correlative coefficient (R^2) of 0.67 (Figure 5). Such correlation suggests that the different weathering mechanisms exert an important influence on the $\sum$REE values. The climate in the sediment provenance area has been warm and humid. Chemical weathering promotes the REEs mobilization and fractionation. While LREE is preferentially adsorbed onto suspended particles [35], HREE is preferentially migrated away in forms of bicarbonate and organic complex in solutions [36]. Compared with LREE, HREE is more mobile for weathering products from granite.

Despite a relatively low abundance in sediments, heavy minerals could considerably contribute to the fraction of bulk REEs [2]. As shown in Table 2, the mean content of heavy minerals in zone II (2.76%) is the highest among the three zones [28]. Nevertheless, the parameters for REEs fractionation, including δCe, δEu, LREE/HREE, $(La/Sm)_N$, and $(Gd/Lu)_N$, do not exhibit similar variations with those of the heavy minerals. As heavy minerals contribute less than 20% of the $\sum$REE in riverine sediments [6], the heavy minerals will not be the dominant factor on $\sum$REE. In general, diagenesis will change the Ce anomalies and make δCe have strong correlations with both $\sum$REE and δEu [37]. However, the $\sum$REE and δEu show weak relationships with δCe in this study area, yielding an R^2 of 0.17 and 0.01, respectively (Figure 6), which implies that diagenesis is not a dominant factor for the $\sum$REE.

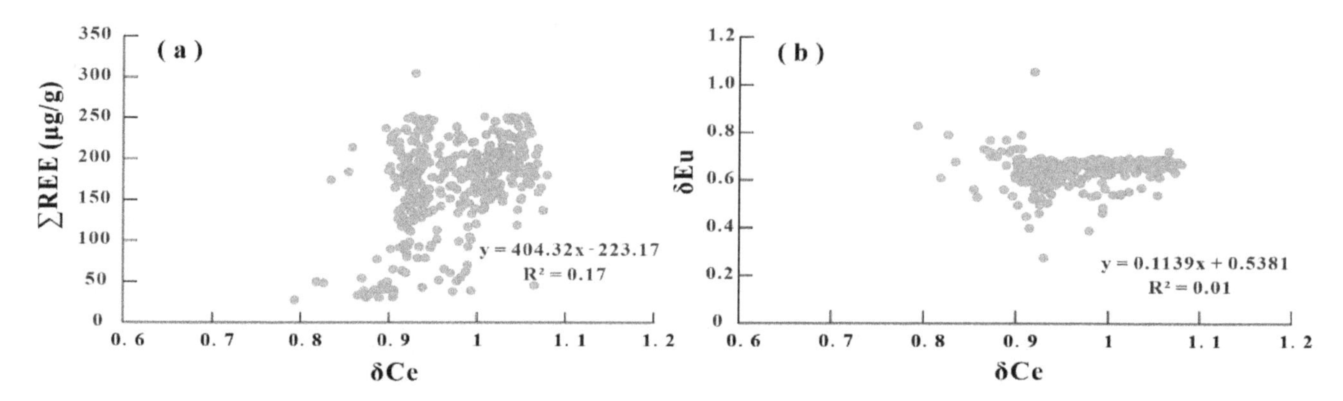

Figure 6. Correlations between cerium anomalies (δCe) with (**a**) ∑REE and (**b**) europium anomalies (δEu).

4.2. Provenance and Transport Route of the Sediments

In general, terrigenous materials transported by rivers are the dominant sediment source in shallow waters [27]. Due to the REEs' conservative behavior in hypergene environments, REEs characteristics can be used to trace the origins of fluvial sediments. Geologically, the South China continent mostly consists of magmatic rocks (mainly granite) and metamorphic rocks [21,38]. If the source rock is granite, the ∑REE value is relatively high and LREE is relatively rich, while the Eu represents an obvious negative anomaly [39]. The average ∑REE value (192.94 μg/g) reported in this study is consistent with the shelf sediments along the Western Pacific (156.00 μg/g, [30]). The REEs abundance of the sediments in this study are also roughly close to the Chinese Loess (171.00 μg/g; [30]), the Taiwanese rivers (193.12 μg/g; [23]), the Hainan Island (124.94 μg/g; [38]), and the Pearl River (255.40 μg/g; [21]), but they are quite different from deep sea clay (411 μg/g; [40]). Chondrite-normalized REEs of sediments are characterized by relatively high LREE and a negative Eu anomaly (Figure 3). The REEs fractionation patterns of sediments in the study area are comparable of those of terrigenous materials, such as the sediments from the Pearl River [21], Hainan Island [24], southwestern Taiwanese rivers [23], Qiongzhou Strait [22], and SCS slope [25]. We also use the discrimination function (DF) to analyze the proximity of the northern SCS sediments to these potential provenances.

$$DF = |(C_{1S}/C_{2S})/(C_{1P}/C_{2P}) - 1| \quad (3)$$

where C_{1S}/C_{2S} and C_{1P}/C_{2P} represent the ratio of two elements, with similar chemical properties in the sediments from the study area and potential provenance. In this study, we choose Sm/Nd to calculate DF [37]. It is generally considered that the chemical properties of the sediments from the study area are close to those of the potential provenance if the value of DF is less than 0.5. All DF values between the study area and Pearl River, Hainan Island, southwestern Taiwanese rivers, Qiongzhou Strait, and SCS slope sediments are very low (Table 3). Both REEs fractionation patterns and DF values (Figure 3, Table 3) show that the REEs have strong terrigenous succession and the surface sediments in the northern SCS shelf are mainly from the surrounding continents.

Table 3. Discrimination function (DF) values of the sediments.

Potential Provenance	DF	Reference
Pearl River	0.02	[21]
Hainan Island	0.05	[24]
Southwestern Taiwanese rivers	0.07	[23]
Qiongzhou Strait	0.03	[22]
SCS slope	0.03	[25]

As a result of their weak correlations with grain size, fractionation parameters, such as LREE/HREE ratio and δEu, are ideal proxies to trace the sources of the fluvial sediments (Figure 5). The LREE/HREE ratio in the study area ranges from 5.91 to 10.93 with an average value of 9.50, which is comparable with those reported for the Pearl River (7.83–11.23; average: 8.98, [21]), Taiwanese rivers (7.48–13.03; average: 8.88, [23]), and Hainan Island (5.16–12.33, average: 9.52, [38]). The diagram of LREE and HREE exhibits a similar positive correlation in all sediment groups (Figure 7). In detail, the Pearl River samples show the highest ∑REE on average, and the Hainan Island sediments have the lowest. The ∑REE values of the study area and Taiwanese riverine samples fall in the middle. Most of the sediments from zone I have relatively uniform and higher LREE and HREE concentrations, indicating a source from the Pearl and Taiwanese rivers. The LREE vs. HREE values in zone II show scattered REEs fractionation, indicating a unique terrigenous source (Figure 7). The distribution patterns indicate enhanced influences from the Hainan Island and Qiongzhou Strait. Sediments in zone III exhibit the highest mean value of LREE/HREE ratio in the study area (Table 2), reflecting the highest fractionation degree. The plots of LREE vs. HREE in zone III shift to the ranges of sediments from the Hainan Island and SCS slope (Figure 7), which indicates the increased influences from these areas.

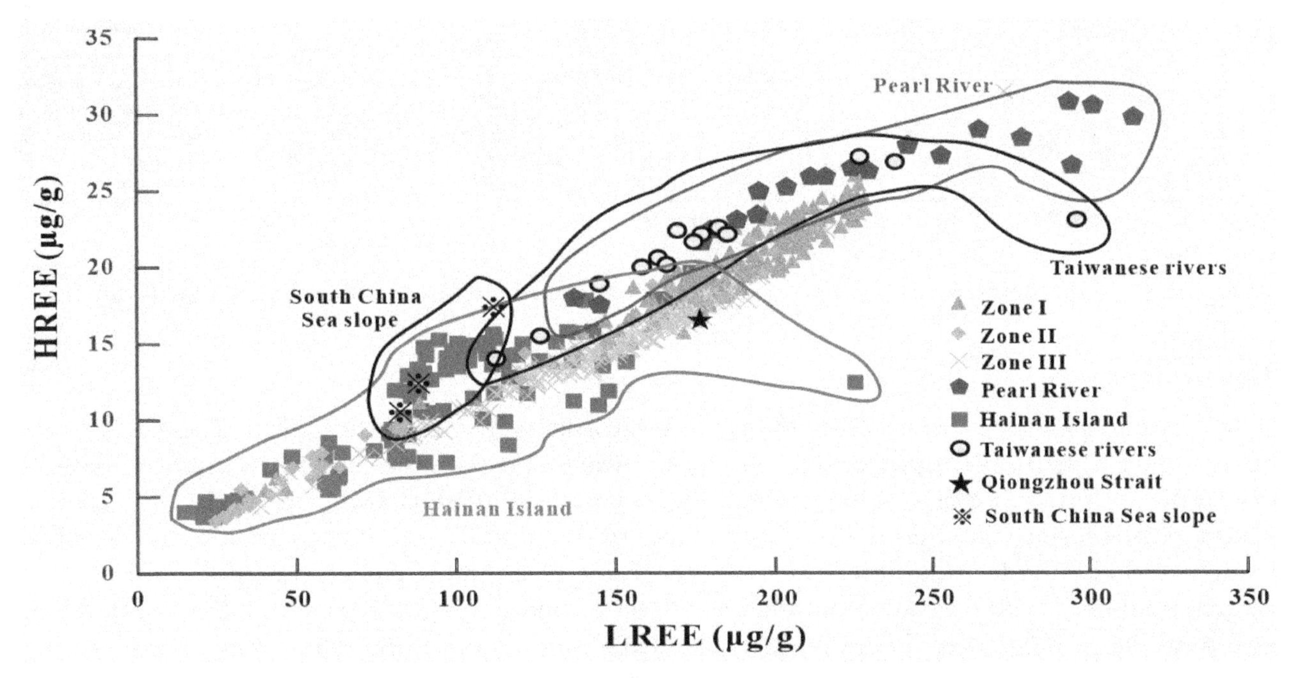

Figure 7. Correlations between LREE and HREE in the study area and potential provenances (the Pearl River data are cited from [21]; the Hainan Island data are cited from [24]; the Taiwanese rivers data are cited from [23]; the Qiongzhou Strait data are cited from [22]; the South China Sea slope data are cited from [25]).

Here, we summarize the modern transportation and depositional processes of sediments on the northern SCS continental shelf (Figure 8) based on the characteristics of the REEs from this study and the grain size, geochemical characteristics [18], sub-bottom profiles [17], and hydrodynamic conditions from previous studies. Under the impact of the East Asian summer monsoon, the weathering and erosion rates in the South Chinese Mainland and Taiwan are high; thus, huge amounts of riverine from the South Chinese Mainland rivers (Pearl River and surrounding small rivers, such as Moyang River, Jian River and so on) and southwestern Taiwanese rivers could flow into the ocean and deposit near the estuary. The strong East Asian winter monsoon strengthens the Chinese Coastal Current, which controls the southwestward longshore transport of re-suspended sediments. During this process, coarse fractions of re-suspended sediments could be transported over a relative short distance and then be deposited firstly, while fine particles could be dispersed over 400 km to the east of Leizhou Peninsula. The coastal current meets with the irregular diurnal tide to form a cyclonic circulation in this area [41,42]. The seawater's carrying capacity is reduced sharply and results in the rapid deposition on the distal depocenter (Figure 8). The sediments distributed in the east end of Qiongzhou Strait have the coarsest grain size and lowest ∑REE in the study area. The coarse materials are mainly re-suspended and transported from the Qiongzhou Strait. With transport southward, the influence of the Hainan Island-derived sediments (such as Wanquan River, Nandu River sediments, and so on) increases. Meanwhile, the sediments from SCS slope could be delivered into the relatively deep-water region during summer.

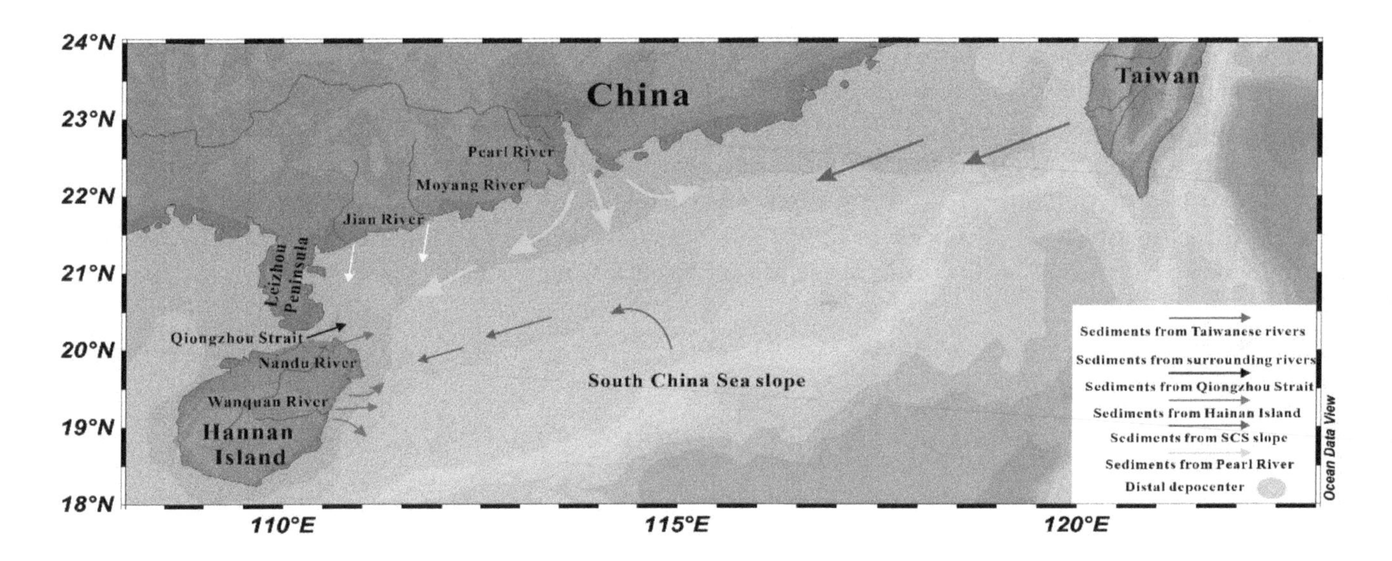

Figure 8. Transport patterns of modern sediments on the northern South China Sea continental shelf.

5. Conclusions

REEs analysis was performed on 388 surface sediments in the northern SCS. The $\sum$REE values range from 32.97 to 349.07 μg/g, with an average value of 192.94 μg/g. The REEs are dominated by LREE, with a mean value of 154.36 μg/g; the mean value of HREE is 16.11 μg/g. The $\sum$REE values are high in the western Guangdong coastal waters. Two accumulation centers are identified near the Xiachuan and Hailing Islands, and the lowest value is located in the eastern Leizhou Peninsula and Qiongzhou Strait. The fractionation parameters reflect a remarkable LREE accumulation, relatively flat HREE patterns, and negative Eu anomalies. The REEs compositions are mainly controlled by source rock properties, hydrodynamic conditions, and intensity of chemical weathering. The study area was divided into three zones according the results of Q-cluster analysis. Zone I is around on the continental shelf off the Guangdong Province, with a mean $\sum$REE value of 233.96 μg/g. The sediments from zone I have relatively uniform and higher LREE and HREE concentrations, indicating sources from the Pearl River and Taiwanese rivers. Zone II is located in the coastal area off Leizhou Peninsula and eastern end of the Qiongzhou Strait, and it has an averaged $\sum$REE value of 107.00 μg/g. The distribution patterns of REEs indicate elevated influences from the Hainan Island and Qiongzhou Strait. Zone III covers the outer shelf waters and eastern Hainan Island. The mean $\sum$REE value in this zone is 180.81 μg/g. The plots of LREE and HREE show that the Hainan Island and SCS slope sediments dominate in this zone. Based on the compositions of REEs, grain size, geochemical characteristics, sub-bottom profiles, and hydrodynamic conditions in the study area, a modern sediment dispersal route can be reconstructed.

Author Contributions: Conceptualization, Q.G., Z.G.X. and F.C.; methodology, Q.G.; software, Q.G.; validation, Q.G., Z.G.X. and F.C.; writing—original draft preparation, Q.G.; writing—review and editing, Z.G.X. and F.C.; project administration, Q.G. and F.C.; funding acquisition, Q.G. and F.C. All authors have read and agreed to the published version of the manuscript.

References

1. Singh, P.; Rajamani, V. REE geochemistry of recent clastic sediments from the Kaveri floodplains, southern India: Implication to source area weathering and sedimentary processes. *Geochim. Cosmochim. Acta* **2001**, *65*, 3093–3108. [CrossRef]
2. Taylor, S.R.; McLennan, S.M. *The Continental Crust: Its Composition and Evolution*; Blackwell: Oxford, UK, 1985; pp. 1–190.
3. Gu, S.; Chen, S.; Wu, B.; Li, S.; Chen, Y. REE geochemistry in surface sediments of South China Sea. *Trop. Oceanol.* **1989**, *8*, 93–101. (In Chinese with English Abstract)
4. Mazumdar, A.; Banerjee, D.; Schidlowski, M.; Balaram, V. Rare-earth elements and Stable Isotope Geochemistry of early Cambrian chert-phosphorite assemblages from the Lower Tal Formation of the Krol Belt (Lesser Himalaya, India). *Chem. Geol.* **1999**, *156*, 275–297. [CrossRef]
5. Munksgaard, N.C.; Lim, K.; Parry, D.L. Rare earth elements as provenance indicators in North Australian estuarine and coastal marine sediments. *Estuar. Coast. Shelf Sci.* **2003**, *57*, 399–409. [CrossRef]
6. Yang, S.; Jung, H.S.; Choi, M.S.; Li, C.X. The rare earth element compositions of the Changjiang (Yangtze) and Huanghe (Yellow) river sediments. *Earth Planet. Sci. Lett.* **2002**, *201*, 407–419. [CrossRef]
7. Li, M.; Ouyang, T.; Zhu, Z.; Tian, C.; Peng, S.; Tang, Z.; Qiu, Y.; Zhong, H.; Peng, X. Rare earth element fractionations of the northwestern South China Sea sediments, and their implications for East Asian monsoon reconstruction during the last 36 kyr. *Quat. Int.* **2019**, *525*, 16–24. [CrossRef]
8. Liu, J.; Chen, Z.; Yan, W.; Chen, M.; Yin, X. Geochemical characteristics of rare earth elements in the fine-grained fraction of surface sediment from South China Sea. *Earth Sci. J. China Univ. Geosci.* **2010**, *35*, 563–571. (In Chinese with English Abstract)
9. Wu, M. REE geochemistry of sea-floor sediments from the Taiwan shallow, China. *Geochimica* **1983**, *3*, 303–313. (In Chinese with English Abstract)
10. Jintu, W. REE geochemistry of surficial sediments from the Yellow Sea of China. *Chin. J. Geochem.* **1991**, *10*, 88–98. [CrossRef]
11. Zhao, Y.; Yan, M. *Geochemistry of Sediments of the China Shelf Sea*; Science Press: Beijing, China, 1994; pp. 5–130. (In Chinese)
12. Wang, L.; Sarnthein, M.; Erlenkeuser, H.; Grimalt, J.O.; Grootes, P.M.; Heilig, S.; Ivanova, E.A.; Kienast, M.; Pelejero, C.; Pflaumann, U. East Asian monsoon climate during the Late Pleistocene: High-resolution sediment records from the South China Sea. *Mar. Geol.* **1999**, *156*, 245–284. [CrossRef]
13. Jian, Z.; Zhao, Q.; Cheng, X.; Wang, J.; Wang, P.; Su, X. Pliocene–Pleistocene stable isotope and paleoceanographic changes in the northern South China Sea. *Palaeogeogr. Palaeoclim. Palaeoecol.* **2003**, *193*, 425–442. [CrossRef]
14. Wang, P.; Li, Q.; Tian, J.; He, J.; Jian, Z.; Ma, W.; Dang, H. Monsoon influence on planktic δ18O records from the South China Sea. *Quat. Sci. Rev.* **2016**, *142*, 26–39. [CrossRef]
15. Gai, C.; Liu, Q.; Roberts, A.P.; Chou, Y.; Zhao, X.; Jiang, Z.; Liu, J. East Asian monsoon evolution since the late Miocene from the South China Sea. *Earth Planet. Sci. Lett.* **2020**, *530*, 115960. [CrossRef]
16. Liu, Z.; Colin, C.; Li, X.; Zhao, Y.; Tuo, S.; Chen, Z.; Siringan, F.P.; Liu, J.T.; Huang, C.-Y.; You, C.-F.; et al. Clay mineral distribution in surface sediments of the northeastern South China Sea and surrounding fluvial drainage basins: Source and transport. *Mar. Geol.* **2010**, *277*, 48–60. [CrossRef]
17. Ge, Q.; Liu, J.P.; Xue, Z.; Chu, F. Dispersal of the Zhujiang River (Pearl River) derived sediment in the Holocene. *Acta Oceanol. Sin.* **2014**, *33*, 1–9. [CrossRef]
18. Ge, Q.; Xue, Z.G.; Ye, L.; Xu, D.; Yao, Z.; Chu, F. Distribution Patterns of Major and Trace Elements and Provenance of Surface Sediments on the Continental Shelf off Western Guangdong Province and Northeastern Hainan Island. *J. Ocean Univ. China* **2019**, *18*, 849–858. [CrossRef]
19. Sun, S.-S.; McDonough, W.F. Chemical and isotopic systematics of oceanic basalts: Implications for mantle composition and processes. *Geol. Soc. Lond. Spéc. Publ.* **1989**, *42*, 313–345. [CrossRef]
20. Boynton, W.V. Geochemistry of the Rare Earth Elements: Meteorite Studies. In *Rare Earth Element Geochemistry*; Henderson, P., Ed.; Elsevier: Amsterdam, The Netherlands, 1984; pp. 63–114.
21. Xu, Z.; Han, G. Rare earth elements (REE) of dissolved and suspended loads in the Xijiang River, South China. *Appl. Geochem.* **2009**, *24*, 1803–1816. [CrossRef]

22. Cui, Z.A.; Hou, Y.M.; Xia, Z.; Lin, J.Q.; Liu, W.T.; Zhang, L. Geochemical characteristics and provenance analysis of Holocene sediments in Beibu Gulf, South China Sea. *Glob. Geol.* **2015**, *34*, 605–614. (In Chinese with English Abstract)
23. Li, C.-S.; Shi, X.-F.; Kao, S.-J.; Liu, Y.-G.; Lyu, H.-H.; Zou, J.-J.; Liu, S.-F.; Qiao, S.-Q. Rare earth elements in fine-grained sediments of major rivers from the high-standing island of Taiwan. *J. Asian Earth Sci.* **2013**, *69*, 39–47. [CrossRef]
24. Sheng, Y. *Geochemical Characteristics of Rare Earth Elements in Soils of Nandu River Basin*; Qingdao University: Qingdao, China, 2011; pp. 16–27. (In Chinese with English Abstract)
25. Zhao, Q.; Gong, J.; Li, S.; He, X.; Fu, S. Geochemical characteristics of rare earth elements of surface sediments in shenhu area of South China Sea. *Mar. Geol. Quat. Geol.* **2010**, *30*, 65–70. [CrossRef]
26. Shi, X.; Liu, S.; Fang, X.; Qiao, S.; Khokiattiwong, S.; Kornkanitnan, N. Distribution of clay minerals in surface sediments of the western Gulf of Thailand: Sources and transport patterns. *J. Asian Earth Sci.* **2015**, *105*, 390–398. [CrossRef]
27. Liu, S.; Zhang, H.; Zhu, A.; Wang, K.; Chen, M.; Khokiattiwong, S.; Kornkanitnan, N.; Shi, X. Distribution of rare earth elements in surface sediments of the western Gulf of Thailand: Constraints from sedimentology and minerlogy. *Quat. Int.* **2019**, *527*, 52–63. [CrossRef]
28. Chu, F. *Report of the Marine Sediment's Characteristics in Zone CJ17*; Second Institute of Oceanography, Ministry of Natural Resources: Hangzhou, China, 2010; pp. 48–57. (In Chinese)
29. Dou, Y.; Yang, S.; Liu, Z.; Clift, P.D.; Shi, X.; Yu, H.; Berne, S. Provenance discrimination of siliciclastic sediments in the middle Okinawa Trough since 30ka: Constraints from rare earth element compositions. *Mar. Geol.* **2010**, *275*, 212–220. [CrossRef]
30. Zhao, Y.; Wang, J.; Qin, C.; Chen, Y.; Wang, X.; Wu, M. Rare-earth elements in continental shelf sediments of the China Sea. *Acta Sedimentol. Sin.* **1990**, *8*, 37–43. (In Chinese with English Abstract)
31. Luo, Y.; Feng, W.; Lin, H. Bottom sediment types and depositional characteristics of sediments of the South China Sea. *Trop. Oceanol.* **1994**, *13*, 47–54. (In Chinese with English Abstract)
32. Wang, W. Propagation of tidal waves and development of sea-bottom sand ridges and sand ripples in northern South China Sea. *Tropic Oceanol.* **2000**, *19*, 1–7. (In Chinese with English Abstract)
33. Elderfield, H.; Upstill-Goddard, R.; Sholkovitz, E. The rare earth elements in rivers, estuaries, and coastal seas and their significance to the composition of ocean waters. *Geochim. Cosmochim. Acta* **1990**, *54*, 971–991. [CrossRef]
34. Liu, Z.; Colin, C.; Huang, W.; Le, K.P.; Tong, S.; Chen, Z.; Trentesaux, A. Climatic and tectonic controls on weathering in south China and Indochina Peninsula: Clay mineralogical and geochemical investigations from the Pearl, Red, and Mekong drainage basins. *Geochem. Geophys. Geosyst.* **2007**, *8*. [CrossRef]
35. Byrne, R.H.; Kim, K.-H. Rare earth element scavenging in seawater. *Geochim. Cosmochim. Acta* **1990**, *54*, 2645–2656. [CrossRef]
36. Song, Y.H.; Shen, L.P.; Wang, X.J. Preliminary discussion on REE in weathering crusts of selected rock types. *Chin. Sci. Bull.* **1987**, *32*, 695–698.
37. Yan, B.; Miao, L.; Huang, W.; Chen, Z.; Lu, J.; Gu, S.; Yan, W. Characteristics of rare earth elements in the surface sediments from the bays along the coast of Guangdong Province and their source tracers. *J. Trop. Oceanogr.* **2012**, *31*, 67–79. (In Chinese with English Abstract)
38. Ma, R.; Yang, Y.; He, Y. Geochemistry of rare earth elements in coastal and estuarial areas of Hainan's Nandu River. *J. Chin. Rare Earth Soc.* **2010**, *28*, 110–114. (In Chinese with English Abstract)
39. Cullers, R.L. The geochemistry of shales, siltstones and sandstones of Pennsylvanian–Permian age, Colorado, USA: Implications for provenance and metamorphic studies. *Lithos* **2000**, *51*, 181–203. [CrossRef]
40. Shen, H. Rare earth elements in deep-sea sediments. *Geochimica* **1990**, *19*, 340–348. (In Chinese with English Abstract)
41. Li, Z.; Ke, X. Preliminary study on tidally-induced sediment fluxes of the Qiongzhou strait. *Mar. Sci. Bull.* **2000**, *19*, 42–49. (In Chinese with English Abstract)
42. Li, K.; Yin, J.; Huang, L.; Zhang, J.; Lian, S.; Liu, C. Distribution and abundance of thaliaceans in the northwest continental shelf of South China Sea, with response to environmental factors driven by monsoon. *Cont. Shelf Res.* **2011**, *31*, 979–989. [CrossRef]

2

Geochemical and Geochronological Constraints on the Genesis of Ion-Adsorption-Type REE Mineralization in the Lincang Pluton, SW China

Lei Lu [1,2], Yan Liu [3,4,*], Huichuan Liu [5,*], Zhi Zhao [2], Chenghui Wang [2] and Xiaochun Xu [1]

1 School of Resources and Environmental Engineering, Hefei University of Technology, Hefei 230009, China; lulei0831@mail.hfut.edu.cn (L.L.); xuxiaochun@hfut.edu.cn (X.X.)
2 MLR Key Laboratory of Metallogeny and Mineral Resource Assessment, Institute of Mineral Resources, Chinese Academy of Geological Sciences, Beijing 100037, China; zhaozhi@cags.ac.cn (Z.Z.); wangchenghui@cags.ac.cn (C.W.)
3 Southern Marine Science and Engineering Guangdong Laboratory (Guangzhou), Guangzhou 511458, China
4 Key Laboratory of Deep-Earth Dynamics of Ministry of Natural Resources, Institute of Geology, Chinese Academy of Geological Science, Beijing 100037, China
5 State Key Laboratory of Petroleum Resources and Prospecting, China University of Petroleum (Beijing), Beijing 102249, China
* Correspondence: ly@cags.ac.cn (Y.L.); lhc@cup.edu.cn (H.L.)

Abstract: Granites are assumed to be the main source of heavy rare-earth elements (HREEs), which have important applications in modern society. However, the geochemical and petrographic characteristics of such granites need to be further constrained, especially as most granitic HREE deposits have undergone heavy weathering. The LC batholith comprises both fresh granite and ion-adsorption-type HREE deposits, and contains four main iRee (ion-adsorption-type REE) deposits: the Quannei (QN), Shangyun (SY), Mengwang (MW), and Menghai (MH) deposits, which provide an opportunity to elucidate these characteristics The four deposits exhibit light REE (LREE) enrichment, and the QN deposit is also enriched in HREEs. The QN and MH deposits were chosen for study of their petrology, mineralogy, geochemistry, and geochronology to improve our understanding of the formation of iRee deposits. The host rock of the QN and MH deposits is granite that includes REE accessory minerals, with monazite, xenotime, and allanite occurring as euhedral inclusions in feldspar and biotite, and thorite, fluorite(–Y), and REE fluorcarbonate occurring as anhedral filling in cavities in quartz and feldspar. Zircon U–Pb dating analysis of the QN (217.8 ± 1.7 Ma, MSWD = 1.06; and 220.3 ± 1.2 Ma, MSWD = 0.71) and MH (232.2 ± 1.7 Ma, MSWD = 0.58) granites indicates they formed in Late Triassic, with this being the upper limit of the REE-mineral formation age. The host rock of the QN and MH iRee deposits is similar to most LC granites, with high A/CNK ratios (>1.1) and strongly peraluminous characteristics similar to S-type granites. The LC granites (including the QN and MH granites) have strongly fractionated REE patterns (LREE/HREE = 1.89–11.97), negative Eu anomalies (Eu/Eu* = 0.06–0.25), and are depleted in Nb, Zr, Hf, P, Ba, and Sr. They have high $^{87}Sr/^{86}Sr$ ratios (0.710194–0.751763) and low $^{143}Nd/^{144}Nd$ ratios (0.511709–0.511975), with initial Sr and Nd isotopic compositions of $(^{87}Sr/^{86}Sr)_i$ = 0.72057–0.72129 and εNd(220 Ma) = −9.57 to −9.75. Their initial Pb isotopic ratios are: $^{206}Pb/^{204}Pb$ = 18.988–19.711; $^{208}Pb/^{204}Pb$ = 39.713–40.216; and $^{207}Pb/^{204}Pb$ = 15.799–15.863. The Sr–Nd–Pb isotopic data and T_{DM2} ages suggest that the LC granitic magma had a predominantly crustal source. The REE minerals are important features of these deposits, with feldspars and micas altering to clay minerals containing Ree^{3+} (exchangeable REE), whose concentration is influenced by the intensity of weathering; the stronger the chemical weathering, the more REE minerals are dissolved. Secondary mineralization is also a decisive factor for Ree^{3+} enrichment. Stable geology within a narrow altitudinal range of 300–600 m enhances Ree^{3+} retention.

Keywords: ion-adsorption-type REE deposit; REE; Lincang batholith; isotopic ages; accessory minerals; Yunan Province

---

1. Introduction

Rare-earth elements (REEs) are currently a focus of global attention because of geopolitical controls on their supply, which have led to them being included in current lists of critical metals. Their importance is due to their use in high-strength magnets, which are fundamental to a range of low-carbon energy-production approaches, and in a wide range of high-technology applications [1]. REE production is currently limited to a small number of large deposits (e.g., Bayan Obo, Baotou, China; Mountain Pass, San Bernardino County, CA, USA; Mount Weld, Laverton, Australia; Lovozero, Russia [2–5]), by-products (e.g., mineral sands in India), or deposits enriched with specific elements of current high demand, such as Dy, Tb, and other heavy REEs (HREEs, e.g., the so-called ion-adsorption-type REE deposits in weathered granite in southern China [6]).

Ion–adsorption-type REE (iRee) deposits supply most of the world's REE requirements, particularly HREEs. Such deposits are often informally referred to as 'South China clays,' but similar systems have been recognized recently in Southeast Asia, Madagascar, and the southeastern USA, where deposits are generally enriched in light REEs (LREEs [6–12]). In China, iRee deposits are distributed mainly in the southeast, although exploration in other areas is ongoing. The Quannei (QN) iRee deposit, which is typically enriched in LREEs and partly enriched in HREEs, is in the Lincang (LC) area of Yunnan Province, SW China, and has an altitudinal range of 1500–2000 m. In this area, four comparable deposits have been identified recently in the LC granite: the QN, Shangyun (SY), Mengwang (MW), and Menghai (MH) iRee deposits. Both fresh and weathered granites occur in the LC area, providing an opportunity to study the effect of weathering on ion-adsorption-type HREE (iHRee) deposits. The elevation of iRee deposits (>1000 m) in SW China are higher than those in SE China, showing that there is an exploring potential of iRee (especially iHRee) deposit in high elevation areas (>1000 m) [10,13–23].

It is generally accepted that REEs are mobilized and fractionated during intense weathering of granite under warm and humid conditions [7,12,24–29]. Rare-earth element-bearing minerals ('REE minerals') are essential to the formation of iRee deposits. Moreover, fluids (organics and inorganics) also display an important role in the mobilization and accumulation of REE [30]. In host rocks, REEs occur isomorphously in the lattices of minerals such as feldspar, biotite, and apatite, or independently in minerals such as allanite, monazite, and xenotime [12]. Dissolution of these minerals is the first stage of the formation of iRee deposits, with clay being the predominant secondary mineral in which REEs are accommodated [7,24–26]. Under natural pH conditions, REEs are adsorbed on clay minerals with negatively charged layers through ion exchange, electrostatic attraction, and surface complexation, migrating into the clay structure [12]. Stable geological and tectonic conditions with a narrow altitudinal range (300–600 m) may be beneficial for retention of REEs in weathering crust [19,24,27,29,31,32].

Despite previous studies of the formation of iRee deposits, the geochemical and petrographic characteristics of granite in such deposits require further constraints. The present study focused on the geochemical characteristics of host granites in the LC area with a view to (i) determine the magma source of the host rock of iRee deposits; (ii) distinguish REE minerals in the host rock; (iii) elucidate the geodynamic mechanism of the host rock generation; and (iv) explain the REE mineralization process in weathering profiles.

2. Study Areas and Sampling

2.1. Geological Background

The LC batholith and the Triassic volcanic belt constitute the southern Lancangjiang zone of SW China. The southern Lancangjiang tectonic zone in the central part of the Sanjiang orogenic belt includes the Jinsha, Lancang, and Nu rivers. The Lanping–Simao Block and Jinshajiang–Ailaoshan suture lie east of the Sanjiang orogenic belt, which represents a Paleo-Tethyan block–arc oceanic basin. The Changning–Menglian suture belt is a remnant of the main Paleo–Tethyan ocean crust in SW China and extends to the south of the Nan and Sr Kaeo sutures.

The LC batholith intrudes the Lancang, Damenglong, and Chongshan groups. The Lancang Group comprises sandstones, mudstones, slates, phyllites, minor chlorite schists, and blueschists. The Danmenglong and Chongshan groups comprise banded gneiss, quartz–feldspathic-mica schist, amphibolite, migmatite, and marble. These three groups are considered to represent Proterozoic basement and have undergone several stages of metamorphism and deformation [33–35].

2.2. *The Lincang Granitic Batholith*

The LC batholith comprises different types of granitic rock with an area of over 7400 km^2, and measures 350 km long by 10–48 km wide. It extends from Yunxian county in the north to Menghai county in the south, with intermittent southward connection to Thailand, the Malay Peninsula, and Sumatra granites (Figure 1a,b). The exposed area represents the largest batholith in Yunnan and is separated into three segments by the Xiaojie–Nadong and Nanling–Chengzi faults (Figure 1c). The altitude of the LC batholith is in the range of 1000–2500 m, with the topography of the region dominated by low hills of 300–600 m in height. Denudation rates are fairly low and weathering profiles are well preserved. The region has a temperate subtropical monsoon climate with an annual-average temperature, rainfall, and humidity of 17.3 °C, 1504.5 mm, and 72.54%, respectively [33].

Four iRee deposits have been recognized in the weathering crust of LC granites, and are distributed in the central and southern areas of the batholith. The QN and SY deposits are in the central part of the batholith, and the MH and MW deposits in the southern area (Figure 1c). The parent rock of the QN deposit (QN granite) is sandwiched between the Tanyao and Jianshitou faults, granodiorite intrudes in the east, and the western area contains light-colored granite. The area of SY, MW, and MH deposits are bounded by the lowest industrial-grade of REE content (REE = 0.05–0.1%) in the weathering crust. The four medium–large-sized deposits have similar mineralization systems with LREE enrichment and host rocks of biotite granite.

2.3. *Sampling and Methods*

2.3.1. Sampling

As shown in Figure 1c, four QN granite samples (lc2-j2, lc2-j3, lc3-j1, and lc4-j1) were taken from the QN iRee deposit in the central LC batholith, and one MH granite sample (mh-j1) from the MH iRee deposit in the southern batholith. The corresponding granite weathering profile was separated to three horizons, A, B, and C. Horizon A, the upper weathering profile, comprises mainly soil and organic matter with some detrital quartz. Horizon B comprises weathered granite, with most minerals being altered to secondary minerals, such as clay, oxides, and phosphates. This sample contains K-feldspar and detrital quartz, and was crushed easily by hand. Horizon C represents a weathering zone where the sample was sandier and stonier than horizon B, and could be crushed by a hammer. Sample numbers in each weathering profile followed Lu et al. (2019, 2020) [28,29].

2.3.2. Methods

Whole rock major and trace element compositions were determined at the National research center of Geoanalysis, Beijing, China (NRCG, CAGS). Major elements of the analyses were determined by X-ray fluorescence spectrometry (XRF) using fused glass discs, with precisions of 1–2%. Trace element analyses were determined by inductively coupled plasma-mass spectrometry (ICP-MS) (PE-300D).

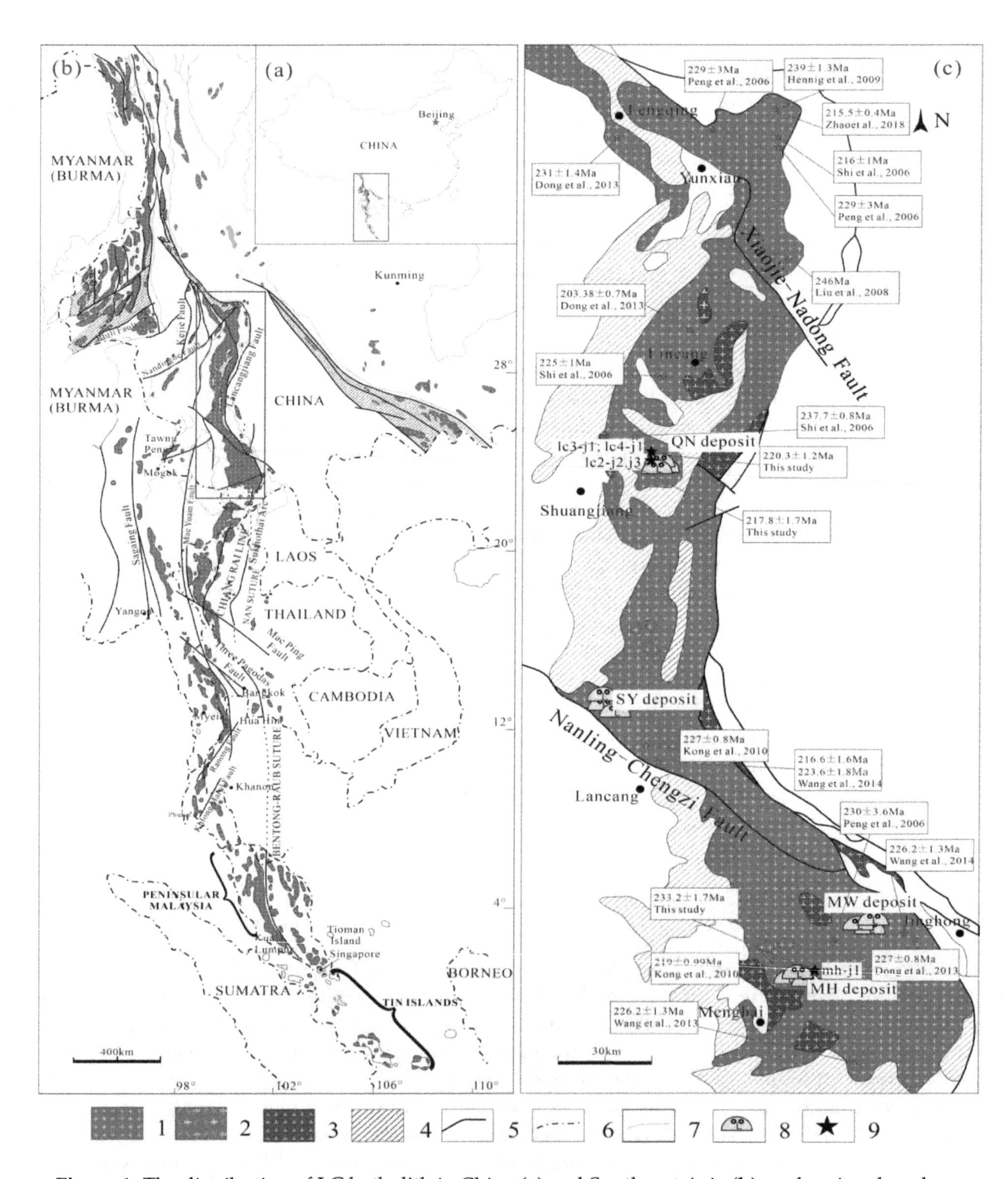

Figure 1. The distribution of LC batholith in China (**a**) and Southeast Asia (**b**), and regional geology of QN, SY, MW and MH iRee deposits showing the location of the samples (**c**) (modified from references [33,36]). 1—granite; 2—porphyritic biotite granite; 3—granodiorite; 4—metamorphic rock; 5—fault; 6—national boundaries; 7—river; 8—iRee deposit; 9—Sampled location.

Chemical compositions of minerals in the granites from the LC granite were determined by a JEOL-8230 electronmicroprobe at the Key Laboratory of Metallogeny and Mineral Assessment, Institute of Mineral Resources, Chinese Academy of Geological Sciences, Beijing. Trace elements of minerals were carried out by fs-LA-ICP-MS, using a femto-second laser ablation system (ASI J200) coupled to an inductively-coupled mass spectrometer (Thermo X series II).

Meanwhile, the zircon U-Pb age and Sr-Nd-Pb isotope of LC granite were examined. Measurements of U, Th, and Pb isotopes of zircon grains were conducted using a Camera IMS–1280HR SIMS at the Institute of Geology and Geophysics, Chinese Academy of Sciences in Beijing. Sr-Nd and Pb isotope analyses were performed on a Neptune Plus MC-ICP-MS (Thermo Fisher Scientific, Dreieich, Germany) at the Wuhan Sample Solution Analytical Technology Co., Ltd, Hubei, China. For the details on the analysis methods, we refer to the Supplementary Materials.

3. Petrography

3.1. Petrographic Characteristics of LC Granite

The granite samples are all biotite granite, barring sample lc2-j2, which is granodiorite. Granodioritic xenoliths occur in the monzonitic granites at several locations (Figure 2A). The biotite granite has either a medium–coarse-grained (samples lc4-j1) or medium–fine-grained (samples lc2-j3, lc3-j1, and mh-j1) texture.

Figure 2. Petrological diagram of QN and MH granites: (**A**) granodiorite was intruded by medium–fine-grained biotite granite; (**B**) medium–coarse-grained biotite granite; (**C**,**D**) granodioritc xenoliths are observed in the monzonitic granites; (**a**) crosshatch twins of microline; (**b**) carlsbad twin of plagioclase; (**c**,**g**) quartz occurs as anhedron fill into the interspace of plagioclase; (**d**,**g**) plagioclase altered to serisite and (**d**) biotite altered to chlorite; (**e**) albite occurs as anhedron; (**f**) biotite was replaced and altered to muscovite; (**h**) muscovite occurs as sheet structure; Qtz—Quartz; Pl—Plagioclase; Ab—Albite; Bt—Biotite; Ms—Muscovite; Chl—Chlorite; Ep—Epidote; Ilm—Ilmenite; Aln—Allanite; Ap—Apatite; Mnz—monazite; Py—Pyrite; Zrn—Zircon.

Granodiorite lc2-j2 (Figure 2A,a) has a porphyritic medium–coarse-grained structure and comprises mainly plagioclase (An_{28-36}; 28–32%), biotite (27–29%), quartz (16–20%), and K-feldspar (8–10%). These rock-forming minerals have similar grain sizes of 0.1–1 mm. Accessory minerals include minor magnetite (0.1–0.3%), apatite (1–2%), and zircon (0.5–1%; Figure 3a).

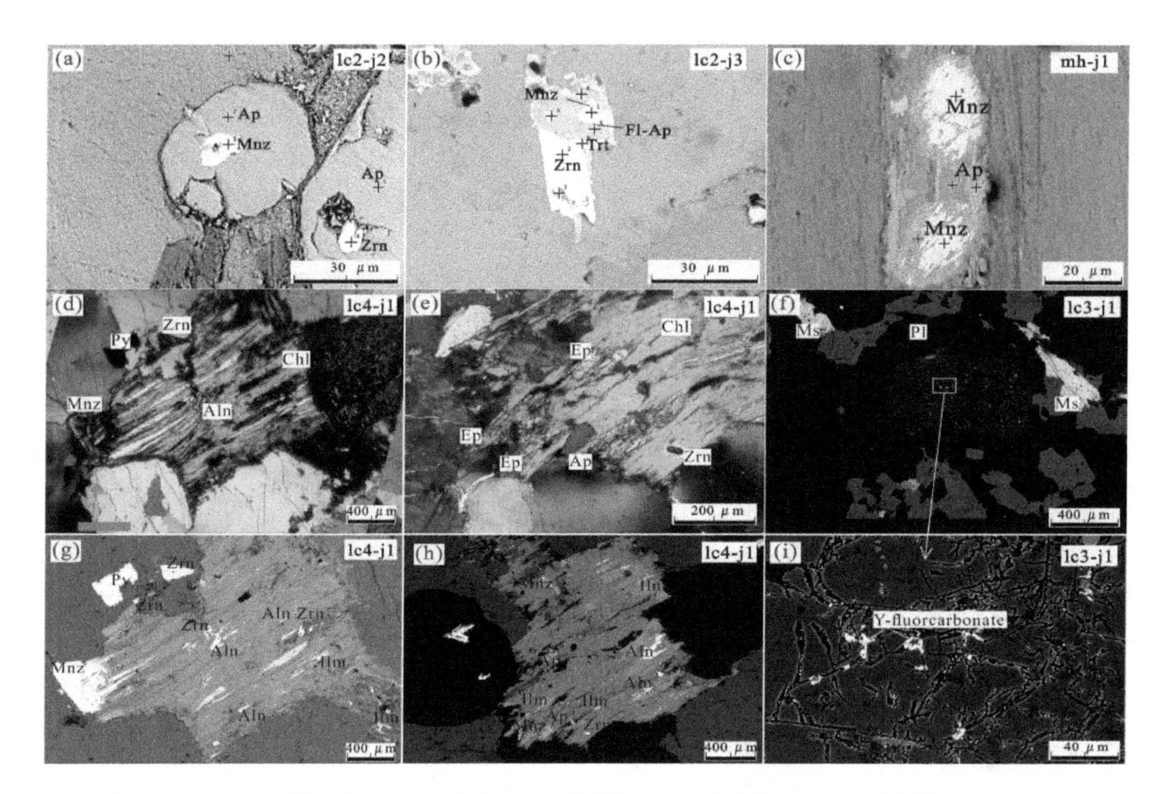

Figure 3. Optical (**d**,**e**) and back scattered electron (BSE, **a**–**c** and **f**–**i**) images of REE minerals in the host rock of QN and MH iRee deposit: (**a**,**c**) monazite and zircon occur as anhedron associated with apatite; (**b**) monazite and thorite occur as anhedron associated with fluorapatite; (**d**,**g**) monazite and allanite zircon occur as subhedral—anhedral associated with biotite, which altered to chlorite; (**e**,**h**) apatite and zircon occur as euhedron associated with biotie, which altered to chlorite; (**f**,**i**) Y-fluorapatite occurs as anhedron filled into the interspace of feldspar; Mnz—Monazite; Xtm—Xenotime; Aln—Allanite; Ap—Apatite; Fl-Ap—Fluorapatite; Trt—Thorite; Zrn—Zircon; Fl—Fluorite.

Biotite granite lc4-j1 (Figure 2B,c,d) has a porphyritic, medium–coarse-grained structure and comprises K-feldspar (20–40%), plagioclase (25–40%), quartz (20–30%), and biotite (4–8%). Phenocrysts include K-feldspar (5–25%) with grain sizes of 0.5–10 cm. The matrix comprises K-feldspar, plagioclase, quartz, and biotite, with grain sizes of 1–5 mm. Mineral crystals are subhedral–anhedral. The K-feldspar comprises microcline and perthite, the former of which has a cross-hatched twin texture and the latter a perthitic texture. Plagioclase is mainly andesine ($An_{28–36}$) with carlsbad twinning and a zoned texture. Alteration in the granite includes chloritization, epidotization, and sericitization, with biotite replaced by chlorite (ink-blue interference color; Figure 2d), and the central part of plagioclase replaced by epidote or sericite (Figure 2d,g). Accessory minerals include ilmenite (0.2–0.5%), sphene (0.2–0.3%), epidote (3–5%), apatite (or fluorapatite, 1–2%), allanite (1–2%), monazite (0.5–1%), zircon (0.5–1%), and pyrite (0.3–0.5%). Accessory minerals occur as euhedral crystals barring allanite, which is anhedral.

The biotite granites lc2-j3, lc3-j1, and mh-j1 (Figure 2A,C,D,b,e–h) comprise plagioclase (31–42%, including oligoclase ($An_{15–20}$) and albite ($An_{0–6}$)); K-feldspar (19–35% microcline); quartz (20–34%), and mica (including 4–6% biotite and 0.2–3% muscovite), all with grain sizes of 0.5–2 mm. Accessory minerals include epidote (0.5–1%), apatite (or fluorapatite, 0.2–0.5%), and zircon (0.6–1%). Quartz, albite, and muscovite contents are higher than those of the granite, and the allanite content lower, but REE minerals such as fluorite(–Y) and REE fluorcarbonate occur. Alteration includes silicification (Figure 2g), albitization (Figure 2e), and muscovitization, with biotite replaced and altered to muscovite (Figure 2f). Fluorapatite is associated with monazite, thorite, and xenotime. Most REE minerals occur as anhedral crystals filling fissures in fluorapatite (Figure 3b,i). Allanite(–Y) is associated with fluorite(–Y) and REE fluorcarbonates that fill fissures between plagioclase and quartz.

3.2. REE Minerals

In the LC granite, REEs are concentrated in accessory minerals such as sphene, apatite, and zircon, or occur as independent REE minerals such as monazite, xenotime, and allanite. Sphene, apatite, zircon, monazite, and xenotime occur as subhedral–euhedral crystals associated with quartz, feldspar, and mica (biotite and muscovite) (Figure 3a,d,e,g,h), whereas allanite, monazite, fluorite(–Y) and REE fluorcarbonate occur as anhedral crystals (Figure 3b,c,g–i). In our samples, different mineral assemblages are observed in each type of granite. For example, in granodiorite lc2-j2, accessory minerals are ilmenite, sphene, apatite, and zircon with minor monazite and xenotime; in the medium–coarse-grained biotite granites lc4-j1 (Figure 3d,e,g,h), allanite is associated with monazite, apatite, and zircon, fluorapatite contents are higher, and monazite, thorite, and zircon occur as anhedral crystals filling fissures (Figure 3b,c); in the medium–fine-grained biotite granite (lc3-j1) of REE mineral are present, such as allanite(–Y), fluorite(–Y), and REE fluorcarbonate (Figure 3f,i).

The REE concentrations for each REE mineral are listed in Table 1. The accessory minerals allanite, monazite, and xenotime have relatively high REE contents (>5 wt.%). Based on their LREE/HREE ratios, REE minerals can be divided into two categories: LREE-type and HREE-type. For example, allanite and monazite are relatively enriched in LREE (LREE/HREE > 1) and are therefore 'LREE minerals,' whereas xenotime and fluorite(–Y) are relatively enriched in HREE, and are therefore 'HREE minerals.'

Allanite, monazite, apatite, and zircon are the main accessory minerals in the LC granite, with minor cerite and xenotime associated with feldspar and quartz. In medium–fine-grained biotite granite (lc3-j1), allanite(–Y) is an HREE mineral with Y_2O_3 contents of up to 30 wt.%. More fluorapatites are found. Monazite and thorite have been re-crystallized and occur as anhedral fillings in fissures in fluorapatite which increasing the REE contents to up to 8.46 wt.%. The REE fluorcarbonates associated with fluorite(–Y) that fill fractures or crystal cleavages in quartz and plagioclase (Figure 3f,i) are LREE minerals, but have high Dy_2O_3 and Y_2O_3 contents of up to 4.6 wt.% and 16.6 wt.%, respectively.

3.3. Weathering Profile of the LC Granite

It is generally considered that weathering crusts are formed by the weathering of granite in a humid environment, with the crust comprising detrital and secondary minerals. Detrital minerals can be subdivided into two groups: (1) quartz, feldspar, and biotite; and (2) heavy minerals that are relatively stable during weathering, such as zircon, monazite, and xenotime. The REE minerals such as allanite, fluorapatite, fluorite, and fluorcarbonate may occur in host rocks, but not in the weathering crust.

Secondary minerals carrying exchangeable REE (Ree^{3+}) include clay minerals and Fe–Mn oxides. Here, the secondary-mineral content of the LC weathering profile ranges from 21% to 33%, from top to bottom. The clay mineral content is highest in horizon A, with most clay minerals finally altering to gibbsite (~7%) in this horizon [28,29]. Horizon B is the most important for Ree^{3+} enrichment. Kaolinite, montmorillonite, and illite are the main clay minerals in the LC granite weathering profile. The proportion of clay minerals varies between horizons based on the intensity of weathering. The horizon A kaolinite content is up to 85%, with 4% illite and 5% vermiculite. Horizon B contains 74–81% kaolinite, 9% illite, and 8–17% montmorillonite (mixed with illite). Horizon C contains 36% kaolinite, 12% illite, and 13% chlorite [29].

Table 1. REE geochemical characteristics in REE minerals from the QN and MH granites (%).

Mineral	Formula	REO	LREO	HREO	LREO/HREO	Nd_2O_3/REO	Ce_2O_3/REO	Eu_2O_3/REO	Dy_2O_3/REO	Y_2O_3/REO
Sphene ($n = 2$)	$CaTi[SiO_4]O$	0.37	0.21	0.39	0.78	0.12	0.25	0.00	0.13	0.19
apatite ($n = 3$)	$Ca_5(PO_4)_3F$	8.46	7.59	1.56	6.49	0.21	0.50	0.00	0.09	0.41
allanite ($n = 17$)	$(Ce,Ca)_2(Al,Fe^{3+})_3(SiO_4)_3(OH)$	16.92	14.53	5.78	8.25	0.22	3.42	0.01	0.03	0.25
Monazite ($n = 21$)	$(Ce,La,Nd,Th)PO_4$	61.03	57.55	17.01	16.80	0.22	12.89	0.00	0.00	1.03
xenotime ($n = 5$)	YPO_4	53.05	0.77	52.28	0.01	0.01	0.00	0.00	0.07	0.74
thorite ($n = 2$)	$(Y,Th,Ca,U)(Ti,Fe^{3+})_3(O,OH)_4$	17.61	16.55	1.07	10.67	0.39	0.31	0.00	0.00	0.00
zircon ($n = 2$)	$(Zr,Y)(Si,P)O_4$	0.67	0.11	0.56	0.26	0.00	0.15	0.00	0.18	0.11
allanite-Y ($n = 7$)	$(Ce,Ca,Y)_2(Al,Fe^{3+})_3(SiO_4)_3(OH)$	31.83	20.89	10.95	1.80	0.34	0.02	0.00	0.09	0.16
flourite-Y ($n = 1$)	$(Ca,Y)F_2$	15.56	9.38	6.18	1.52	0.30	0.03	0.00	0.06	0.21
REE-flourcarbonate ($n = 1$)		59.20	31.94	27.26	1.17	0.27	0.06	0.00	0.08	0.28

4. Results

4.1. Geochemistry of the QN Granite

Results of geochemical analysis of representative samples of the QN and MH granites are listed in Table 2. The average loss on ignition was <1.5 wt.%. The samples have high SiO_2 (62.67–75.12 wt.%) and K_2O (4.45–4.85 wt.%) contents; low MgO (0.39–0.87 wt.%) and FeOt (total Fe oxides; 1.71–6.44 wt.%) contents; and MnO and P_2O_5 contents of 0.05–0.07 wt.% and 0.05–0.07 wt.%, respectively. Their K_2O/Na_2O ratios of 0.88–2.36 indicate that the QN and MH granites are high-K calc-alkaline series rocks (Figure 4a, b). Their peraluminous nature is indicated by A/CNK ratios (mol. $Al_2O_3/(CaO + K_2O + Na_2O)$) of generally >1.1 (Figure 4c). In the ternary classification diagram (Figure 5), the samples plot within the S-type granite field (with one exception in the I-type field), close to the boundary of the A-type field. In the Harker diagrams (Figure 6), FeO, MgO, TiO_2, CaO, MnO, and P_2O_5 contents exhibit strong negative correlations with SiO_2 content. Five samples (lc2-j2, lc2-j3, lc3-j1, lc4-j1, and mh-j1) have REE contents of 212–749 ppm with similar chondrite-normalized REE patterns that are relatively enriched in LREEs (LREE/HREE = 1.89–7.74), and obvious negative Eu anomalies (Eu/Eu* = 0.14–0.2). The Eu/Eu* ratios are negatively related to HREE contents. The sample lc3-j1 with the highest REE content exhibits a notable negative Ce anomaly (Figure 7a). Five samples are enriched in large-ion lithophile elements, and display negative Ba, Nb, Sr, Zr, Hf, and Ti anomalies in primitive-mantle-normalized trace-element spidergrams (Figure 7b).

Table 2. Major (wt.%), trace elements ($\times 10^{-6}$), and rare earth elements ($\times 10^{-6}$) of QN and MH granites.

Sample	lc2-j2	lc2-j3	lc3-j1	lc4-j1	mh1-j1
			(wt.%)		
SiO_2	62.7	75.4	72.7	74.7	67.7
Al_2O_3	15.9	12.1	13.7	12.9	0.53
CaO	2.37	1.44	1.46	1.59	14.7
Fe_2O_3	6.44	2.77	0.12	0.09	4.29
FeO	5.28	2.14	2.18	1.62	3.53
K_2O	2.27	0.87	0.68	0.39	0.09
MgO	0.08	0.05	0.07	0.05	1.94
MnO	2.12	1.88	2.66	2.35	1.76
Na_2O	5.40	4.45	4.52	4.85	2.42
P_2O_5	0.85	0.32	0.27	0.20	4.61
TiO_2	0.25	0.05	0.07	0.05	0.19
CO_2	0.36	0.20	0.23	0.20	0.48
H_2O^+	0.87	0.65	0.70	0.44	0.92
LOI	0.83	0.50	0.63	0.49	0.90
			($\times 10^{-6}$)		
Li	37.8	14.8	36.0	16.2	39.1
Be	2.27	1.46	5.00	1.40	6.03
Cr	41.4	41.4	19.1	17.6	64.6
Co	15.4	5.68	4.09	3.00	10.2
Ni	16.0	10.8	6.71	2.61	21.9
Cu	17.2	11.1	10.2	4.25	13.2
Zn	93.4	38.4	36.0	23.0	63.4
Ga	26.1	15.9	16.6	13.3	19.0
Rb	277	163	257	159	268
Sr	181	136	85.5	119	126
Mo	1.06	0.84	0.39	0.68	0.82
Cd	0.07	<0.05	0.07	0.03	0.10
In	0.13	<0.05	0.05	0.02	0.07
Cs	8.66	3.77	9.51	4.47	16.4

Table 2. *Cont.*

Sample	lc2-j2	lc2-j3	lc3-j1	lc4-j1	mh1-j1
Ba	1972	1322	638	792	754
Tl	1.47	0.82	1.38	0.90	1.50
Pb	43.3	55.2	41.1	37.6	42.4
Bi	0.19	0.17	1.13	0.11	1.10
Th	25.2	22.9	28.6	40.5	24.2
U	3.45	3.53	6.03	3.53	7.90
Nb	19.7	8.66	10.9	3.96	13.6
Ta	1.70	0.91	1.47	0.48	1.54
Zr	339	153	96.1	92.9	172
Hf	9.21	4.74	3.71	3.27	5.21
Sn	5.06	2.73	10.5	3.89	10.9
Sb	0.16	0.09	0.09	0.07	0.07
W	2.82	0.68	2.48	0.95	2.44
As	2.51	1.03	1.01	1.09	11.5
V	116	36.1	20.8	16.4	73.5
Sc	16.1	6.60	5.42	3.68	10.8
La	56.5	41.5	154	59.5	41.5
Ce	115	83.8	81.1	112	87.1
Pr	12.4	9.35	41.3	13.5	9.89
Nd	45.6	34.1	166	50.1	36.3
Sm	9.65	6.52	41.7	7.93	7.52
Eu	1.45	1.27	6.18	1.03	1.14
Gd	9.11	5.17	38.9	6.28	6.11
Tb	1.79	0.80	6.38	0.78	0.95
Dy	10.5	4.24	33.5	3.68	5.07
Ho	1.92	0.70	5.78	0.70	0.83
Er	5.29	1.87	14.2	1.89	2.28
Tm	0.64	0.26	1.87	0.23	0.31
Yb	3.53	1.65	13.6	1.71	2.02
Lu	0.51	0.25	1.47	0.20	0.29
Y	54.3	21.1	143	16.0	24.0
ΣREE	328	212	749	275	225
LREE	240	176	490	244	183
HREE	87.6	36.0	258	31.5	41.8
LREE/HREE	2.75	4.90	1.89	7.74	4.38
δEu	0.15	0.22	0.15	0.15	0.17

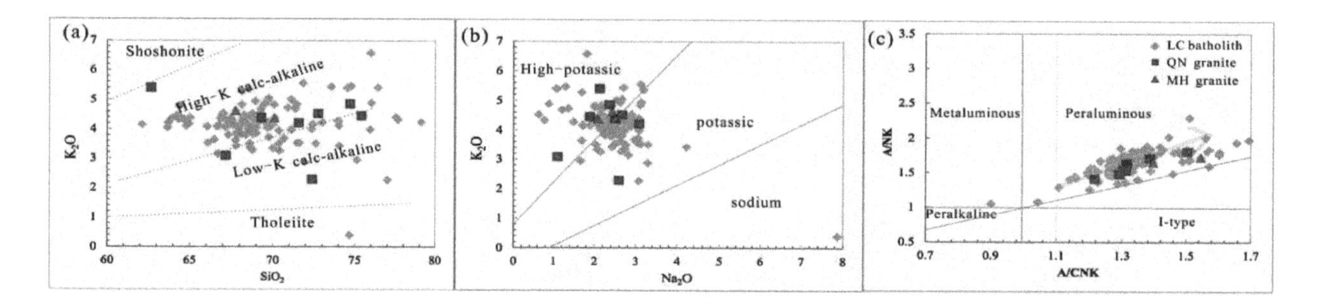

Figure 4. Plots of SiO_2 *versus* K_2O (**a**), Na_2O *versus* K_2O (**b**) and A/CNK *versus* A/NK (**c**) for the LC granite (modified from references [37–39]; the LC batholith data after e.g., [40–45]).

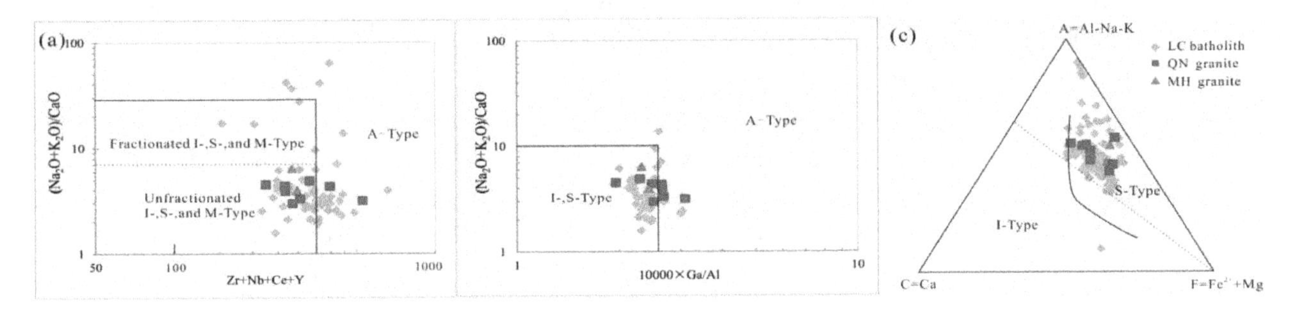

Figure 5. Plots of Zr+Nb+Ce+Y *versus* $(Na_2O+K_2O)/CaO$ (**a**), 1000×Ga/Al *versus* $(Na_2O+K_2O)/CaO$ (**b**) and A(Al—Na—K)-C(Ca)-F(Fe^{3+}+Mg) (**c**) for the LC granite (modified from reference [46]).

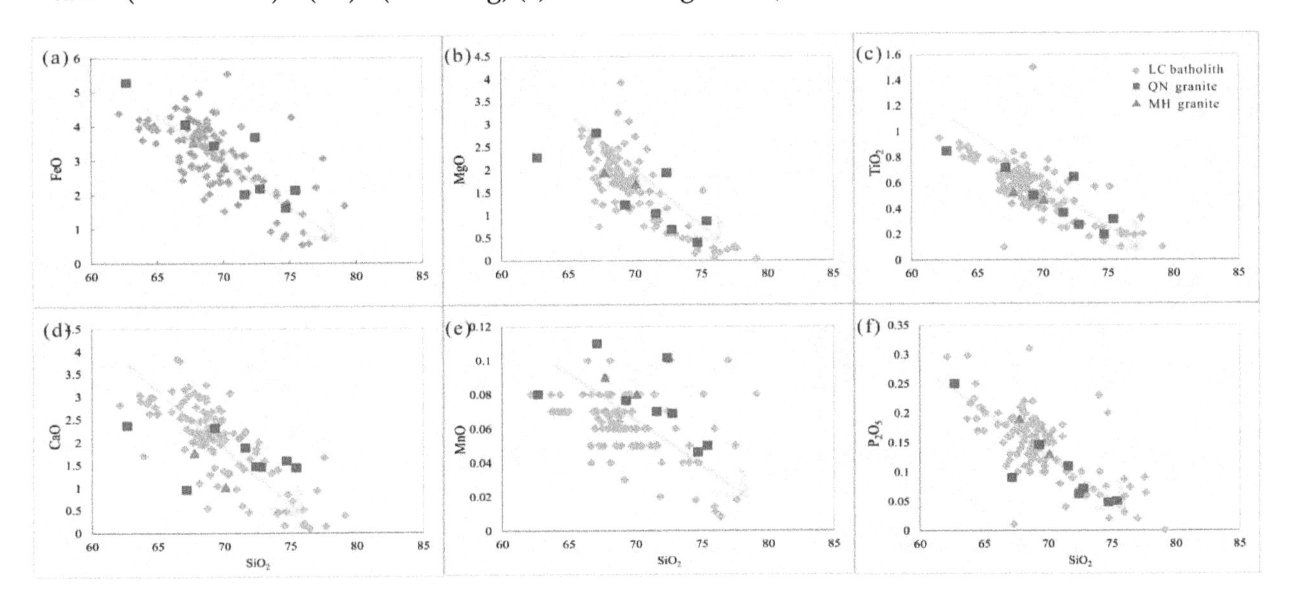

Figure 6. Harker diagrams (SiO_2 *versus* FeO (**a**), MgO (**b**), TiO_2(**c**), CaO (**d**), MnO (**e**), P_2O_5 (**f**)) for the major elements in the LC granite.

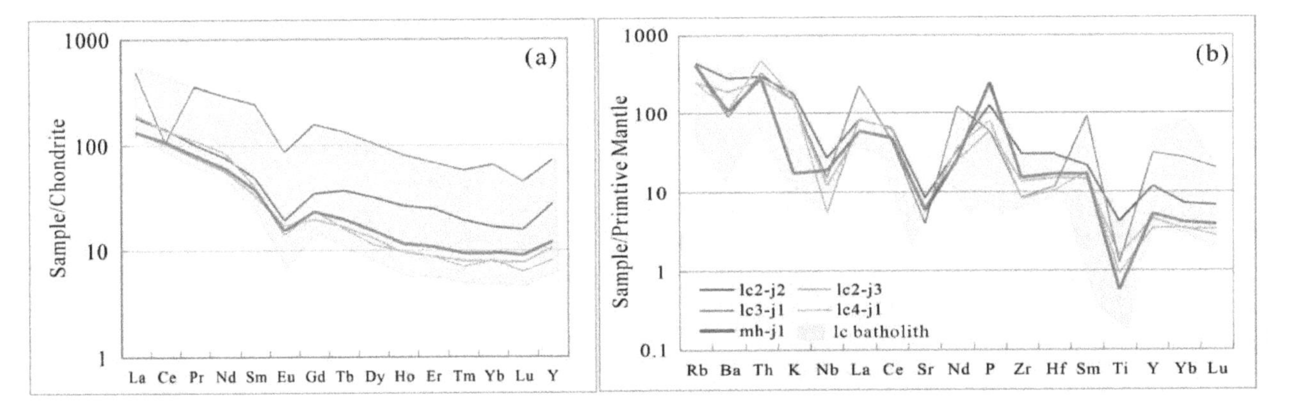

Figure 7. Chondrite-normalized REE patterns (**a**), primitive mantle-normalized elements diagrams (**b**) of the LC granite (normalization values after [47]; the LC batholith data after, e.g., [40–45]).

4.2. *Zircon U–Pb Dating*

Zircons from the QN porphyritic biotite granite sample (lc4-j1) are predominantly yellow to colorless euhedral crystals, 50–150 μm long with aspect ratios of 1–3, and exhibit oscillatory magmatic zoning in cathodoluminescence (CL) images. Their U and Th contents are 117–21,152.8 and 69–28,873 ppm, respectively, with Th/U ratios of 0.19–1.36, which is typical of magmatic zircons. Twenty analyses yielded concordant $^{206}Pb/^{238}U$ ages of 231.4–215.2 Ma, with a weighted-mean age of 220.3 ± 1.2 Ma (mean squared weighted deviation, MSWD = 0.71), which is interpreted as the crystallization age of the central part of the LC batholith (Table 3; Figure 8a,b).

Zircons from biotite granite lc2-j3 are a similar color to those of sample lc4-j1, with lengths of 50–200 μm and aspect ratios of 1.5–2.5, and magmatic oscillatory zoning in CL images. Their U and Th contents are 107–4335 and 48–1146 ppm, respectively, with Th/U ratios of 0.05–1.59. The 12 youngest ages are concordant with a weighted-mean $^{206}Pb/^{238}U$ age of 217.8 ± 1.7 Ma (MSWD = 1.06). Older ages were obtained from cores and xenocrysts. Three analysis yielded Concordia $^{206}Pb/^{238}U$ ages of 443.1 ± 13.9 to 384.7 ± 12.1 Ma, and 10 analyses yield ages of 1575.1 ± 45.0 to 509.0 ± 15.8 Ma (Table 3; Figure 8c,d).

Zircons from the MH biotite granite mh-j1 are similar to the QN zircons and are yellow to colorless, with lengths of 50–250 μm and aspect ratios of 1–4, and magmatic oscillatory zoning in CL images. Their U and Th contents are 98–709 and 16–294 ppm, respectively, with Th/U ratios of 0.18–0.85. One xenocryst analysis yielded a $^{206}Pb/^{238}U$ age of 410.6 ± 13.2 Ma. Four other xenocryst analyses yielded $^{206}Pb/^{238}U$ ages of 1015.1 ± 30.0 to 902.4 ± 26.8 Ma, possibly indicating the existence of early Proterozoic basement in the region, similar to that in the Yangtze Block [48,49]. These ages suggest that the petrogenesis of the LC granite was associated with anatexis of crustal materials. The 19 youngest analyses are concordant, with a weighted-mean age of 232.2 ± 1.7 Ma (MSWD = 0.58; Table 3; Figure 8e,f).

Table 3. LA-ICP-MS zircon U-Pb analytical data of the QN and MH granites.

Sample/No.	U	Th	Pb(t)	Th/U	$^{207}Pb/^{235}U$		$^{206}Pb/^{238}U$		$^{207}Pb/^{235}U$		$^{208}Pb/^{232}Th$	
		($\times10^{-6}$)			Ratio	1σ	Ratio	1σ	Ma	1σ	Ma	1σ
					lc4-j1, N = 25							
1	21153	28873	1703	1.36	0.2496	0.0081	0.0365	0.0012	226.2	6.60	230.9	7.30
2	1341	467	44.2	0.35	0.2516	0.0082	0.0364	0.0012	227.9	6.70	230.5	7.30
3	285	125	10.8	0.44	0.2599	0.0093	0.0366	0.0012	234.6	7.50	231.4	7.40
4	326	182	13.8	0.56	0.2552	0.0088	0.0361	0.0012	230.8	7.20	228.6	7.30
5	285	163	11.9	0.57	0.2362	0.0083	0.0342	0.0011	215.3	6.80	216.6	6.90
6	967	443	35.4	0.46	0.2367	0.0078	0.0342	0.0011	215.7	6.40	216.9	6.90
7	475	295	21.9	0.62	0.2361	0.0079	0.0344	0.0011	215.2	6.50	217.8	6.90
8	1076	660	48.5	0.61	0.2366	0.0078	0.0343	0.0011	215.6	6.40	217.2	6.90
9	242	132	10.0	0.54	0.2376	0.0084	0.0343	0.0011	216.4	6.90	217.6	6.90
10	358	189	14.4	0.53	0.2337	0.0118	0.0341	0.0012	213.3	9.70	216.1	7.30
11	334	342	21.5	1.02	0.2519	0.0085	0.0361	0.0012	228.1	6.90	228.3	7.20
12	1248	232	30.2	0.19	0.2506	0.0082	0.0365	0.0012	227.1	6.70	230.9	7.30
13	117	72.0	5.1	0.62	0.2563	0.0154	0.0361	0.0013	231.7	12.50	228.9	8.10
14	395	251	18.4	0.64	0.2545	0.0086	0.0362	0.0012	230.2	6.90	229.1	7.20
15	784	494	36.2	0.63	0.2375	0.0081	0.0344	0.0011	216.4	6.70	218.2	6.90
16	520	278	20.7	0.53	0.2356	0.0079	0.0341	0.0011	214.8	6.50	216.3	6.80
17	276	139	10.6	0.50	0.2348	0.0087	0.0344	0.0011	214.2	7.20	218.0	7.00
18	137	69.4	5.3	0.51	0.2382	0.0106	0.0343	0.0011	217.0	8.70	217.6	7.10
19	165	92.6	7.2	0.56	0.2402	0.0107	0.034	0.0011	218.6	8.80	215.2	7.10
20	196	116	9.2	0.60	0.2403	0.0135	0.0344	0.0012	218.7	11.00	218.1	7.50
21	694	254	21.9	0.37	0.2364	0.0078	0.0343	0.0011	215.5	6.40	217.5	6.90
22	785	336	28.1	0.43	0.2341	0.0077	0.0344	0.0011	213.5	6.30	218.0	6.90
23	182	78.1	6.5	0.43	0.2399	0.0178	0.034	0.0013	218.3	14.60	215.4	8.10
24	133	74.6	5.6	0.56	0.2364	0.0089	0.0343	0.0011	215.5	7.30	217.2	6.90
25	577	167	16.6	0.29	0.2352	0.0082	0.0343	0.0011	214.5	6.80	217.2	6.90

Table 3. *Cont.*

Sample/No.	U	Th	Pb(t)	Th/U	$^{207}Pb/^{235}U$		$^{206}Pb/^{238}U$		$^{207}Pb/^{235}U$		$^{208}Pb/^{232}Th$	
		($\times 10^{-6}$)			Ratio	1σ	Ratio	1σ	Ma	1σ	Ma	1σ
					lc2-j3, N = 25							
1	1027	309	32.1	0.30	0.217	0.0073	0.0321	0.0010	199	6.10	204.50	4.50
2	974	331	32.2	0.34	0.2237	0.0075	0.0322	0.0011	205	6.20	196.60	4.30
3	689	256	22.3	0.37	0.239	0.0080	0.0348	0.0011	218	6.50	205.40	4.50
4	271	136	10.5	0.50	0.2407	0.0084	0.0348	0.0011	219	6.80	203.40	4.70
5	1159	75.0	21.4	0.06	0.2471	0.0082	0.0349	0.0011	224	6.70	213.30	5.80
6	597	48.2	11.1	0.08	0.2418	0.0081	0.0349	0.0011	220	6.60	226.10	6.20
7	449	23.4	7.9	0.05	0.247	0.0094	0.0349	0.0012	224	7.70	252.00	13.9
8	2237	1146	104	0.51	0.2393	0.0078	0.0350	0.0011	218	6.40	212.00	4.40
9	171	81.0	6.2	0.47	0.2496	0.0089	0.0350	0.0011	226	7.30	192.40	4.80
10	425	161	13.6	0.38	0.2452	0.0083	0.0351	0.0011	223	6.70	194.60	4.40
11	383	166	13.8	0.43	0.2444	0.0083	0.0351	0.0011	222	6.80	207.00	4.80
12	4335	1110	142.7	0.26	0.2443	0.0080	0.0351	0.0011	222	6.50	243.20	5.10
13	1931	381	91.5	0.20	0.452	0.0147	0.0615	0.0020	379	10.3	368.50	7.70
14	1036	529	79.2	0.51	0.48	0.0157	0.0644	0.0021	398	10.8	402.80	8.40
15	1038	94.1	51.0	0.09	0.6739	0.0223	0.0712	0.0023	523	13.5	588.20	14.1
16	1808	813	150	0.45	0.5983	0.0195	0.0822	0.0027	476	12.4	450.40	9.40
17	1153	115	82.8	0.10	0.9613	0.0319	0.1055	0.0034	684	16.5	784.80	18.8
18	58.0	80.0	16.5	1.37	1.0337	0.0491	0.1180	0.0041	721	24.5	692.60	19.4
19	517	136	58.6	0.26	1.3384	0.0438	0.1415	0.0046	863	19.0	745.60	16.0
20	464	263	84.9	0.57	1.4514	0.0475	0.1499	0.0048	911	19.7	817.10	17.1
21	629	109	68.2	0.17	1.45	0.0471	0.1504	0.0048	910	19.5	898.80	18.9
22	976	169	157	0.17	2.2981	0.0751	0.1733	0.0056	1212	23.1	1240.80	26.3
23	356	238	118	0.67	2.2211	0.0731	0.1963	0.0064	1188	23.0	1136.30	23.5
24	107	170	52.7	1.59	2.2998	0.0760	0.2039	0.0066	1212	23.4	1004.40	20.8
25	377	122	97.8	0.32	3.7222	0.1205	0.2768	0.0089	1576	25.9	1329.70	27.1
					mh1-j1, N = 25							
1	286	163	13.8	0.57	0.2632	0.0243	0.0364	0.0015	237.2	19.5	230.5	9.6
2	283	215	15.4	0.76	0.2627	0.0158	0.0365	0.0013	236.8	12.7	231.2	8.1
3	403	267	23.1	0.66	0.2593	0.0101	0.0366	0.0012	234.1	8.1	231.6	7.4
4	136	80.8	6.20	0.60	0.2606	0.0111	0.0366	0.0012	235.1	9.0	231.9	7.5
5	520	271	24.7	0.52	0.2593	0.0144	0.0366	0.0013	234.1	11.6	232.0	7.9
6	459	294	25.4	0.64	0.2611	0.0103	0.0367	0.0012	235.5	8.3	232.1	7.4
7	297	196	16.0	0.66	0.2638	0.0101	0.0367	0.0012	237.7	8.1	232.3	7.4
8	295	15.9	5.40	0.05	0.2622	0.0135	0.0367	0.0013	236.5	10.8	232.3	7.8
9	142	99.4	7.00	0.70	0.2568	0.0122	0.0367	0.0012	232.0	9.8	232.5	7.6
10	297	159	12.7	0.53	0.2611	0.0094	0.0367	0.0012	235.6	7.6	232.5	7.3
11	98.0	55.7	4.4	0.57	0.2577	0.0114	0.0367	0.0012	232.8	9.2	232.6	7.5
12	511	180	19.6	0.35	0.2535	0.0090	0.0368	0.0012	229.4	7.3	232.9	7.3
13	497	217	20.7	0.44	0.2533	0.0102	0.0368	0.0012	229.3	8.2	233.1	7.5
14	238	81.8	7.40	0.34	0.2627	0.0134	0.0369	0.0013	236.8	10.8	233.4	7.8
15	108	50.7	4.10	0.47	0.2579	0.0117	0.0369	0.0012	233.0	9.5	233.7	7.6
16	179	49.6	6.60	0.28	0.2705	0.0204	0.0369	0.0014	243.1	16.3	233.9	8.9
17	445	184	18.6	0.41	0.2604	0.0097	0.037	0.0012	235.0	7.8	234.1	7.4
18	510	251	25.9	0.49	0.2618	0.0137	0.0371	0.0013	236.1	11.0	234.6	7.9
19	181	154	10.5	0.85	0.2598	0.0122	0.0373	0.0013	234.5	9.8	235.8	7.7
20	284	52.5	8.80	0.18	0.2976	0.0244	0.0405	0.0016	264.6	19.1	256.2	10.0
21	709	127	35.5	0.18	0.5302	0.0230	0.0658	0.0022	432.0	15.2	410.6	13.2
22	332	265	79.7	0.80	1.4443	0.0485	0.1503	0.0048	907.5	20.2	902.4	26.8
23	540	253	109	0.47	1.4412	0.0462	0.1507	0.0048	906.2	19.2	904.7	26.7
24	167	44.0	28.4	0.26	1.9706	0.0636	0.1509	0.0048	1105.5	21.8	905.8	26.8
25	136	40.6	22.8	0.30	2.1645	0.0730	0.1706	0.0055	1169.7	23.4	1015.1	30.0

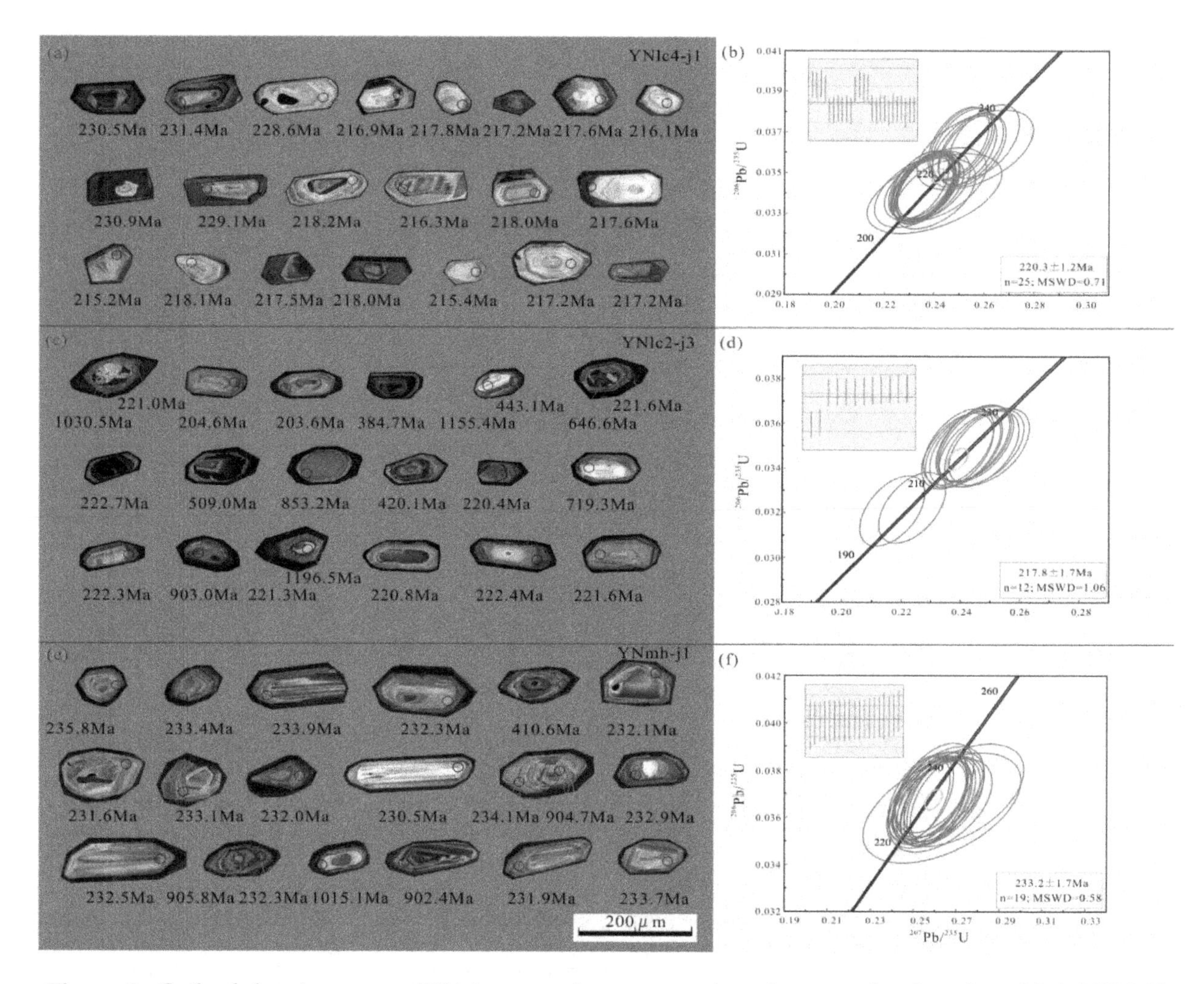

Figure 8. Cathodoluminescence (CL) images of representative zircon grains (**a**,**c**,**e**) and LA-ICP-MS zircon U-Pb weighted age diagram for the QN and MH granites (**b**,**d**,**f**).

4.3. Sr–Nd–Pb Isotopic Compositions

Results of whole-rock Sr–Nd isotope analyses of the QN and MH granites are shown in Table 4 and plotted in the εNd(t)–($^{87}Sr/^{86}Sr$) diagram (Figure 9a). The QN and MH granites have variable $^{87}Sr/^{86}Sr$ ratios of 0.733459–0.733626, and $^{143}Nd/^{144}Nd$ ratios of 0.511982–0.511990. Their initial Sr and Nd isotopic compositions are indicated by $(^{87}Sr/^{86}Sr)_i$ = 0.72057–0.72129 and εNd(232 Ma) = −9.57 to −9.75.

Table 4. Sr-Nd isotope ratios of the QN and MH granites.

Sample	t (Ma)	Rb (ppm)	Sr (ppm)	$^{87}Rb/^{86}Sr$	$^{87}Sr/^{86}Sr$	2σ	I_{Sr}	ε_{Sr} (t)	$^{147}Sm/^{144}Nd$	$^{143}Nd/^{144}Nd$	2σ	I_{Nd}	εNd (0)	εNd (t)	$f_{Sm/Nd}$	T_{2DM}
YNlc3-j1	220	257	85.5	3.919	0.733514	0.000007	0.72058	232.2	0.09349	0.511982	0.000003	0.51184	−12.8	−9.75	−0.52	1801
YNlc4-j1	220	159	119	3.905	0.733459	0.000008	0.72057	232.1	0.09267	0.51199	0.000007	0.511849	−12.64	−9.57	−0.53	1787
YNmh-j1	220	246	146	3.918	0.733616	0.000006	0.72069	233.8	0.09349	0.511982	0.000006	0.51184	−12.8	−9.75	−0.52	1801

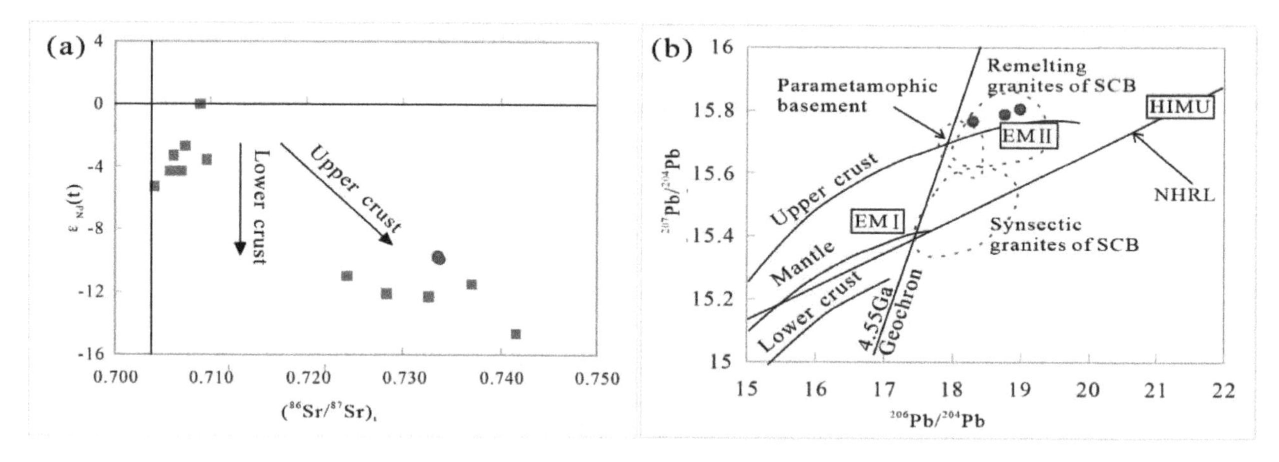

Figure 9. Diagrams of $(^{87}Sr/^{86}Sr)_i$-εNd(t) (**a**) and Pb-isotopic composition (**b**) of biotite granite obtained from the LC granite (the LC batholith data after e.g., [41,50]).

The initial Pb isotopic compositions of the QN and MH granites are moderately radiogenic, with isotopic ratios of $^{206}Pb/^{204}Pb$ = 18.988–19.711, $^{208}Pb/^{204}Pb$ = 39.713–40.216, and $^{207}Pb/^{204}Pb$ = 15.799–15.863. In $(^{207}Pb/^{204}Pb)$–$(^{206}Pb/^{204}Pb)$ diagrams (Table 5; Figure 9b), the LC granites are clustered towards the upper-crust field.

Table 5. Pb isotope ratios of the QN and MH granites.

Sample	$^{206}Pb/^{204}Pb$	2σ	$^{207}Pb/^{204}Pb$	2σ	$^{208}Pb/^{204}Pb$	2σ	$(^{206}Pb/^{204}Pb)_i$	$(^{207}Pb/^{204}Pb)_i$	$(^{208}Pb/^{204}Pb)_i$
YNlc3-j1	19.346	0.0000	15.823	0.00001	39.690	0.001	18.993	15.805	39.145
YNlc4-j1	18.988	0.0000	15.799	0.00001	39.819	0.001	18.763	15.788	38.979
YNmh-j1	19.326	0.0000	15.817	0.00001	39.713	0.001	18.293	15.765	39.107

5. Discussion

5.1. The Timing of REE Mineralization in the LC Granite

Previously published ages for the LC granite have generally been in the range of 252–199 Ma (e.g., [17–19,33,40–45,50,51]; Figure 10). Our zircon U–Pb dating in the central LC batholith, which contains HREE mineralization, yielded an age of ca. 220.3 Ma, and zircons from the southern area yielded an age of ca. 233.2 Ma, both of which are within this range, suggesting that the main LC granitic body intruded during the middle–late Triassic.

Zircon usually crystallizes during the early stages of rock formation and can survive temperatures as high as 3000 °C. Zircon U–Pb ages may, therefore, reflect those of their host granite [52]. However, REE minerals such as monazite, allanite, and xenotime crystallize later in granite than zircon. Furthermore, the types and amounts of REE mineral (including HREE minerals) increase with hydrothermal alteration after crystallization being [30,53]. Hence, zircon U–Pb ages should be considered upper limits of the REE-mineralization age; i.e., 220.3 Ma in the QN granite, and 233.2 Ma in the MH granite.

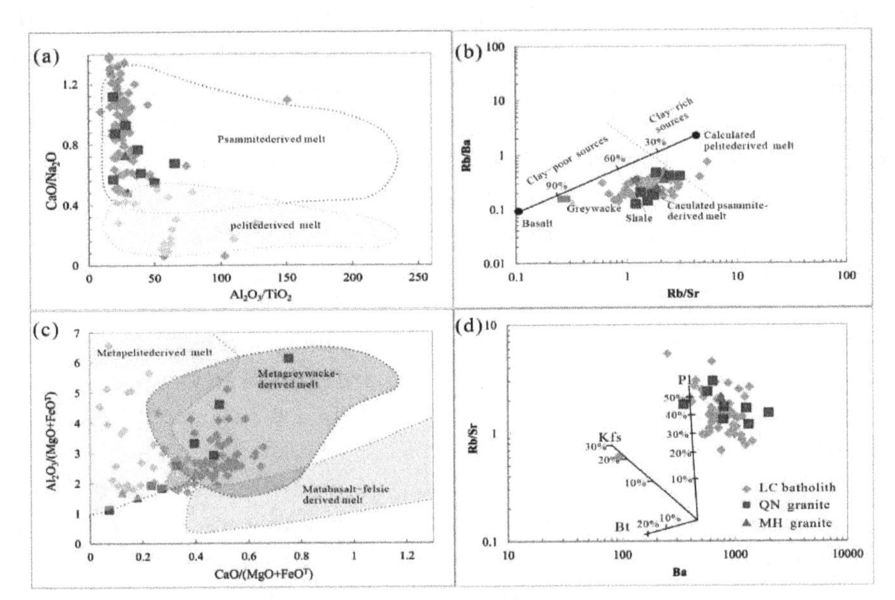

Figure 10. Plots of Al_2O_3/TiO_2 *versus* CaO/Na_2O (**a**), Rb/Sr *versus* Rb/Ba (**b**), $Ca/(MgO+FeO^T)$ *versus* $Al_2O_3/(MgO+FeO^T)$ (**c**), Ba *versus* Rb/Sr (**d**) for the potential magma source of the LC granite (modified from references [54,55]).

5.2. Petrogenesis and Tectonic Setting of the LC Pluton

The decrease in FeOt, MgO, TiO_2, CaO, MnO, and P_2O_5 contents of the LC granites with increasing SiO_2 content suggests that the magmas underwent continuous crystallization during magmatic evolution, and their peraluminous nature (A/CNK > 1.1) indicates S-type affinity (Section 4.1; Figure 4, Figure 5c, and Figure 6; [54–58]). Geochemical, isotopic, and petrological constraints confirm that most large-volume peraluminous granites originate from the melting of crustal rocks (e.g., [54]). Here, the strong negative Eu, Ba, Nb, Sr, P, and Ti anomalies, and positive Rb, Th, and U anomalies of the LC granite are similar to those of magma derived from the partial melting of crustal rocks [56,59].

S-type granites are produced mainly from the melting of metasediments such as psammite and pelite [60], and their CaO/Na_2O ratios are influenced by their magma source. The CaO/Na_2O ratios in psammite-derived melt are generally > 0.3 (average 0.8), whereas in pelite-derived melt the ratios are <0.5 [54]. Here, most LC granite samples have CaO/Na_2O ratios of 0.06–2.67 (n = 147), with only 21 samples having ratios of <0.5, and Al_2O_3/TiO_2 ratios of 9.29–150.9, with only four samples having ratios of >79. In the (CaO/Na_2O)–(Al_2O_3/TiO_2) diagram (Figure 10a), most samples plot in the field of psammite-derived melt. The Rb/Ba and Rb/Sr ratios in psammite-derived melt are lower than those in pelite-derived melt [56]. In the (Rb/Ba)–(Rb/Sr) diagram ([61]; Figure 10b), most LC granite samples plot close to the psammite-derived melt field, with such melt having a clay-poor source, including minor shale. Some samples plot near the clay-rich field with a mixed source dominated by quartz-feldspathic psammites. In the (Al_2O_3/(MgO + FeOt))–(CaO/(MgO + FeO^T)) diagram (Figure 10c), most samples plot in the field of psammite-derived melt. Furthermore, Sr and Eu are enriched mainly in plagioclase, whereas K-feldspar is enriched in Ba. The strong negative Eu, Ba, and Sr anomalies in LC granite samples indicate that plagioclase and K-feldspar are the main residual phase in partial melting. In the (Rb/Sr)–Ba diagram (Figure 10d), most LC granite samples plot with 20–60% plagioclase, suggesting that at least 20–60% plagioclase was present in the magma source [54,55]. It follows that the parental magma source for the LC granites involved mainly quartz-feldspathic materials such as psammite-dominated clastic metasediments, with minor shales.

The high initial $^{87}Sr/^{86}Sr$ ratios, low ε_{Nd}(231 Ma) values and Pb isotopic compositions, consistent with the isotopic composition of the field of remelting granites of SCB (South China Block; [62]; Figure 9), indicates that the LC granitic magma was derived from partial melting of ancient crystallized basement.

The compositions of the Changning–Menglian suture belt in the Lancangjiang zone of SW China, the Chiang Mai–Inthanon suture belt in northern Thailand, and the Bentong–Raub suture belt in the Malay Peninsula indicate that the Paleo-Tethys Ocean opened in the early Devonian and closed in the Middle–Late Triassic [33,34,40–43,63–65]. The LC batholith, exposed to the east of the Changning–Menglian suture zone, represents a tectono-magmatic zone that records the evolution of the Paleo-Tethys Ocean from subduction to post-collisional regimes [41,64].

The LC granite exhibits peraluminous and S-type characteristics and, with psammite-derived upper-crustal melt being the main source of S-type granites, some studies have concluded that the LC batholith formed in a syn-collisional setting [33,51,55–57]. This is supported by most LC granites being peraluminous (Figure 5c) and plotting in the volcanic arc and syn-collision fields in the Nb–Y diagram (Figure 11a). Peralkaline and alkaline granites are commonly associated with post-tectonic within-plate extension [55]. In the SiO_2–K_2O diagram (Figure 5a), most LC granite samples plot in the peralkaline and alkaline fields, while in the Rb–(Y+Nb) diagram (Figure 11c), all but one plot in the post-collision field, consistent with a post-collision tectonic setting. Our zircon U–Pb ages of 233.2–217.8 Ma from the LC granite are consistent with those found in most other studies (252–199 Ma), and the common absence of Early Triassic strata in the Lancangjiang zone reflects pronounced uplift and erosion at that time [33,40]. Moreover, in the Rb–(Y+Ta) diagram (Figure 11b), all LC granites plot in the within-plate granite field, indicating a within-plate extensional environment. Experimental studies have demonstrated that bimodal igneous suites comprising mafic members of tholeiitic gabbros and basalts, and felsic members displaying A-type characteristics, are common in such a setting [37,40,66]. We, therefore, conclude that closure of the Paleo-Tethys Ocean at ca. 295 Ma was followed by syn-collisional extension accompanied by crustal melting and a degree of mantle-material ascent, with magmatism occurring mainly at ca. 199 Ma within a post-collision extensional environment. The LC granite, thus, originated from the partial melting of early Paleozoic crustal basement.

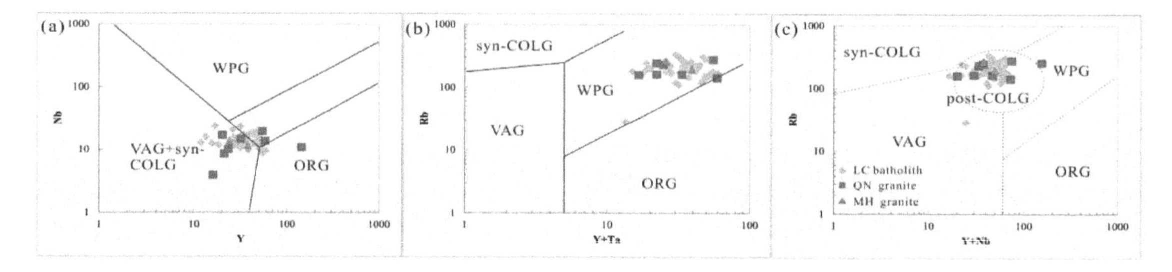

Figure 11. Plots of Y *versus* Nb (**a**), Y+Ta *versus* Rb (**b**) and Y+Nb *versus* Rb (**a**) for the LC granite (modified from references [67,68]).

5.3. *Key Factors in REE Mineralization*

5.3.1. Rare-Earth Element Minerals

Surface water and groundwater may contain little Ree^{3+}, and bedrock is considered the main source of iRee deposits [24–26]. The Nangling (NL) granites are the host rocks for most iRee deposits in SE China, and in both the NL and LC granites, accessory rare-earth element (REE) minerals are critical for the formation of iRee deposits [7,24–27]. However, fewer types and lower quantities of REE minerals occur in the LC granite than in, for example, the Zudong and Longnan granites that are enriched in prisite (–Y), bastnaesite, samarskite, eschynite, monazite, and allanite. The behavior of REE minerals during chemical weathering is a critical factor that affects the accumulation and differentiation of REEs in the weathering profile. Minerals in weathering crust can be divided into three groups: (1) strongly stable during chemical weathering; e.g., quartz, monazite, xenotime, and zircon; (2) moderately stable during weathering; e.g., feldspar, biotite, apatite, and allanite; and (3) weakly stable during weathering; e.g., fluorite(–Y) and REE fluorcarbonate [7,24–27].

REE minerals are the main source of iRee deposits in the LC granite. Although feldspar, biotite, sphene, and apatite have low REE contents, weathering cycles may result in these minerals contributing to the deposit. Minerals in granites enriched in LREEs contribute to iLRee deposits when they are dissolved in the weathering crust. Allanite(–Y), fluorite(–Y), and REE fluorcarbonate are relatively enriched in HREEs, and are the main HREE source for iRee deposits. However, HREE minerals are not common in the LC granite. The QN granites, for example, are mainly enriched in LREEs, with relatively few exhibiting HREE enrichment, leading to QN deposits being rich in LREEs, but with a degree of HREE enrichment. The HREE minerals are scarce in MH granite, so the MH deposit is enriched with LREEs only.

5.3.2. Secondary Minerals

Rock-forming minerals such as feldspar and mica (biotite and muscovite) are altered to clay minerals during weathering and become carriers of Ree^{3+} [7,24–29,69]. Clay minerals such as those of the kaolin-group (kaolinite, dickite, nacrite, and halloysite), illite, and montmorillonite, which adsorb Ree^{3+}, can also be considered as iRee ores. This study found that kaolinite, illite, and montmorillonite are the main clay minerals in the LC granite weathering profile, while halloysite also occurs in the NL weathering profile.

The REE adsorption capacity is influenced by surface structure, composition, and surface charge of clay minerals [31]. Illite and montmorillite occur in 2:1 clays, with higher adsorption capacities than kaolin-group minerals in 1:1 clays. The high adsorption capacity of montmorillonite is due to negative charges generated by isomorphism in the lattice, whereas kaolinite and halloysite have negative charges where –OH in the crystal lattice releases H^+. The higher the pH, the higher the adsorption capacity [31,68,69]. The pH points of zero charge (pH_{pzc}) of kaolinite, illite, and montmorillonite are <3.7, ~2.5, and 7–9, respectively [70], so under natural pH conditions (4–7), kaolin-group minerals (especially kaolinite and halloysite) and illite are more capable of surface complexation of Ree^{3+} than montmorillonite [32,69]. In both the LC and NL areas, the kaolin-group represents the predominant clay minerals in the weathering profile, providing enrichment in Ree^{3+}. In the appropriate pH range, the deposit's Ree^{3+} content, thus, depends on the amount of kaolin-group minerals in the weathering profile.

Horizon B of the granite weathering profile has the highest Ree^{3+} content, owing to its pH range of 4–6.8 and higher kaolinite and/or halloysite content. In the LC area, horizon B contains an average of 21–26% clay minerals [29], whereas in the NL area the clay-mineral content is up to 50% [24–26,32]. In the NL weathering profile, the content of kaolinite decreases but that of halloysite increases with depth, with both clay minerals contributing to the accumulation and fractionation of REEs at pH 5.5–6.3 [32]. The higher Ree^{3+} content of the NL weathering profile than that of the LC profile is, therefore, likely related to its higher kaolinite and halloysite contents.

5.3.3. Intensity of Granite Weathering

The Ree^{3+} content of the weathering profile is influenced by the intensity of granite weathering; the stronger the weathering, the more REE minerals are dissolved, and the more Ree^{3+} is released to the profile. The LC weathering profile has a lower REE content than the NL profile: 134–1111 ppm (mean 447 ppm; $n = 77$) and 168–2347 ppm (mean 572 ppm; $n = 70$), respectively, and the difference may be due to the relative intensities of weathering. In the NL area, the chemical index of alteration (CIA; $100 \times Al_2O_3/(Al_2O_3 + CaO + Na_2O + K_2O)$) of granite and the weathering profile are in the ranges of 45–60% and 65–95%, respectively, with the REE content increasing with CIA in the range 45–60%, but decreasing in the CIA range of 65–95% [27]. The CIA of LC granite and weathering profile are similar to the NL profile (Figure 12a, b).

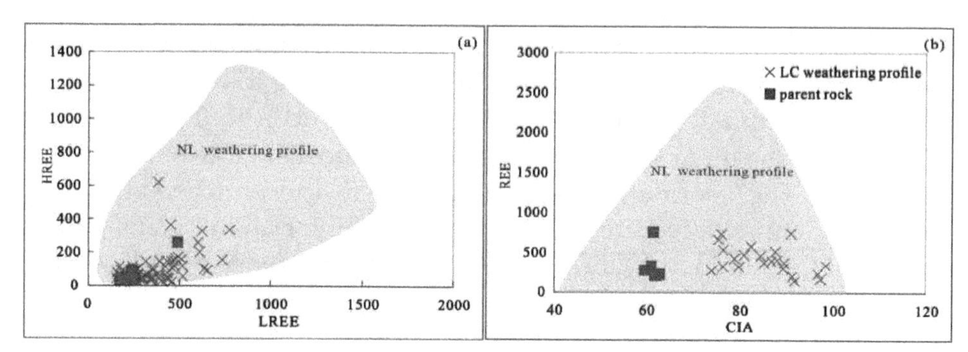

Figure 12. Diagrams of LREE *versus* HREE (**a**) and CIA *versus* REE (**b**) of the LC granite and weathering profile.

The intensity of weathering can also be reflected in the amounts and types of secondary minerals in the horizon. In natural chemical weathering, feldspar and mica (muscovite and biotite) are altered to clay through several stages, with each stage corresponding to a different clay mineral:

- Feldspar→kaolinite→gibbsite;
- Muscovite→illite→montmorillonite→kaolinite→gibbsite;
- Biotite→chlorite→vermiculite→montmorillonite→beidellite→kaolinite→gibbsite.

Halloysite is in the kaolin group, and normally coexists with kaolinite in modern soils and sediments. The higher the intensity of weathering, the higher the content of clay minerals, such as those of the kaolin group (and possibly gibbsite). In NL granitc weathering profile, halloysie is in common and play an important role in the carrier of Ree^{3+} [32,69]. But in the most of LC granite weathering profile, there is kaolinite only [28,29], which could be the reason why the concentration of Ree^{3+} in NL area are higher than LC area.

Weathering cycles are key to the secondary mineralization of Ree^{3+} in the weathering profile, with re-enrichment and re-fractionation and with Ree^{3+} being re-migrated and re-accumulated. As soil pH changes from acidic to alkalescent with increasing depth in the weathering profile, the Ree^{3+} concentration increases with the increasing adsorption capacity of clays. Previous studies have shown that a pH range of 5.4–6.8 is optimum for the accumulation of Ree^{3+} [32,70,71]. The HREEs of higher atomic mass are more soluble than LREEs in water, and accumulate at higher pH [32,70,71], with the depth of HREE enrichment, thus, being greater than that of LREE enrichment.

6. Conclusions

(1) REE minerals, including allanite, fluorapatite, and monazite, and HREE minerals such as xenotime, thorite, allanite(–Y), fluorite(–Y), and REE fluorcarbonate, make the greatest contribution to the REE content of the LC granite.The QN and MH granites are host rocks for Ree deposits with zircon U–Pb ages of ca. 217 Ma and 232 Ma, respectively, with these being the earliest times of REE mineral formation.

(2) The host rocks of the Ree deposits are strongly peraluminous with S-type granite affinities, suggesting they originated from partly melted crustal basement. Sr–Nd–Pb isotopic systematics and geochemical signatures of the LC granites indicate that they were derived from partial melting of the upper crust. Paleo-Tethyan continent–continent collision occurred during the early Indosinian, followed by post-collisional extension. Middle Indosinian magmatism was generated in a post-collisional tectonic setting related to early–middle Indosinian slab break-off.

(3) REE minerals such as allanite(–Y), fluorapatite, fluorite(–Y), and REE fluorcarbonate dissolved in the weathering crust are the main sources of Ree deposits. Dissolution of LREE and HREE minerals lead to LREE and HREE enrichment, respectively. Secondary minerals, such as kaolin-group minerals and illite (from altered feldspars and micas), are carriers of Ree^{3+}, and its enrichment is influenced by the intensity of weathering; the stronger the weathering, the more REE minerals are dissolved

in the weathering profile. Weathering cycles are key to the secondary mineralization of Ree^{3+}, and stable geological conditions over a narrow altitudinal range promote retention of Ree^{3+} in the weathering profile.

Author Contributions: Writing—original draft preparation, L.L.; writing—review and editing, Y.L. and H.L.; methodology, Z.Z.; investigation, C.W.; resources, C.W. and X.X.; funding acquisition, Y.L. All authors have read and agreed to the published version of the manuscript.

Acknowledgments: We thank Yang Yue-qing, Wang Rui-jiang, Wang Denghong, Qin Jinghua, and other geologists for their help during the field investigations and sample analysis.

References

1. Chakhmouradian, A.R.; Wall, F. Rare Earth Elements: Minerals, Mines, Magnets (and More). *Elements* **2012**, *8*, 333–340. [CrossRef]
2. Faris, N.; Ram, R.; Tardio, J.; Bhargava, S.; McMaster, S.A.; Pownceby, M.I. Application of ferrous pyrometallurgy to the beneficiation of rare earth bearing iron ores—A review. *Miner. Eng.* **2017**, *110*, 20–30. [CrossRef]
3. Faris, N.; Ram, R.; Tardio, J.; Bhargava, S.; Pownceby, M.I. Characterisation of a ferruginous rare earth bearing lateritic ore and implications for rare earth mineral processing. *Miner. Eng.* **2019**, *134*, 23–36. [CrossRef]
4. Soltani, F.; Abdollahy, M.; Petersen, J.; Ram, R.; Becker, M.; Koleini, S.J.; Moradkhani, D. Leaching and recovery of phosphate and rare earth elements from an iron-rich fluorapatite concentrate: Part I: Direct baking of the concentrate. *Hydrometallurgy* **2018**, *177*, 66–78. [CrossRef]
5. Soltani, F.; Abdollahy, M.; Petersen, J.; Ram, R.; Koleini, S.J.; Moradkhani, D. Leaching and recovery of phosphate and rare earth elements from an iron-rich fluorapatite concentrate: Part II: Selective leaching of calcium and phosphate and acid baking of the residue. *Hydrometallurgy* **2019**, *184*, 29–38. [CrossRef]
6. Kanazawa, Y.; Kamitani, M. Rare earth minerals and resources in the world. *J. Alloy. Compd.* **2006**, *37*, 1339–1343. [CrossRef]
7. Yang, Y.Q.; Hu, C.S.; Luo, Z.M. Geological characteristic of mineralization of rare earth deposit of the ion–absorption type and their prospecting direction. *Bull. Chin. Acad. Geol. Sci.* **1981**, *2*, 102–118.
8. Sanematsu, K.; Murakami, H.; Watanabe, Y.; Duangsurigna, S.; Siphandone, V. Enrichment of rare earth elements (REE) in granitic rocks and their weathered crusts in central and southern Laos. *Bull. Geol. Surv. Jpn.* **2009**, *60*, 527–558. [CrossRef]
9. Sanematsu, K.; Kon, Y.; Imai, A.; Watanabe, K.; Watanabe, Y. Geochemical and mineralogical characteristics of ion-adsorption type REE mineralization in Phuket, Thailand. *Miner. Depos.* **2013**, *48*, 437–451. [CrossRef]
10. Yuan, Z.X.; He, H.H.; Liu, L.J.; Wang, D.H.; Zhao, Z. *Rare Metal and Rare Earth Deposit Abroad*; Beijing Science Press: Beijing, China, 2016; pp. 1–110.
11. Padrones, J.T.; Imai, A.; Takahashi, R. Geochemical Behavior of Rare Earth Elements in Weathered Granitic Rocks in Northern Palawan, Philippines. *Resour. Geol.* **2017**, *67*, 231–253. [CrossRef]
12. Ram, R.; Becker, M.; Brugger, J.; Etschmann, B.; Burcher-Jones, C.; Howard, D.; Kooyman, P.J.; Petersen, J. Characterisation of a rare earth element- and zirconium-bearing ion-adsorption clay deposit in Madagascar. *Chem. Geol.* **2019**, *522*, 93–107. [CrossRef]
13. Wang, D.H.; Wang, R.J.; Li, J.K.; Zhao, Z.; Yu, Y.; Dai, J.J.; Chen, Z.H.; Li, D.X.; Qu, W.J.; Deng, M.C.; et al. The progress in the strategic research and survey of rare earth, rare metal and rare–scattered elements mineral resources. *Geol. China* **2013**, *40*, 361–370.
14. Wang, D.H.; Zhao, Z.; Yu, Y.; Zhao, T.; Li, J.K.; Dai, J.J.; Liu, X.X. Progress, problems and research orientation of ion–adsorpotion type rare earth resources. *Rock Miner. Anal.* **2013**, *32*, 796–802.
15. Wang, F.; Liu, F.L.; Liu, P.H.; Shi, J.R.; Cai, J. Petrogenesis of Lincang granites in the south of Lancangjiang area: Constrain from geochemistry and zircon U–Pb geochronology. *Acta Petrol. Sinica* **2014**, *30*, 3034–3050.
16. Wang, D.H.; Zhao, Z.; Yu, Y.; Wang, C.H.; Dai, J.J.; Sun, Y.; Zhao, T.; Li, J.K.; Huang, F.; Chen, Z.Y.; et al. A Review of the Achievements in the Survey and Study of Ion-absorption Type REE Deposits in China. *Acta Geosci. Sinica* **2017**, *38*, 317–325.

17. Zhao, Z.; Wang, D.H.; Chen, Z.Y.; Guo, N.X.; Liu, X.X.; He, H.H. Metallogenic specialization of rare earth mineralized igneous rocks in the Eastern Nanling Region. *Geotecton. Metallog.* **2014**, *38*, 255–263.
18. Zhao, Z.; Wang, D.H.; Liu, X.X.; Zhang, Q.W.; Yao, M.; Gu, W.N. Geochemical features of rare earth elements in different weathering stage of the Guangxi Huashan granite and its influence factors. *Chin. Rare Earth* **2015**, *3*, 14–20.
19. Zhao, Z.; Wang, D.H.; Chen, Z.H.; Chen, Z.Y. Progress of research on metallogenic regularity of ion-adsorption type REE deposit in the Nanling Range. *Acta Geol. Sinica* **2017**, *91*, 2814–2827.
20. Liu, Y.; Chakhmouradian, A.R.; Hou, Z.; Song, W.; Kynický, J. Development of REE mineralization in the giant Maoniuping deposit (Sichuan, China): Insights from mineralogy, fluid inclusions, and trace-element geochemistry. *Miner. Depos.* **2018**, *54*, 701–718. [CrossRef]
21. Liu, Y.; Hou, Z. A synthesis of mineralization styles with an integrated genetic model of carbonatite-syenite-hosted REE deposits in the Cenozoic Mianning-Dechang REE metallogenic belt, the eastern Tibetan Plateau, southwestern China. *J. Asian Earth Sci.* **2017**, *137*, 35–79. [CrossRef]
22. Liu, Y.; Chen, Z.Y.; Yang, Z.S.; Sun, X.; Zhu, Z.M.; Zhang, Q.C. Mineralogical and geochemical studies of brecciated ores in the Dalucao REE deposit, Sichuan Province, southwestern China. *Ore Geol. Rev.* **2015**, *70*, 613–636. [CrossRef]
23. Liu, Y.; Zhu, Z.; Chen, C.; Zhang, S.; Sun, X.; Yang, Z.; Liang, W. Geochemical and mineralogical characteristics of weathered ore in the Dalucao REE deposit, Mianning—Dechang REE Belt, western Sichuan Province, southwestern China. *Ore Geol. Rev.* **2015**, *71*, 437–456. [CrossRef]
24. Wu, C.Y.; Huang, D.H.; Bai, G.; Ding, X.S. Differentiation of rare earth elements and origin of granitic rocks,Nanling Mountain area. *Acta Petrol. Mineral.* **1990**, *9*, 106–116.
25. Wu, C.Y.; Bai, G.; Huang, D.H.; Zhu, Z.S. Characteristics and significance of HREE-rich granitoids of the Nanling mountain area. *Bull. Chin. Acad. Geol. Sci.* **1992**, *25*, 43–58.
26. Wu, C.Y.; Huang, D.H.; Guo, Z.X. REE geochemistry in the weathering process of granites in Longnan County, Jiangxi Province. *Acta Geol. Sinica* **1992**, *63*, 349–362.
27. Bao, Z.; Zhao, Z. Geochemistry of mineralization with exchangeable REY in the weathering crusts of granitic rocks in South China. *Ore Geol. Rev.* **2008**, *33*, 519–535. [CrossRef]
28. Lu, L.; Wang, D.H.; Wang, C.H.; Zhao, Z.; Feng, W.J.; Xu, X.C.; Yu, F. Mineralization regularity of ion–adsorption type REE deposits on Lincang granite in Yunnan Province. *Acta Geol.* **2019**, *96*, 1466–1478.
29. Lu, L.; Wang, D.H.; Wang, C.H.; Zhao, Z.; Feng, W.J.; Xu, X.C.; Chen, C.; Zhong, H.R. The metallogenic regularity of ion-adsorption type REE deposit in Yunnan Province. *Acta Geol.* **2020**, *94*, 179–191.
30. Migdisov, A.; Williams-Jones, A.; Brugger, J.; Caporuscio, F. Hydrothermal transport, deposition, and fractionation of the REE: Experimental data and thermodynamic calculations. *Chem. Geol.* **2016**, *439*, 13–42. [CrossRef]
31. Yusoff, Z.M.; Ngwenya, B.T.; Parsons, I. Mobility and fractionation of REEs during deep weathering of geochemically contrasting granites in a tropical setting, Malaysia. *Chem. Geol.* **2013**, *349*, 71–86. [CrossRef]
32. Yang, M.; Liang, X.; Ma, L.; Huang, J.; He, H.; Zhu, J. Adsorption of REEs on kaolinite and halloysite: A link to the REE distribution on clays in the weathering crust of granite. *Chem. Geol.* **2019**, *525*, 210–217. [CrossRef]
33. YNBGMR (Yunnan Bureau Geological Mineral Resource). *Regional Geology of Guizhou Province*; Geology Publish House: Beijing, China, 1990; pp. 1–729.
34. Zhang, Y.F. An approach to the characeristics of the Indo–China movement in western Yunnan area. *Yunnan Geol.* **1985**, *1*, 59–68.
35. Heppe, K.; Helmcke, D.; Wemmer, K. The Lancang River zone of southwestern Yunnan, China: A questionable location for the active continental margin of Paleo-Tethys. *J. Asian Earth Sci.* **2007**, *30*, 706–720. [CrossRef]
36. Liu, H.; Wang, Y.; Fan, W.; Zi, J.; Cai, Y.; Yang, G. Petrogenesis and tectonic implications of Late-Triassic high ε Nd(t)-ε Hf(t) granites in the Ailaoshan tectonic zone (SW China). *Sci. China Earth Sci.* **2014**, *57*, 2181–2194. [CrossRef]
37. Turner, S.; Foden, J.; Morrison, R. Derivation of some A-type magmas by fractionation of basaltic magma: An example from the Padthaway Ridge, South Australia. *Lithos* **1992**, *28*, 151–179. [CrossRef]
38. Peccerillo, A.; Taylor, S.R. Geochemistry of eocene calc-alkaline volcanic rocks from the Kastamonu area, Northern Turkey. *Contrib. Miner. Pet.* **1976**, *58*, 63–81. [CrossRef]
39. Kemp, A.; Hawkesworth, C. Granitic Perspectives on the Generation and Secular Evolution of the Continental Crust. *Treatise Geochem.* **2003**, *3*, 349–410. [CrossRef]

40. Peng, T.P.; Wang, Y.J.; Fan, W.M.; Liu, D.Y.; Shi, Y.R.; Miao, L.C. SHRIMP zircon U–Pb geochronology of early Mesozoic felsic igneous rocks from the southern Lancangjiang and its tectonic implications. *Sci. China Ser. D Earth Sci.* **2006**, *49*, 1032–1042. [CrossRef]
41. Peng, T.; Wilde, S.A.; Wang, Y.; Fan, W.; Peng, B. Mid-Triassic felsic igneous rocks from the southern Lancangjiang Zone, SW China: Petrogenesis and implications for the evolution of Paleo-Tethys. *Lithos* **2013**, 15–32. [CrossRef]
42. Jian, P.; Liu, D.; Kröner, A.; Zhang, Q.; Wang, Y.; Sun, X.; Zhang, W. Devonian to Permian plate tectonic cycle of the Paleo-Tethys Orogen in southwest China (I): Geochemistry of ophiolites, arc/back-arc assemblages and within-plate igneous rocks. *Lithos* **2009**, *113*, 748–766. [CrossRef]
43. Jian, P.; Liu, D.; Kröner, A.; Zhang, Q.; Wang, Y.; Sun, X.; Zhang, W. Devonian to Permian plate tectonic cycle of the Paleo-Tethys Orogen in southwest China (II): Insights from zircon ages of ophiolites, arc/back-arc assemblages and within-plate igneous rocks and generation of the Emeishan CFB province. *Lithos* **2009**, *113*, 767–784. [CrossRef]
44. Kong, H.L. Genchemistty, Geochronology and Petrogenisis of Lincang Granites in Southern Lancangjiang Zone of Sanjiang Area. Master's Thesis, China University of Geosciences, Beijing, China, 2011.
45. Kong, H.L.; Dong, G.C.; Mo, X.X.; Zhao, Z.D.; Zhu, D.C.; Wang, S.; Li, R.; Wang, Q.L. Petrogenesis of Lincang granites in Sanjiang area of western Yunnan Province: Constraints from geochemistry, zircon U-Pb geochronology and Hf isotope. *Acta Petrol. Sinica* **2012**, *28*, 1438–1452.
46. Whalen, J.B.; Currie, K.L.; Chappell, B.W. A-type granites: Geochemical characteristics, discrimination and petrogenesis. *Contrib. Miner. Pet.* **1987**, *95*, 407–419. [CrossRef]
47. Sun, S.-S.; McDonough, W.F. Chemical and Isotopic Systematics of Oceanic Basalts: Implications for Mantle Composition and Processes. In *Magmatism in the Ocean Basins*; Geological Society: London, UK, 1989; Volume 42, pp. 313–345.
48. Cong, B.L.; Wu, G.Y.; Zhang, Q. Petrotectonic evolution of the Tethys zone in western Yunnan, China. *Chin. Sci. Bull.* **1993**, *23*, 1201–1207.
49. Qiu, Y.M.; Gao, S.; McNaughton, N.J.; Groves, D.I.; Ling, W.L. First evidence of >3.2 Ga continental crust in the Yangtze craton of south China and its implications for Archean crustal evolution and Phanerozoic tectonics. *Geology* **2000**, *33*, 309–314.
50. Shi, X.B.; Qiu, X.L.; Liu, H.L.; Chu, Z.Y.; Xia, B. Thermochronological analyses on the cooling history of the Lincang granitoid batholith, Western Yunnan. *Acta Petrol. Sinica* **2006**, *22*, 465–479.
51. Li, X.L. Basic characteristic and formation structural environment of Lincang composite granite batholith. *Yunnan Geol.* **1996**, *1*, 1–18.
52. Pupin, J.P. Zircon and granite petrology. *Contrib. Miner. Pet.* **1980**, *73*, 207–220. [CrossRef]
53. Zapata, A.; Botelho, N.F. Mineralogical and geochemical characterization of rare-earth occurrences in the Serra do Mendes massif, Goiás, Brazil. *J. Geochem. Explor.* **2018**, *188*, 398–412. [CrossRef]
54. Sylvester, P.J. Post-collisional strongly peraluminous granites. *Lithos* **1998**, *45*, 29–44. [CrossRef]
55. Altherr, R.; Holl, A.; Hegner, E.; Langer, C.; Kreuzer, H. High-potassium, calc-alkaline I-type plutonism in the European Variscides: Northern Vosges (France) and northern Schwarzwald (Germany). *Lithos* **2000**, *50*, 51–73. [CrossRef]
56. Chappell, B.W.; White, A.J.R. I- and S-type granites in the Lachlan Fold Belt. *Earth Environ. Sci. Trans. R. Soc. Edinb.* **1992**, *83*, 1–26. [CrossRef]
57. Chappell, B.W.; White, A.J.R. Two contrasting granite types: 25 years later. *Aust. J. Earth Sci.* **2001**, *48*, 488–489. [CrossRef]
58. Chappell, B.W.; White, A.J.R. Two contrasting granite types. *Pacific Geol.* **1974**, *8*, 173–174.
59. Harris, N.B.W.; Pearce, J.A.; Tindle, A.G. Geochemical Characteristics of Collision-Zone Magmatism. In *Collision Tectonics*; Geological Society: London, UK, 1986; pp. 67–81.
60. Brown, M. Granite: From genesis to emplacement. *Bull. Geol. Soc. Am.* **2013**, *125*, 1079–1113. [CrossRef]
61. Harris, N.B.W.; Inger, S. Trace element modelling of pelite-derived granites. *Contrib. Miner. Pet.* **1992**, *110*, 46–56. [CrossRef]
62. Zindler, A.; Hart, S. Chemical geodynamics. *Annu. Rev. Earth Planet. Sci.* **1986**, *14*, 493–571. [CrossRef]
63. Mo, X.X.; Shen, S.Y.; Zhu, Q.W. Volcanics-Ophiolite and Mineralization of Middle and Southern Part. In *Sanjiang, Southern China*; Geological Publishing House: Beijing, China, 1998; pp. 1–128.

64. Sone, M.; Metcalfe, I. Parallel Tethyan sutures in mainland Southeast Asia: New insights for Palaeo-Tethys closure and implications for the Indosinian orogeny. *Comptes Rendus Geosci.* **2008**, *340*, 166–179. [CrossRef]
65. Fan, W.M.; Wang, Y.J.; Zhang, A.M.; Zhang, F.F.; Zhang, Y.Z. Permian arc-backarc basin development along the Ailaoshan tectonic zone: Geochemical, isotopic and geochronological evidence from the Mojiang volcanic rocks, Southwest China. *Lithos* **2010**, *119*, 553–568. [CrossRef]
66. Bonin, B. A-type granites and related rocks: Evolution of a concept, problems and prospects. *Lithos* **2007**, *97*, 1–29. [CrossRef]
67. Pearce, J.A.; Harris, N.B.W.; Tindle, A.G. Trace Element Discrimination Diagrams for the Tectonic Interpretation of Granitic Rocks. *J. Petrol.* **1984**, *25*, 956–983. [CrossRef]
68. Pearce, J.A. Sources and settings of granitic rocks. *Episodes* **1996**, *19*, 120–125. [CrossRef]
69. Li, M.Y.H.; Zhou, M.-F.; Williams-Jones, A.E. The Genesis of Regolith-Hosted Heavy Rare Earth Element Deposits: Insights from the World-Class Zudong Deposit in Jiangxi Province, South China. *Econ. Geol.* **2019**, *114*, 541–568. [CrossRef]
70. Henderson, P. *Rare Earth Element Geochemistry*; Elsevier: Amsterdam, The Netherlands, 1984; pp. 180–213.
71. Chen, D.Q.; Wu, J.S. The mineralization mechanism of ion-adsorbed REE deposit. *J. Chin. Rare Earth Soc.* **1990**, *8*, 175–179.
72. Qi, L.; Hu, J.; Gregoire, D.C. Determination of trace elements in granites by inductively coupled plasma mass spectrometry. *Talanta* **2000**, *51*, 507–513.
73. Liu, Y.S.; Gao, S.; Hu, Z.C.; Gao, C.G.; Zong, K.Q.; Wang, D.B. Continental and oceanic crust recycling-induced melt-peridotite interactions in the Trans-North China Orogen: U-Pb dating, Hf isotopes and trace elements in zircons from mantle xenoliths. *J. Petrol.* **2010**, *51*, 537–571. [CrossRef]
74. Li, X.H.; Liu, Y.; Li, Q.L.; Guo, C.H.; Chamberlain, K.R. Precise determination of Phanerozoic zircon Pb/Pb age by multi-collector SIMS without external standardi- zation. *Geochem. Geophys. Geosyst.* **2009**, *10*. [CrossRef]
75. Sláma, J.; Košler, J.; Condon, D.J. Plešovice zircon—A new natural reference material for U-Pb and Hf isotopic microanalysis. *Chem. Geol.* **2008**, *249*, 1–353. [CrossRef]
76. Wiedenbeck, M.; Alle, P.; Corfu, F.; Griffin, W.L.; Meier, M.; Oberli, F.; Vonquadt, A.; Roddick, J.C.; Speigel, W. Three natural zircon standards for U-Th–Pb, Lu–Hf, trace–element and REE analyses. *Geostandard Newslett.* **1995**, *19*, 1–23. [CrossRef]
77. Ludwig, K.R. Users Manual for Isoplot/Ex rev. 2.49. Berkeley Geochronology Centre Special Publication. 2001. Available online: http://www.geo.cornell.edu/geology/classes/Geo656/Isoplot%20Manual.pdf (accessed on 21 February 2020).

3

Zoned Laurite from the Merensky Reef, Bushveld Complex, South Africa: "Hydrothermal" in Origin?

Federica Zaccarini * and Giorgio Garuti

Department of Applied Geological Sciences and Geophysics, University of Leoben, A-8700 Leoben, Austria; giorgio.garuti1945@gmail.com
* Correspondence: federica.zaccarini@unileoben.ac.at

Abstract: Laurite, ideally (Ru,Os)S_2, is a common accessory mineral in podiform and stratiform chromitites and, to a lesser extent, it also occurs in placer deposits and is associated with Ni-Cu magmatic sulfides. In this paper, we report on the occurrence of zoned laurite found in the Merensky Reef of the Bushveld layered intrusion, South Africa. The zoned laurite forms relatively large crystals of up to more than 100 μm, and occurs in contact between serpentine and sulfides, such as pyrrhotite, chalcopyrite, and pentlandite, that contain small phases containing Pb and Cl. Some zoned crystals of laurite show a slight enrichment in Os in the rim, as typical of laurite that crystallized at magmatic stage, under decreasing temperature and increasing sulfur fugacity, in a thermal range of about 1300–1000 °C. However, most of the laurite from the Merensky Reef are characterized by an unusual zoning that involves local enrichment of As, Pt, Ir, and Fe. Comparison in terms of Ru-Os-Ir of the Merensky Reef zoned laurite with those found in the layered chromitites of the Bushveld and podiform chromitites reveals that they are enriched in Ir. The Merensky Reef zoned laurite also contain high amount of As (up to 9.72 wt%), Pt (up to 9.72 wt%) and Fe (up to 14.19 wt%). On the basis of its textural position, composition, and zoning, we can suggest that the zoned laurite of the Merensky Reef is "hydrothermal" in origin, having crystallized in the presence of a Cl- and As-rich hydrous solution, at temperatures much lower than those typical of the precipitation of magmatic laurite. Although, it remains to be seen whether the "hydrothermal" laurite precipitated directly from the hydrothermal fluid, or it represents the alteration product of a pre-existing laurite reacting with the hydrothermal solution.

Keywords: laurite; sulfides; fluids; platinum group elements (PGE); platinum group minerals (PGM); Merensky Reef; Bushveld Complex; South Africa

---

1. Introduction

Minerals of ruthenium are very rare and only five of them, namely anduoite (Ru,Os)As_2, laurite (Ru,Os)S_2, ruarsite RuAsS, ruthenarsenite (Ru,Ni)As, and ruthenium (Ru,Ir,Os), have been approved by the International Mineralogical Association (IMA). They occur as accessory minerals associated with mafic–ultramafic rocks, especially with chromitite, and as nuggets in placer deposits. Among the minerals of ruthenium, laurite is the most common. It was discovered in 1866 in a placer from Laut, Banjar, South Kalimantan Province, Borneo, Indonesia [1]. Laurite is a common constituent of the suite of platinum group minerals (PGM) inclusions (usually less than 20 μm) in podiform and stratiform chromitites [2–4]. Less frequently, laurite has been reported from placers and Ni-Cu magmatic sulfide deposits [5–7]. Laurite forms a complete solid solution with erlichmanite (OsS_2) [5], and their typical mode of occurrence, i.e., included in chromite grains, indicate that they crystallized at high temperatures, in a thermal range of about 1300–1000 °C prior to, or coeval with, the precipitation of the host chromitite [2–4]. The reciprocal stability of laurite and erlichmanite is strongly controlled by sulfur fugacity and temperature. In particular, laurite precipitates at a higher temperature and lower sulfur fugacity, compared to erlichmanite. This order of crystallization can be observed in the zoning of the small crystals of laurite and erlichmanite enclosed in fresh chromite grains that, typically, show an Os-poor core, grading into a high-Os rim [2–4]. This magmatic zoning can be obliterated by low temperature processes such as serpentinization and weathering, as documented in laurite associated with podiform chromitites [8–10]. During alteration processes at low temperature, laurite and erlichmanite lose their original S and release part of Os and Ir to form secondary Ru-Os-Ir alloys, in which the lost S may be replaced by Fe-oxide [8–10]. The occurrence of laurite in the Bushveld Complex of South Africa has been documented by several authors [2,7,11–17]. Most commonly, the mineral occurs as small polygonal grains enclosed in chromite grains of the Critical Zone chromitite layers and has only occasionally been found as part of the sulfide ore of the Merensky Reef. In this contribution, we have investigated in detail the mineral chemistry of the laurite associated with the sulfide-rich zone of the Merensky Reef. The grains are characterized by an unusual zoning and composition compared with laurite inclusions in the Bushveld chromitites, suggesting that the mineral was generated under different thermodynamic conditions in the two cases.

2. Sample Provenance and Petrographic Description

The Bushveld layered intrusion is located in the central part of the Transvaal province, north of Pretoria, South Africa (Figure 1A), and it is divided into Eastern, Western, and Northern limbs (Figure 1B). The Bushveld intrusion is well known among economic geologists because it contains the world's largest deposits of platinum group elements (PGE), namely: the UG-2 chromitite and the Merensky Reef [18]. The noun Merensky Reef refers to a sulfide-bearing pegmatoidal feldspathic pyroxenite enriched in PGE, marked at the bottom and top, by two centimetric layers of chromitite. The Reef can be traced for a total strike of about 280 km, marking the limit between the Critical and Main Zone [18,19]. The investigated samples were collected by one of the authors (G.G.) in the Rustenburg underground mine, during the third International Platinum Symposium held in Pretoria from 6 to 10 July 1981 [20,21]. The Rustenburg mine is located in the Western limb of the Bushveld Complex, about 100 km west of Pretoria (Figure 1B,C). Here, four different zones of Bushveld (undifferentiated in Figure 1C) are intruded by the Pilanesberg Alkaline Complex [22]. Four square polished blocks, about 2.5 × 2.5 cm (Figure 2), were prepared for petrographic and mineralogical investigation. The blocks consist of a thin layer of chromitite, about 0.2 cm thick, in contact with pegmatoidal feldspathic pyroxenite and large blebs of sulfide. In agreement with observations made by several authors [21,23,24], the sulfide-rich zone contains accessory actinolite, micas, talc, chlorite, and a serpentine subgroup mineral.

3. Methods

The polished blocks were previously studied by reflected-light microscope. Quantitative chemical analyses of laurite were performed with a JEOL JXA-8200 electron microprobe (JEOL, Tokyo, Japan), installed in the E. F. Stumpfl laboratory, Leoben University, Leoben, Austria, operated in WDS (wavelength dispersive spectrometry) mode. Major and minor elements were determined at 20 kV accelerating voltage and 10 nA beam current, with 20 s as counting time for the peak and 10 s for the backgrounds. The beam diameter was about 1 μm in size. The Kα lines were used for S, As, Fe and Ni; Lα for Ir, Ru, Rh, Pd, and Pt, and Mα for Os. The reference materials were pure metals for the six PGE (Ru, Rh, Pd, Os, Ir and Pt), synthetic NiS, natural pyrite and niccolite for Ni, Fe, S and As. The following diffracting crystals were selected: PETJ for S; PETH for Ru, Os, Pd and Rh; LIFH for Fe, Ni, Ir and Pt; and TAP for As. Automatic correction was performed for the Ru-Rh and Rh-Pd interferences. The detection limits were calculated by the software and are: Os (0.07 wt%), Ir (0.06 wt%), Ru, Pd, and Pt (0.04 wt%), Rh (0.03 wt%), Fe, Ni, As and S (0.02 wt%). The grains smaller than 10 μm were analyzed by EDS. The same instrument was used to obtain back-scattered electron images (BSE) and X-ray elemental distribution maps.

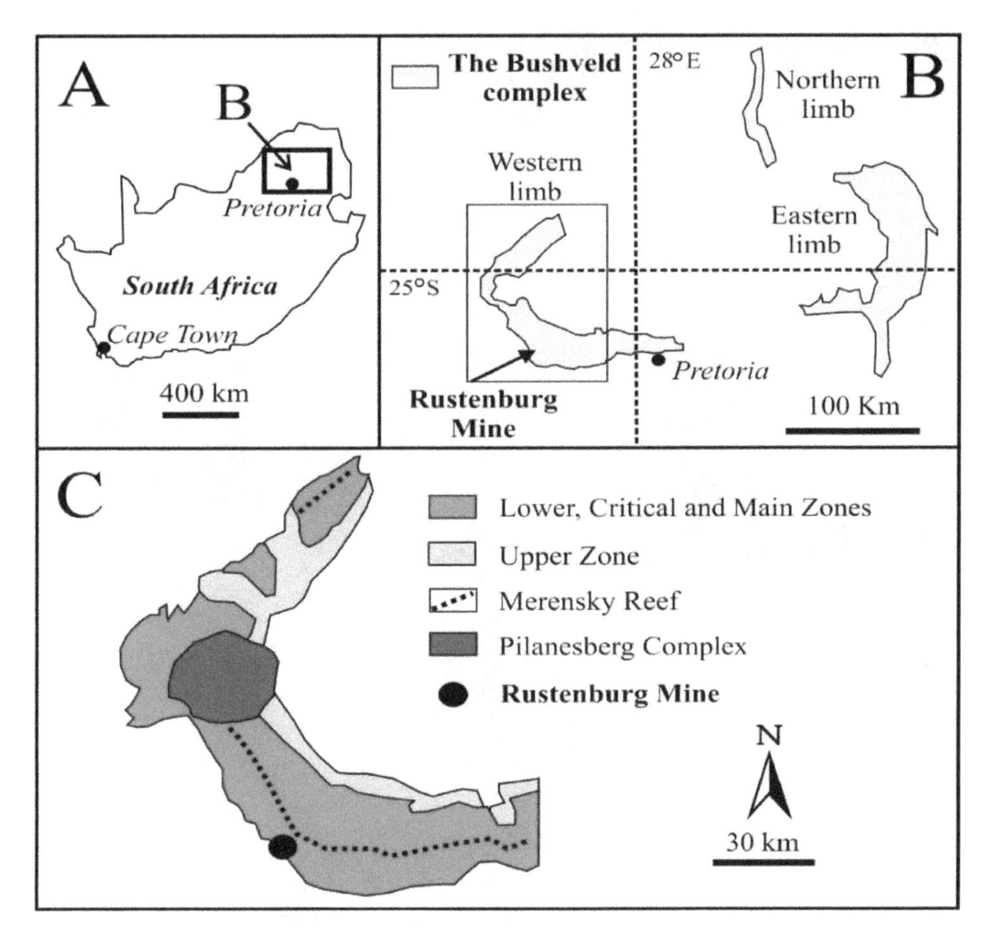

Figure 1. The Bushveld Complex, South Africa (**A**) and locations of the Rustenburg mine and the Merensky Reef (**B**,**C**) in the Western limb (modified after [15,18]).

Figure 2. Example of the studied polished blocks from the Merensky Reef, showing the sulfide blebs (creamy–yellow) and the cumulitic chromitite (small polygonal dark grey grains) in the pegmatoidal feldspathic pyroxenite.

4. Laurite: Morphology, Texture, and Composition

The investigated samples contain laurite in two different textural positions, either included in fresh chromite of the thin chromitite layer (Figure 3A), or at the contact between sulfide patches (pyrrhotite, chalcopyrite, pentlandite) and serpentine (Figure 3B–D).

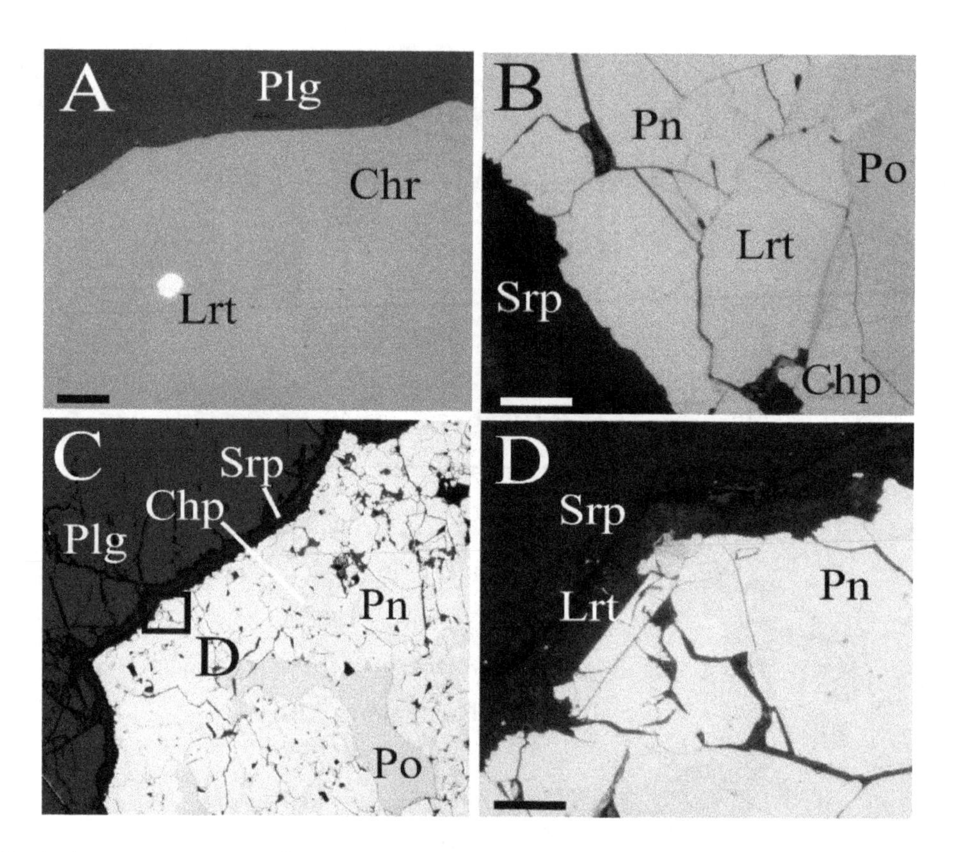

Figure 3. Digital image in reflected plane polarized light showing laurite from the Merensky Reef. (**A**) Laurite enclosed in fresh chromite. (**B**,**C**) Zoned laurite and (**D**) enlargement of (**C**). Abbreviations: Lrt = laurite, Plg = plagioclase, Chr = chromite, Srp = serpentine, Pn = pentlandite, Po = pyrrhotite, Chp = chalcopyrite. Scale bar = 20 μm.

Quantitative analyses of laurite enclosed in chromite and associated with sulfides are listed in Tables 1 and 2, respectively. Laurite included in chromite forms tiny crystals, usually not exceeding 10 μm in size, characterized by euhedral to subeuhedral morphology and homogenous composition. Laurite associated with sulfides and serpentine is bigger, up to more than 100 μm, and may occur as single crystals or clusters of grains (Figure 3B–D), characterized by subeuhedral to anhedral shape. The BSE images of large laurite display remarkable zoning emphasized by marked contrast in the electronic reflectivity (Figure 4A–D). Laurite in the sulfide assemblage is accompanied by a complex association of precious minerals comprising: cooperite (PtS), moncheite ($PtTe_2$), platarsite (PtAsS), rustenburgite (Pt_3Sn), Pt-Fe alloy, undetermined Pt-Te-Bi and Pd-Te-Bi compounds, Au-Ag alloy, and the recently discovered PGM bowlesite PtSnS [21].

Table 1. Selected wavelength dispersive spectrometry (WDS) electron microprobe analyses of Merensky Reef laurite enclosed in chromite.

Sample	As	S	Ru	Os	Ir	Rh	Pt	Pd	Ni	Fe	Total
					wt%						
mr18a	1.69	36.81	50.36	5.42	2.85	1.24	0.00	1.79	0.03	1.00	101.18
mr8a	1.85	36.93	49.99	4.97	3.03	1.06	0.00	1.68	0.08	0.90	100.48
Sample	**As**	**S**	**Ru**	**Os**	**Ir**	**Rh**	**Pt**	**Pd**	**Ni**	**Fe**	
					at%						
mr18a	1.28	65.25	28.32	1.62	0.84	0.69	0.00	0.96	0.03	1.02	
mr8a	1.40	65.58	28.16	1.49	0.90	0.59	0.00	0.90	0.08	0.91	

Table 2. WDS electron microprobe analyses of Merensky Reef zoned laurite.

Sample	As	S	Ru	Os	Ir	Rh	Pt	Pd	Ni	Fe	Total
					wt%						
MR2a1	1.08	37.50	39.32	4.79	7.23	1.09	0.00	1.32	0.26	6.54	99.13
MR2a2	1.28	36.97	37.20	7.97	9.49	0.90	0.00	1.18	0.06	5.76	100.81
MR2a3	1.63	37.63	43.23	2.53	6.22	1.38	0.00	1.35	0.42	5.87	100.27
MR2a4	1.51	37.58	45.24	3.15	5.75	1.32	0.00	1.55	0.81	4.23	101.15
MR2a5	1.38	37.68	43.61	3.59	6.02	1.35	0.00	1.47	0.90	4.73	100.73
MR2a6	1.74	37.32	42.99	2.78	6.46	1.37	0.00	1.41	0.65	5.22	99.94
MR2a7	1.44	38.69	44.54	1.88	5.27	1.19	0.00	1.47	0.49	5.24	100.21
MR2a8	1.42	39.13	44.65	2.00	4.50	1.63	0.00	1.58	0.39	5.44	100.73
MR2a9	1.50	38.46	43.86	1.86	6.06	1.05	0.10	1.42	0.53	5.28	100.13
MR2a10	1.54	38.98	43.03	2.77	6.25	1.16	0.00	1.45	0.31	6.17	101.66
MR2a11	2.22	38.17	44.63	1.57	5.35	1.53	0.00	1.63	0.57	4.57	100.23
MR2a12	9.72	28.45	28.24	5.90	13.43	1.75	7.20	1.05	0.80	3.05	99.57
MR2a13	3.66	36.78	42.64	1.90	5.85	2.02	1.78	1.58	0.63	4.35	101.20
MR2a14	2.90	37.25	43.43	1.64	5.83	1.98	1.04	1.55	0.54	4.75	100.91
MR2a15	2.74	36.97	41.61	2.55	6.73	1.77	0.85	1.48	0.76	5.57	101.02
MR2a16	1.39	39.25	44.22	2.38	5.75	1.08	0.00	1.39	0.42	5.26	101.14
MR2a17	1.41	39.21	44.41	1.92	5.11	1.12	0.00	1.37	0.34	5.74	100.63
MR2a18	1.49	39.68	44.34	1.70	4.52	1.18	0.00	1.75	0.31	6.26	101.24
MR2a19	1.40	39.70	43.20	1.73	5.22	1.38	0.00	1.42	0.35	6.38	100.80
MR2a20	1.42	38.14	41.29	3.31	6.79	1.36	0.00	1.49	0.27	5.78	99.84
MR2a21	1.42	39.38	44.15	2.30	5.31	1.11	0.00	1.38	0.41	5.51	100.97
MR2a22	1.49	39.29	45.22	1.47	4.26	1.23	0.00	1.44	0.22	5.84	100.46
MR2a23	1.50	39.63	43.90	1.38	4.63	1.32	0.00	1.48	0.39	6.40	100.63
MR2a24	1.54	37.92	43.50	3.35	7.10	0.82	0.00	1.55	0.82	4.61	101.20
MR4a1	2.06	41.46	33.93	0.94	2.68	3.03	0.57	1.49	0.23	14.19	100.58
MR4a2	3.90	38.45	35.37	1.51	2.98	3.50	2.53	1.61	0.30	10.13	100.27
MR4a3	8.19	32.52	33.93	3.15	2.83	2.69	8.69	1.37	0.36	5.66	99.38
MR4a4	4.32	37.85	37.63	2.52	2.25	2.75	3.79	1.58	0.35	7.79	100.83
MR4a5	5.78	34.99	34.15	3.55	3.08	2.80	5.61	1.50	0.31	7.78	99.56
MR4a6	6.88	34.58	36.72	2.47	2.64	2.68	7.33	1.36	0.35	5.90	100.90
MR4a7	7.94	32.54	33.50	3.92	4.15	2.51	8.20	1.41	0.31	5.97	100.44
MR4a8	4.26	37.60	36.69	3.21	3.26	2.54	2.95	1.62	0.27	8.77	101.17
MR1a1	1.38	37.10	42.11	4.44	9.00	1.18	0.00	1.40	0.56	4.03	101.19
MR1a2	1.36	37.54	42.06	4.85	8.61	1.22	0.00	1.35	0.56	3.82	101.37
MR1a3	1.44	36.34	42.08	4.46	8.24	1.09	0.00	1.45	0.71	3.70	99.51
MR1a4	1.43	37.43	41.91	4.47	8.97	1.10	0.00	1.43	0.52	3.94	101.21
MR1a5	1.45	37.60	39.58	6.03	7.48	1.14	0.00	1.22	0.12	5.98	100.59
MR1a6	1.46	38.23	37.79	5.25	6.84	1.31	0.00	1.24	0.13	8.68	100.92
MR1a7	1.45	36.94	41.04	7.93	7.14	1.09	0.00	1.31	0.73	3.75	101.38
MR1a8	2.03	36.96	41.34	2.77	9.50	1.56	0.00	1.45	0.67	4.13	100.41
MR1a9	1.51	37.75	36.32	7.72	8.20	1.21	0.00	1.10	1.39	4.95	100.16
MR1a10	1.69	36.81	50.36	5.42	2.85	1.24	0.00	1.79	0.03	1.00	101.18
MR1a11	1.85	36.93	49.99	4.97	3.03	1.06	0.00	1.68	0.08	0.90	100.48

Table 2. *Cont.*

Sample	As	S	Ru	Os	Ir	Rh	Pt	Pd	Ni	Fe
					at%					
MR2a1	0.81	65.69	21.85	1.41	2.11	0.59	0.00	0.70	0.25	6.58
MR2a2	0.97	65.76	20.99	2.39	2.82	0.50	0.00	0.63	0.05	5.88
MR2a3	1.20	64.94	23.67	0.74	1.79	0.74	0.00	0.70	0.39	5.82
MR2a4	1.12	64.99	24.82	0.92	1.66	0.71	0.00	0.81	0.77	4.20
MR2a5	1.02	65.21	23.94	1.05	1.74	0.73	0.00	0.76	0.85	4.70
MR2a6	1.29	64.96	23.74	0.82	1.88	0.74	0.00	0.74	0.62	5.21
MR2a7	1.05	65.89	24.06	0.54	1.50	0.63	0.00	0.75	0.45	5.13
MR2a8	1.02	65.97	23.89	0.57	1.26	0.86	0.00	0.80	0.36	5.26
MR2a9	1.10	65.81	23.81	0.54	1.73	0.56	0.03	0.73	0.50	5.19
MR2a10	1.11	65.72	23.02	0.79	1.76	0.61	0.00	0.73	0.28	5.98
MR2a11	1.63	65.43	24.27	0.45	1.53	0.82	0.00	0.84	0.53	4.50
MR2a12	8.48	58.03	18.27	2.03	4.57	1.11	2.41	0.64	0.89	3.57
MR2a13	2.73	64.07	23.56	0.56	1.70	1.10	0.51	0.83	0.60	4.35
MR2a14	2.15	64.45	23.84	0.48	1.68	1.07	0.30	0.81	0.51	4.72
MR2a15	2.03	64.14	22.90	0.75	1.95	0.96	0.24	0.77	0.72	5.55
MR2a16	1.01	66.26	23.68	0.68	1.62	0.57	0.00	0.71	0.39	5.10
MR2a17	1.02	66.10	23.75	0.55	1.44	0.59	0.00	0.69	0.31	5.56
MR2a18	1.06	66.04	23.41	0.48	1.25	0.61	0.00	0.88	0.28	5.98
MR2a19	1.00	66.30	22.89	0.49	1.45	0.72	0.00	0.72	0.32	6.11
MR2a20	1.05	65.91	22.63	0.96	1.96	0.73	0.00	0.78	0.25	5.73
MR2a21	1.03	66.27	23.58	0.65	1.49	0.58	0.00	0.70	0.38	5.32
MR2a22	1.07	66.01	24.10	0.42	1.19	0.64	0.00	0.73	0.20	5.63
MR2a23	1.07	66.11	23.23	0.39	1.29	0.68	0.00	0.74	0.35	6.13
MR2a24	1.14	65.45	23.82	0.97	2.04	0.44	0.00	0.80	0.78	4.56
MR4a1	1.39	65.32	16.96	0.25	0.71	1.49	0.15	0.71	0.20	12.84
MR4a2	2.78	64.02	18.68	0.42	0.83	1.82	0.69	0.81	0.28	9.68
MR4a3	6.50	60.32	19.97	0.98	0.88	1.55	2.65	0.77	0.37	6.02
MR4a4	3.13	64.09	20.21	0.72	0.63	1.45	1.06	0.81	0.32	7.58
MR4a5	4.40	62.15	19.24	1.06	0.91	1.55	1.64	0.80	0.30	7.94
MR4a6	5.25	61.69	20.78	0.74	0.78	1.49	2.15	0.73	0.34	6.04
MR4a7	6.29	60.18	19.66	1.22	1.28	1.45	2.49	0.79	0.31	6.34
MR4a8	3.09	63.62	19.69	0.91	0.92	1.34	0.82	0.82	0.25	8.52
MR1a1	1.04	65.42	23.56	1.32	2.65	0.65	0.00	0.74	0.54	4.08
MR1a2	1.02	65.85	23.41	1.43	2.52	0.66	0.00	0.72	0.54	3.85
MR1a3	1.11	65.22	23.96	1.35	2.47	0.61	0.00	0.78	0.70	3.81
MR1a4	1.08	65.77	23.36	1.32	2.63	0.60	0.00	0.76	0.50	3.98
MR1a5	1.08	65.67	21.93	1.77	2.18	0.62	0.00	0.64	0.11	6.00
MR1a6	1.07	65.12	20.43	1.51	1.94	0.70	0.00	0.63	0.12	8.49
MR1a7	1.10	65.50	23.09	2.37	2.11	0.60	0.00	0.70	0.71	3.81
MR1a8	1.53	65.24	23.15	0.82	2.80	0.86	0.00	0.77	0.65	4.19
MR1a9	1.14	66.34	20.25	2.29	2.41	0.66	0.00	0.58	1.33	5.00
MR1a10	1.28	65.25	28.32	1.62	0.84	0.69	0.00	0.96	0.03	1.02
MR1a11	1.40	65.58	28.16	1.49	0.90	0.59	0.00	0.90	0.08	0.91

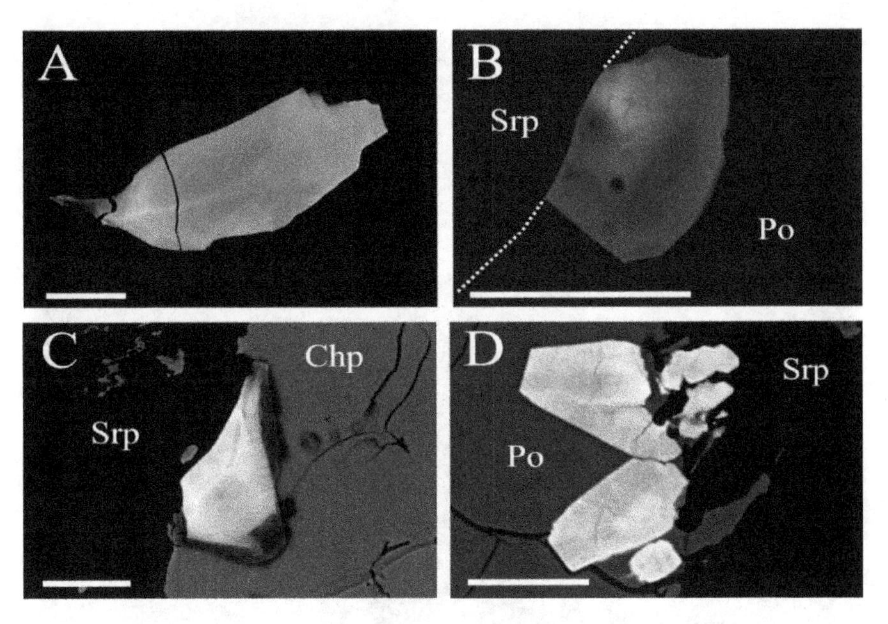

Figure 4. Back-scattered electron (BSE) images of zoned laurite. (**A**) See reflected-light image of Figure 3B for the mineralogical assemblage of the grain. (**B**) Laurite in contact with pyrrhotite and serpentine, (**C**) laurite in contact with chalcopyrite and serpentine and (**D**) laurite grains in contact with pyrrhotite and serpentine Abbreviations as in Figure 3. Scale bar = 20 μm.

As previously reported by [14], abundant Pb-Cl minerals, less than 10 μm in size, were also observed enclosed in the sulfides (Figure 5A), and qualitatively identified by EDS (Figure 5B). The EDS overlap between Pb and S was checked by a WDS semi-quantitative analysis that gave a composition (wt%) of 77.8 for Pb and 18.9 for Cl, very similar to the mineral analyzed by [14].

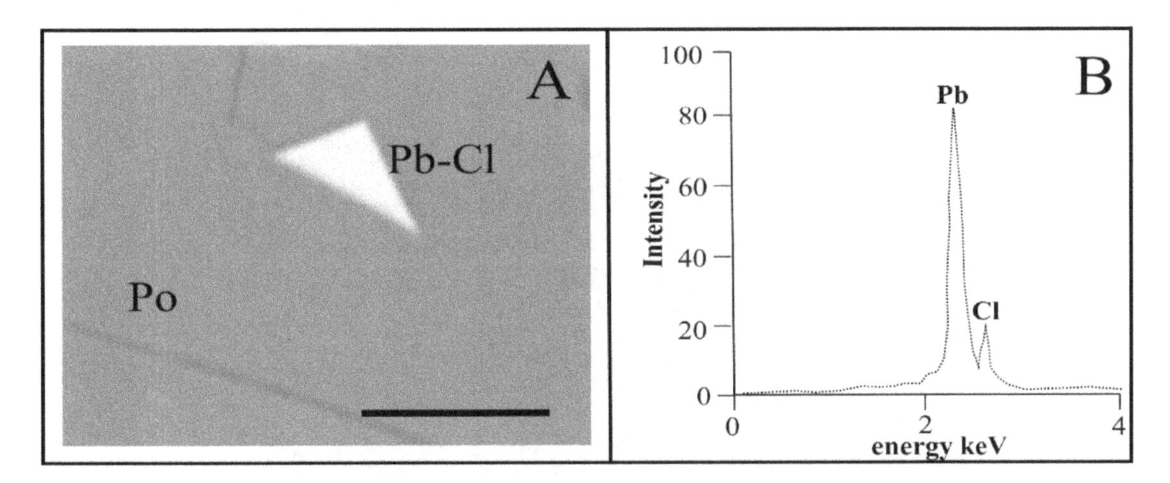

Figure 5. (**A**) BSE image of a Pb-Cl mineral (Pb-Cl) enclosed in pyrrhotite (Po), scale bar = 10 μm, and (**B**) its EDS spectrum (see the text for the Pb-S overlap).

Electron microprobe analyses of the zoned laurite (Table 2) and elemental distribution maps (Figures 6–8) showed unusual enrichments in As, Ir, Os, Pt, and Fe. Distribution of Rh, Pd, and Ni was not visible in the X-ray maps because of the low concentrations, while Cu (not analyzed, but visible in Figure 7) was due to a Cu-phase filling fissures in laurite. Substitution of As for S occurs systematically from a homogeneous background of about 1.00–1.50 wt% (Figure 6) up to patchy enrichment of 3.66–9.7 wt% (Figures 7 and 8). The enrichments of Os and Ir are closely related and may occur either at the rim of grains as described by [14] (Figure 6), or as irregular patches (Figures 7 and 8). The Pt appears to be particularly concentrated, up to 8.69 wt%, in the As-rich zones (Figures 7 and 8).

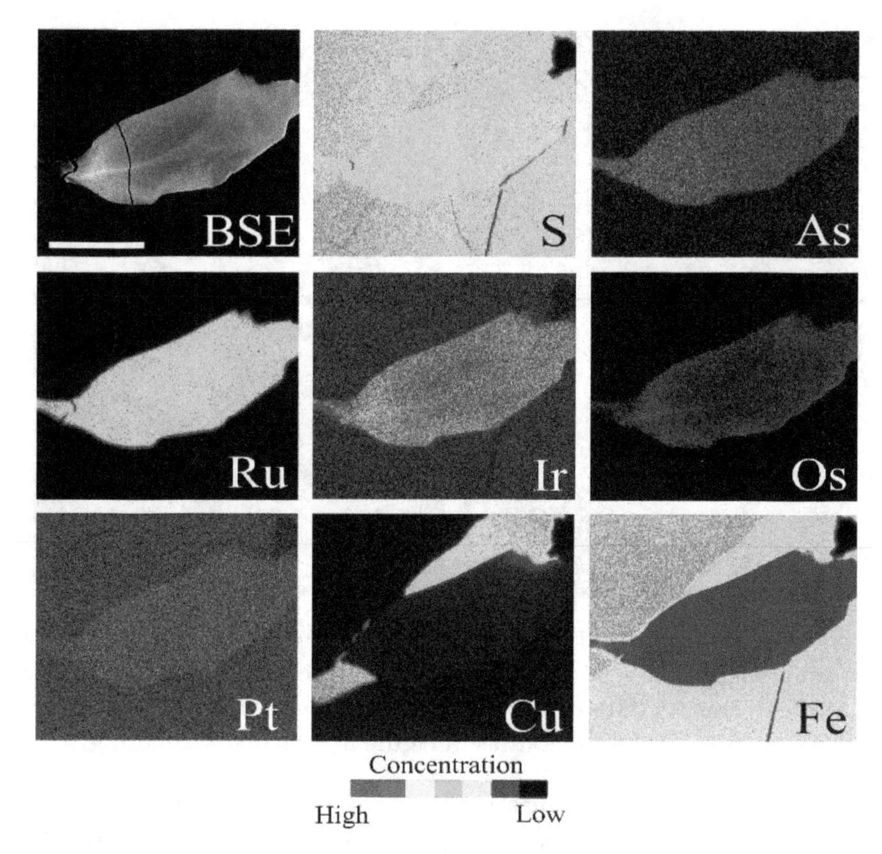

Figure 6. BSE image and X-ray element-distribution maps of S, As, Ru, Ir, Os, Pt, Cu, and Fe in zoned laurite from the Merensky Reef. See Figures 3B and 4A for the paragenetic assemblage. Scale bar = 20 µm.

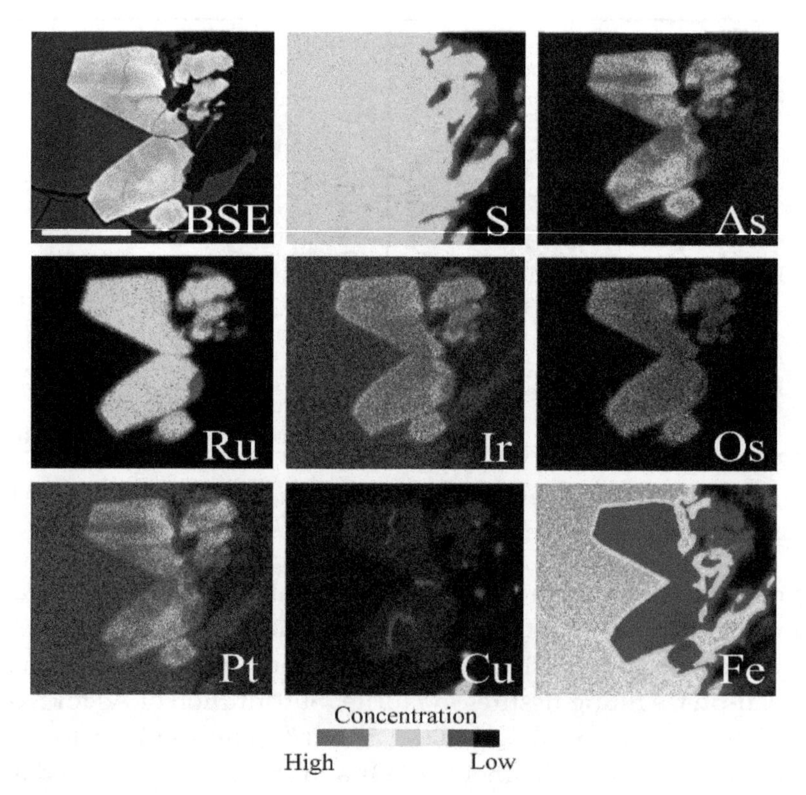

Figure 7. BSE image and X-ray element-distribution maps of S, As, Ru, Ir, Os, Pt, Cu, and Fe, showing the zoning of the laurite from the Merensky Reef, see reflected-light image of Figure 4D for the mineralogical assemblage. Scale bar is 20 µm.

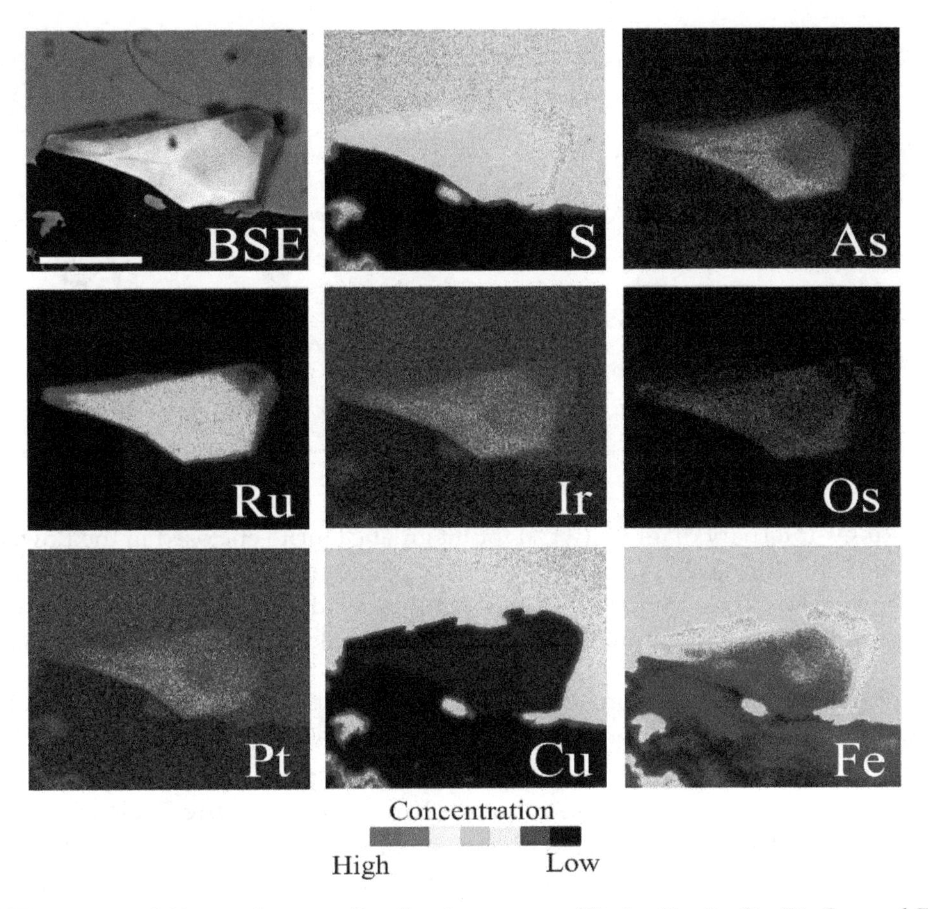

Figure 8. BSE image and X-ray element-distribution maps of S, As, Ru, Ir, Os, Pt, Cu, and Fe, showing the zoning of the laurite from the Merensky Reef, see reflected-light image of Figure 4C for the mineralogical assemblage. Scale bar is 20 μm.

The Ru-Os-Ir (wt%) ternary diagram (Figure 9) shows that the zoned laurite of the Merensky Reef are significantly enriched in Ir, compared with laurite enclosed in the chromitite of the same Reef, and other chromitite layers of the Bushveld. They also do not display the Ru-Os negative correlation inferred by the Ru-Os substitution trend due to the laurite-erlichmanite solid solution trend (Figure 8). Based on more than 1000 published analyses, and unpublished data of the authors, laurite associated with ophiolitic, stratiform, and Alaskan-type magmatic chromitites exhibit a pronounced negative correlation between Ru and Os (R = −0.97). In contrast, the correlation matrix calculated from our electron microprobe analyses (Table 3) indicates that zoned laurite of the Merensky Reef are characterized by the absence of Ru-Os correlation (R = −0.07).

Table 3. Element correlation in the zoned laurite for the Merensky Reef.

at%	As	S	Ru	Os	Ir	Rh	Pt	Pz	Ni	Fe
As	1.00									
S	**−0.98**	1.00								
Ru	−0.56	0.53	1.00							
Os	0.10	−0.14	−0.07	1.00						
Ir	0.04	−0.09	−0.04	0.56	1.00					
Rh	0.72	−0.70	−0.68	−0.25	−0.42	1.00				
Pt	**0.97**	**−0.95**	−0.58	0.06	−0.10	0.75	1.00			
Pd	0.03	−0.05	0.45	−0.37	−0.50	0.20	0.03	1.00		
Ni	0.07	−0.09	−0.02	0.20	0.50	−0.12	−0.01	−0.21	1.00	
Fe	0.11	−0.10	**−0.78**	−0.35	−0.37	0.56	0.18	−0.27	−0.29	1.00

Note: the relevant correlations are highlighted in bold.

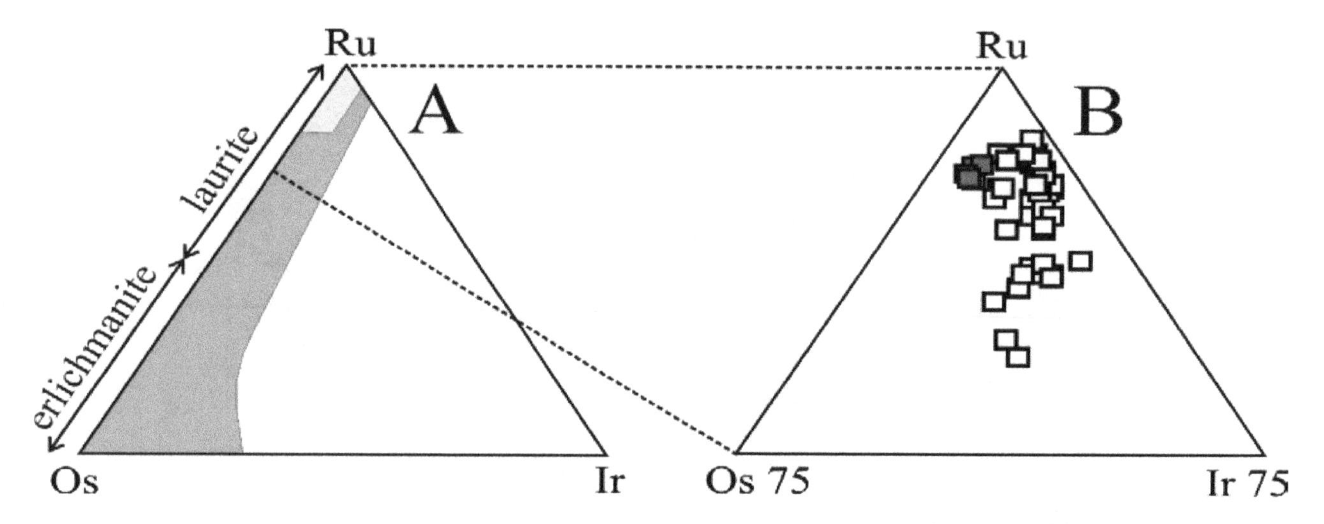

Figure 9. (**A**) Ru-Os-Ir ternary diagram (wt%) for magmatic laurite enclosed in the Lower-, Middle-, and Upper-group chromitite layers: from the Bushveld Complex (yellow field) and podiform chromitite (blue field). Compositional fields after [11,12] and unpublished data of the authors. (**B**) Merensky Reef zoned laurite (open square) and laurite enclosed in the Merensky Reef chromite (red square).

In addition, the high concentrations of As (up to 9.72 wt%), Pt (up to 9.72 wt%), and Fe (up to 14.19%) (Table 2), distinguish the zoned laurite of the Merensky Reef from the laurite inclusions in different types of chromitite (ophiolitic, stratiform, Alaskan-type).

The correlation matrix (Table 3) and distribution X-ray maps (Figure 7) clearly support a positive correlation between Pt-As ($R = +0.97$), and a negative correlation of both elements with S ($R = -0.95$ and -0.98, respectively). The possible existence of submicroscopic inclusions of sperrylite ($PtAs_2$) or platarsite (PtAsS) in laurite was carefully checked, and discarded.

The fact that the sum of S+As atoms is consistently close to stoichiometry ($S + As = 2.00$) supports that Pt and As are parts of the laurite structure. Notably also the high concentrations of Fe are not due to exotic inclusions, but Fe appears to be the major substitute for Ru, showing a negative correlation of $R = -0.78$ (Table 3), that is a clear discrepancy with common magmatic laurite in general.

5. Origin of the Zoned Laurite in the Merensky Reef

Several theories for the origin of the Merensky Reef have been proposed, and they have been recently summarized by [25]. The proposed genetic models include: (i) gravitational settling of crystals that precipitated in the magma chamber during the orthomagmatic stage; (ii) hydrodynamic sorting of a mobilized cumulate slurry in a large magma chamber, under slow cooling; (iii) crystallization at the crystal mush-magma interface caused by a replenishment event; (iv) interaction of a hydrous melt with a partially molten cumulate assemblage; (v) intrusion of magmas as sills into undifferentiated norite; and (vi) intrusion of magma into a pre-existing cumulate pile ([25] and references therein).

To explain the formation of the coarse-grained feldspathic orthopyroxenite enriched in PGE, and associated chromitite in the Merensky Reef, several authors have invoked the reaction between a late-stage hydrous melt with an unconsolidated cumulate assemblage [23–27]. On the basis of natural observations supported by experimental results, formation of tiny Os-Ir-Ru PGM inclusions in chromitite can be modeled by a sequence of crystallization events controlled by sulfur fugacity and temperature (T) [2–4]. The sulfur fugacity is expected to increase with decreasing T in magmatic systems between about 1300–1000 °C, and a consistent order of crystallization can be observed. Alloys in the system Os-Ir-Ru are the first to precipitate, followed by laurite, usually characterized by a core-to-rim increment of Os content, and finally, erlichmanite. Incorporation of IrS_2 molecules in the laurite structure is generally low, controlled by Ir activity in the system. However, the systematic substitution of Os for Ru can be remarkable, and the composition can enter the field of erlichmanite if sulfur fugacity increases sufficiently during magmatic crystallization.

At a first instance, the Os enrichment observed at the rim of some zoned laurite of the Merensky Reef may correspond to such a magmatic crystallization trend. However, other characteristics of the Merensky Reef zoned laurite, such as the unusual zoning that involves local enrichment of As, Pt, Ir, and Fe (Figures 7 and 8), and the absence of Ru-Os substitution, are in apparent contrast with this conclusion. The observed zoning also does not reflect a fluctuation of the sulfur fugacity, suggested to explain the oscillatory zoning of Ru and Os, described in the laurite from the Penikat Layered Complex of Finland [28]. The presence of abundant minerals containing Pb and Cl and occurring enclosed in the sulfides associated with the zoned laurite indicates the presence of Cl in the system [14]. According to [14], the Cl-rich phase precipitated from a late-stage solution or formed as a result of replacement of a precursor galena by an aqueous hydrochloric solution in the final stage of hydrothermal alteration, at low-temperature. Theoretical and experimental work coupled with natural observations suggest that both Cl and As may be important for the transport and mobilization of the PGE during metasomatic and hydrothermal events [23,28–32].

The textural position and the coarse grain size of laurite crystals, as well as their paragenesis including hydrous silicates, suggest crystallization at a late stage from a volatile-rich melt enriched in As and Cl, after coalescence of an immiscible sulfide liquid. The close stabilization of serpentine points to relatively low temperatures for the precipitation of zoned laurite, certainly much lower than those required for the crystallization of tiny laurite included in the chromite seams of the Merensky Reef. Although it is not possible to provide a precise temperature for the genesis of the zoned laurite in the Merensky Reef, we suggest they were in the range of 400–200 °C, similar to temperatures calculated for PGM deposition in the hydrothermal Waterberg platinum deposit of Mookgophong, South Africa [31]. Therefore, we can suggest that the zoned laurite of the Merensky Reef is "hydrothermal" in origin, having crystallized in the presence of a Cl- and As-rich hydrous solution, at temperatures much lower than those typical of the precipitation of magmatic laurite. Although, it remains to be seen whether the "hydrothermal" laurite precipitated directly from the hydrothermal fluid, or it represents the alteration product of a pre-existing laurite reacting with the hydrothermal solution.

Author Contributions: F.Z. and G.G. wrote the manuscript and provided support in the data interpretation. G.G. collected the studied sample, and F.Z. performed the chemical analyses. All authors have read and agreed to the published version of the manuscript.

Acknowledgments: Many thanks are due to the editorial staff of Minerals, to the guest editor Maria Economou-Eliopoulos, and two referees for their useful comments. We are honored to dedicate this manuscript to the memory of our friend and colleague, Demetrios G. Eliopoulos.

References

1. Blackburn, W.H.; Dennen, W.H. *Encyclopedia of Minerals Names, The Canadian Mineralogist Special Publication*; Robert, M., Ed.; Mineralogical Society of Canada: Ottawa, ON, Canada, 1997; Volume 1, p. 360. ISBN 0-921294-45-x.
2. Prichard, H.M.; Barnes, S.J.; Fisher, P.C.; Zientek, M.L. Laurite and associated PGM in the Stillwater chromitites: Implications for processes on formation, and comparisons with laurite in the Bushveld and ophiolitic chromitites. *Can. Mineral.* **2017**, *55*, 121–144. [CrossRef]
3. Zaccarini, F.; Garuti, G.; Pushkarev, E.; Thalhammer, O. Origin of Platinum Group Minerals (PGM) Inclusions in Chromite Deposits of the Ural. *Minerals* **2018**, *8*, 379. [CrossRef]
4. Garuti, G.; Proenza, J.; Zaccarini, F. Distribution and mineralogy of platinum-group elements in altered chromitites of the Campo Formoso layered intrusion (Bahia State, Brazil): Control by magmatic and hydrothermal processes. *Mineral. Pet.* **2007**, *89*, 159–188. [CrossRef]
5. Bowles, J.F.W.; Tkin, D.A.; Lambert, J.L.M.; Deans, T.; Phillips, R. The chemistry, reflectance, and cell size of the erlichmanite (OsS_2)-laurite (RuS_2) series. *Mineral. Mag.* **1983**, *47*, 465–471. [CrossRef]

6. Bowles, J.F.W.; Suárez, S.; Prichard, H.M.; Fisher, P.C. The mineralogy, geochemistry and genesis of the alluvial platinum-group minerals of the Freetown Layered Complex, Sierra Leone. *Mineral. Mag.* **2018**, *82*, 223–246. [CrossRef]
7. Oberthür, T. The Fate of Platinum-Group Minerals in the Exogenic Environment—From Sulfide Ores via Oxidized Ores into Placers: Case Studies Bushveld Complex, South Africa, and Great Dyke, Zimbabwe. *Minerals* **2019**, *9*, 581. [CrossRef]
8. Garuti, G.; Zaccarini, F. In-situ alteration of platinum-group minerals at low temperature: Evidence from chromitites of the Vourinos complex (Greece). *Can. Mineral.* **1997**, *35*, 611–626.
9. Zaccarini, F.; Proenza, J.A.; Ortega-Gutierrez, F.; Garuti, G. Platinum Group Minerals in ophiolitic chromitites from Tehuitzingo (Acatlan Complex, Southern Mexico): Implications for postmagmatic modification. *Mineral. Petrol.* **2005**, *84*, 147–168. [CrossRef]
10. Zaccarini, F.; Bindi, L.; Garuti, G.; Proenza, J. Ruthenium and magnetite intergrowths from the Loma Peguera chromitite, Dominican Republic, and relevance to the debate over the existence of platinum-group element oxides and hydroxides. *Can. Mineral.* **2015**, *52*, 617–624. [CrossRef]
11. Maier, W.D.; Prichard, H.M.; Fisher, P.C.; Barnes, S.J. Compositional variation of laurite at Union Section in the Western Bushveld Complex. *S. Afr. J. Geol.* **1999**, *102*, 286–292.
12. Zaccarini, F.; Garuti, G.; Cawthorn, G. Platinum group minerals in chromitites xenoliths from the ultramafic pipes of Onverwacht and Tweefontein (Bushveld Complex). *Can. Mineral.* **2002**, *40*, 481–497. [CrossRef]
13. Kaufmann, F.E.D.; Hoffmann, M.C.; Bachmann, K.; Veksler, I.V.; Trumbull, R.B.; Hecht, L. Variations in Composition, Texture, and Platinum Group Element Mineralization in the Lower Group and Middle Group Chromitites of the Northwestern Bushveld Complex, South Africa. *Econ. Geol.* **2019**, *14*, 569–590. [CrossRef]
14. Barkov, A.Y.; Martin, R.F.; Kaukonen, R.J.; Alapieti, T.T. The occurrence of Pb–Cl–(OH) and Pt–Sn–S compounds in the Merensky Reef, Bushveld layered complex, South Africa. *Can. Mineral.* **2001**, *39*, 1397–1403. [CrossRef]
15. Prichard, H.M.; Barnes, S.J.; Maier, W.D.; Fisher, P.C. Variations in the nature of the platinum-group minerals in a cross-section through the Merensky Reef at Impala platinum: Implications for the mode of formation of the Reef. *Can. Mineral.* **2004**, *42*, 423–437. [CrossRef]
16. Hutchinson, D.; Foster, J.; Prichard, H.; Gilbertm, S. Concentration of particulate Platinum-Group Minerals during magma emplacement; a case study from the Merensky Reef, Bushveld Complex. *J. Petrol.* **2015**, *56*, 113–159. [CrossRef]
17. Junge, M.; Oberthür, T.; Melcher, F. Cryptic variation of chromite chemistry, platinum group element and platinum group mineral distribution in the UG-2 chromitite: An example from the Karee mine, Western Bushveld complex, South Africa. *Econ. Geol.* **2014**, *109*, 795–810. [CrossRef]
18. Cawthorn, R.G. The Platinum Group Element Deposits of the Bushveld Complex in South Africa. *Plat. Met. Rev.* **2010**, *54*, 205–215. [CrossRef]
19. Chistyakova, S.; Latypov, R.; Youlton, K. Multiple Merensky Reef of the Bushveld Complex, South Africa. *Contrib. Mineral. Petrol.* **2019**, *174*, 26. [CrossRef]
20. Vermaak, C.F.; Von Gruenewaldt, G. Third international platinum symposium, excursion guidebook. *Geol. Soc. S. Afr.* **1981**, *62*, 5.
21. Vymazalová, A.; Zaccarini, F.; Garuti, G.; Laufek, F.; Mauro, D.; Stanley, C.J.; Biagioni, C. Bowlesite, IMA 2019-079. CNMNC Newsletter No. 52. *Mineral. Mag.* **2019**, *83*. [CrossRef]
22. Cawthorn, R.G. The geometry and emplacement of the Pilanesberg Complex, South Africa. *Geol. Mag.* **2015**, *152*, 1–11. [CrossRef]
23. Ballhaus, C.G.; Stumpfl, E.F. Sulfide and platinum mineralization in the Merensky Reef: Evidence from hydrous silicates and fluid inclusions. *Contrib. Mineral. Petrol.* **1986**, *94*, 193–204. [CrossRef]
24. Nicholson, D.M.; Mathez, E.A. Petrogenesis of the Merensky Reef in the Rustenburg section of the Bushveld Complex. *Contrib. Mineral. Petrol.* **1991**, *107*, 293–309. [CrossRef]
25. Hunt, E.J.; Latypov, R.; Horváth, P. The Merensky Cyclic Unit, Bushveld Complex, South Africa: Reality or Myth? *Minerals* **2018**, *8*, 144. [CrossRef]
26. Mathez, E.A. Magmatic metasomatism and formation of the Merensky Reef, Bushveld Complex. *Contrib. Mineral. Petrol.* **1995**, *119*, 277–286. [CrossRef]
27. Boudreau, A.E. Modeling the Merensky Reef, Busvheld Complex, Republic of South Africa. *Contrib. Mineral. Petrol.* **2008**, *156*, 431–437. [CrossRef]

28. Barkov, A.; Fleet, M.E.; Martin, R.F.; Alapieti, T.T. Zoned sulfides and sulfarsenides of the platinum-group elements from the Penikat layered complex, Finland. *Can. Mineral.* **2004**, *42*, 515–537. [CrossRef]
29. Boudreau, A.E. Chlorine as an exploration guide for the platinum-group elements in layered intrusions. *J. Geochem. Explor.* **1993**, *48*, 21–37. [CrossRef]
30. Kislov, E.V.; Konnikov, E.G.; Orsoev, D.; Pushkarev, E.; Voronina, L.K. Chlorine in the Genesis of the Low-Sulfide PGE Mineralization in the Ooko-Dovyrenskii Layered Massif. *Geokhimiya* **1997**, *5*, 521–528.
31. Oberthür, T.; Melcher, F.; Fusswinkel, T.; van den Kerkhof, A.M.; Sosa, G.M. The hydrothermal Waterberg platinum deposit, Mookgophong (Naboomspruit), South Africa. Part 1: Geochemistry and ore mineralogy. *Mineral. Mag.* **2018**, *82*, 725–749. [CrossRef]
32. Le Vaillant, M.; Barnes, S.J.; Fiorentini, M.; Miller, J.; Mccuaig, C.; Mucilli, P. A hydrothermal Ni-As-PGE geochemical halo around the Miitel komatiite-hosted nickel sulfide deposit, Yilgarn craton, Western Australia. *Econ. Geol.* **2015**, *110*, 505–530. [CrossRef]

4

A Comprehensive Review of Rare Earth Elements Recovery from Coal-Related Materials

Wencai Zhang [1,*], Aaron Noble [1], Xinbo Yang [2] and Rick Honaker [2]

1 Department of Mining and Minerals Engineering, Virginia Polytechnic Institute and State University, Blacksburg, VA 24061, USA; aaron.noble@vt.edu

2 Department of Mining Engineering, University of Kentucky, Lexington, KY 40506, USA; xinbo.yang@uky.edu (X.Y.); rick.honaker@uky.edu (R.H.)

* Correspondence: wencaizhang@vt.edu

Abstract: Many studies have been published in recent years focusing on the recovery of rare earth elements (REEs) from coal-related materials, including coal, coal refuse, coal mine drainage, and coal combustion byproducts particularly fly ash. The scientific basis and technology development have been supported by coal geologists and extractive metallurgists, and through these efforts, the concept has progressed from feasibility assessment to pilot-scale production over the last five years. Physical beneficiation, acid leaching, ion-exchange leaching, bio-leaching, thermal treatment, alkali treatment, solvent extraction, and other recovery technologies have been evaluated with varying degrees of success depending on the feedstock properties. In general, physical beneficiation can be a suitable low-cost option for preliminary upgrading; however, most studies showed exceedingly low recovery values unless ultrafine grinding was first performed. This finding is largely attributed to the combination of small RE-bearing mineral particle size and complex REE mineralogy in coal-based resources. Alternatively, direct chemical extraction by acid was able to produce moderate recovery values, and the inclusion of leaching additives, alkaline pretreatment, and/or thermal pretreatment considerably improved the process performance. The studies reviewed in this article revealed two major pilot plants where these processes have been successfully deployed along with suitable solution purification technologies to continuously produce high-grade mixed rare earth products (as high as +95%) from coal-based resources. This article presents a systematic review of the recovery methods, testing outcomes, and separation mechanisms that are involved in REE extraction from coal-related materials. The most recent findings regarding the modes of occurrence of REEs in coal-related materials are also included.

Keywords: rare earth elements; recovery; coal; acid mine drainage; coal combustion byproducts

---

1. Introduction

Rare earth elements (REEs) including the 15 lanthanides plus yttrium and scandium have been identified as critical commodities by several international agencies and national governments due to their crucial roles in clean energy, high tech, and national defense industries [1–3]. REEs scarcely form natural ore depositsthat are economically recoverable. As a result, only a few commercial deposits are currently being extracted worldwide. This combination of restricted supply and critical need has prompted many public and private entities to evaluate alternative REE sources. Coal-related materials, including coal refuse, coal fly ash, and coal mine drainage have been identified as a potentially promising resource. The average REE contents in lignite and bituminous coals as well as lignite and bituminous coal ashes worldwide have been estimated to be around 69, 72, 378, and 469 ppm, respectively [4]. Many coal deposits with elevated REE contents have been found and reported in the literature [5–8]. REE concentration in coal mine drainage varies from site-to-site, ranging from several ppb to ppm levels [9,10].

Recovering REEs from coal-related materials has several advantages over commercial rare earth ores: (1) coal-related materials contain more heavy and critical REEs (HREEs and CREEs) relative to the light REEs (LREEs) [5,10]; (2) the mining costs are negligible since REEs can be produced as byproducts from the coal production and utilization processes [11]; (3) coal-based materials, particularly coal mine drainage, tend to have lower concentrations of radionuclide (e.g., U and Th) when compared to traditional ore deposits [12]; and (4) REE recovery from coal materials has the potential to mitigate or eliminate legacy environmental issues [13]. The concept has been discussed in several review articles [5,14–16]; however, these articles have primarily addressed geologic considerations, such as depositional settings and resource characteristics.

To this end, the U.S. Department of Energy initiated research and development efforts in 2014 to evaluate the technical and economicviability of extracting REEs from coal-related materials with a particular focus on technology maturation and process development [17]. During the last five years, the extraction of REEs from coal-related materials has progressed through feasibility assessment, field sampling and characterization, bench-scale REE separation, and pilot-scale REE production stages. During this period, many innovative findings concerning the recovery of REEs from coal-related materials have been published, which significantly contributed to the scientific knowledge in this area [11,12,18–26]. A review article focused on promising methods for REE recovery from coal and coal byproducts was published by the authors of [27] in 2015. However, experimental results and findings were limited at that time. Considerable developments in this area since 2015 led to a need for another review in order to cover the latest findings.

This review article summarizes technical information regarding process flowsheets, metallurgical performance, and economical metrics of the various methods that have been developed for REE extraction from coal-related materials. This review will provide comparative data to prompt further study and analysis in this area. The review is divided into three sections based on the types of the coal-related materials that were investigated as reported in the literature, i.e., coal and coal refuse, coal combustion ash, and acid mine drainage. Recent findings regarding the modes of occurrence of REEs in coal-related materials and how these findings inform process design are also covered.

2. REEs in Coal and Coal Refuse

The geological aspects such as accumulation mechanisms and modes of occurrence of REEs in coal and coal refuse have been extensively investigated by coal geologists, and these topics have been systematically reviewed in several prior publications [5,14,15,28,29]. Therefore, this article only focuses on the recovery aspect of REEs from coal and coal refuse.

2.1. Physical Beneficiation of REEs from Coal and Coal Refuse

Several studies have been performed to concentrate rare earth (RE)-bearing mineral particles from coal and coal refuse using physical beneficiation techniques such as gravity, magnetic, and flotation separations [11,20,30–34]. Table 1 summarizes some of the beneficiation performances. It is worth noting that decarbonization is normally conducted prior to REE beneficiation, which enables the production of clean coal. A systematic study was performed by Honaker et al. [35] to evaluate the viability of using physical separation methods (riffle table, multi-gravity, and wet high intensity magnetic separations as well as froth flotation) to concentrate rare earth minerals from coal and coal refuse, and representative test results are shown in Table 1. With respect to gravity-based separations, the REEs were only enriched by a factor of 1.1 using a riffle table, and the separation performance was still unsatisfactory when using a multi-gravity separator. This finding has been corroborated by a separate study performed by another group of researchers [24]. Physical separations (size, density, and magnetic) performed on a clean coal sample collected from Kentucky, USA showed that a maximum enrichment ratio of 1.21 was obtained. As such, a high degree of enrichment of rare earth minerals from coal and coal refuse is difficult using physical separation methods.

Table 1. A summary of physical beneficiation of rare earth minerals from coal and coal refuse.

Separation Method	Sample	Sources	Separation Method	REE (ppm)	*ER*	*Re* (%)	Reference
Gravity Separation	Coarse refuse (28 × 100 mesh fraction)	Fire Clay	Riffle table	252 [w]	1.1	16.8	[35]
	Coarse refuse (28 × 100 mesh fraction)	Eagle Seam	Riffle table	213 [w]	1.1	16.1	
	Coarse refuse (28 × 100 mesh fraction)	Fire Clay Rider	Riffle table	234 [w]	1.1	24.75	
	Coarse refuse (<100 mesh fraction)	Eagle Seam	Multi-gravity separation	257 [a]	1.2	90	
	Coarse refuse (<100 mesh fraction)	Fire Clay	Multi-gravity separation	290 [a]	1.2	85	
	Coarse refuse (<100 mesh fraction)	Fire Clay Rider	Multi-gravity separation	254 [a]	1.1	87	
Flotation	Decarbonized thickener underflow	Fire Clay	Multi-stage flotation using a conventional cell with sodium oleate as the collector	2300 [a]	5.3	<5	[33]
	Decarbonized thickener underflow	Fire Clay	Multi-stage flotation using a column with sodium oleate as the collector	4700 [a]	10.9	<5	
	Decarbonized thickener underflow	Fire Clay	Single-stage conventional cell flotation using oleic acid as the collector	386 [w]	1.4	23	[35]
	Decarbonized thickener underflow	Eagle Seam	Single-stage flotation using a conventional cell with oleic acid as the collector	367 [w]	1.8	31	
	Decarbonized thickener underflow	Fire Clay Rider	Single-stage conventional cell flotation using oleic acid as the collector	377 [w]	1.3	13	
HHS	Decarbonized thickener underflow	Fire Clay	Potassium octylhydroxamate and sorbitan monooleate were used as the hydrophobizing agent	17,428 [a]	53	5.9	[11]
Magnetic Separation	Decarbonized middling	Fire Clay	Three-stage wet high intensity magnetic separation (1.4 T)	7000 [w]	14	<5	[20]

Note: [w] and [a] represent dry whole sample basis and dry ash basis, respectively; *ER* and *Re* represent enrichment ratio and recovery, respectively.

Froth flotation is normally utilized to treat fine particles based on their varying degrees of surface hydrophobicity. Rare earth mineral particles present in decarbonized thickener underflows of coal preparation plants can be selectivily recovered by froth flotation. As shown in Table 1, concentrates containing 2300 and 4700 ppm of REEs were obtained by using multiple treatment stages involving conventional flotation cells and column flotation, respectively. A limited concentration was also obtained when using a single-stage of flotation. For example, REEs in the decarbonized thickener underflow derived from treating Eagle seam coal was concentated by nearly two times with the flotation product containing 367 ppm of REEs on a dry whole sample basis (see Table 1). Oleic acid was used as the collector for the flotation test results shown in Table 1, which has been widely used to recover rare earth minerals (e.g., monazite and xenotime) from heavy mineral sands [36,37].

As an alternative to flotation, the hydrophobic–hydrophilic separation (HHS) process is a novel ultrafine particle concentrator that provides improved recovery of micron-size material while also providing a dewatered product [38]. The process uses hydrocarbon oils to agglomerate hydrophobic particles and recover the particles through a phase separation. As shown in Table 1, a concentrate containing 17,428 ppm of REEs was obtained from the Fire Clay decarbonized thickener underflow using HHS. Octylhydroxamate was used as the collector, which is another commonly used reagent for rare earth mineral beneficiation [39,40]. The enrichment ratio of the HHS test reached as high as 53:1. In addition to the aforementioned gravity separation and flotation studies, a concentrate containing 7,000 ppm of REEs was also obtained from the decarbonzied Fire Clay middlings using high-intensity magnetic separation [20].

Overall, rare earth minerals in the decarbonized materials can be concentrated by using flotation, HHS, and magnetic separations. However, the recovery values obtained using these technologies are often too low to be economically viable. One explanation for this consistent finding is that RE-bearing particles in coal and coal refuse are extremely fine and often interlocked within the host-particles. SEM characterizations showed that RE-enriched particles occurring in coal refuse and middlings normally have a particle size of <10 μm [20,33]. In this case, extensive grinding is required to liberate the encapsulated rare earth particles, thereby making the operating process cost prohibitive [23]. However, physical separations without significant size reduction can be employed to generate a higher-grade feedstock for downstream recovery and purification processes.

2.2. Chemical Extraction

Given the low recovery and subsequent high production costs associated with physical beneficiation, direct chemical extraction of REEs from coal refuse has been evaluated by a number of researchers [11,20,21,25,41–43]. Moreover, thermal and alkaline treatments prior to acid leaching have also been used to improve the REE leachability [21,44–47]. These technologies have included acid leaching, salt/ion exchange leaching, and leaching with pretreatment, which are reviewed in the following sections.

2.2.1. Salt and Acid Leaching

In the commerical production of REEs from ion-adsorbed clays, ammonium sulfate is commonly used as a salt lixiviant due to the relatively low hydration energy of ammonium ions. Rozelle et al. [48] collected two high-ash content samples from the overlying strata of the Upper Kittanning bed. The samples were crushed and screened to obtain the 595 μm × 150 μm fraction for salt leaching tests. It was found that around 80% of the total REEs were extracted from the solid, which provided promise as an economically viable option. However, in subsequent investigations performed by other researchers, only a small fraction (e.g., 10%) of the total REEs were determined to be ion-exchangeable in the components of the other coal deposits [41,49]. Given these disparate findings, successful extraction of REEs from coal and coal refuse using salt leaching largely depends on the nature of coal deposit.

REEs can be efficiently extracted using acid solutions of relatively low concentration from some coal sources. Laudal et al. successfully extracted nearly 90% of total REEs from a lignite using 0.5 M H_2SO_4 [25]. The high recovery was explained by the fact that the REEs in the lignite are primarily complexed with organic acids, and the complexation can be destroyed under mild acidic conditions. For bituminous coals, sytematic acid leaching studies have been condcuted on decarbonized middlings, fine refuse, and coarse refuse produced from the treatment of coals originating from different seams [11,20,43,50,51]. It was found that REEs in the mineral matter of coal middlings obtained by grinding and flotation were more leachable than those in coal refuse. For example, nearly 83% of REEs were leached from the decarbonized Fire Clay middlings using a nitric acid solution of pH 0 at 75 °C, whereas less than 30% of REEs were extracted under the same conditions from the decarbonized thickener underflow [11]. However, liberation of the mineral matter from coal middlings consumes a significant amount of energy, which requires consideration for determining economic viability [23].

Improvements in the REE leaching efficiency from coal refuse have been achieved by optimizing the operation parameters such as particle size, temperature, and leaching duration. REE recovery values were relatively low (<30%) for most of the investigated sources [11]. A summary of salt and acid leaching of REEs from coal and coal refuse is shown in Table 2. Overall, direct leaching with salt or acid failed to provide satisfactory recovery from high-rank coal and coal refuse, whereas, direct leaching is a promising choice for recovering REEs from low-rank coal. Therefore, thermal and/or alkaline pretreatment has been utilized in subsequent studies to improve the leaching performance.

Table 2. A summary of salt and acid leaching of rare earth elements (REEs) from coal and coal refuse.

Sample	Coal Seam	Extraction Condition	Leaching Recovery	Reference
Decarbonized thickener underflow	West Kentucky No. 13	0.1 M $(NH_4)_2SO_4$, pH 5	Around 10% of total REEs, 7% of LREEs, and 18% of HREEs	[41]
		0.1 M $(NH_4)_2SO_4$, pH 3	Around 12% of total REEs, 10% of LREEs, and 21% of HREEs	
Roof material, 595 μm × 150 μm	Upper Kittanning	1 M $(NH_4)_2SO_4$, 1/2 solid/solution mass ratio, room temperature	Nearly 90% of the total REEs were extracted after 1 h of reaction	[48]
Lignite	Fort Union	0.5 M H_2SO_4, 40 °C, 48 h	Nearly 90% of the total REEs	[25]
Decarbonized middlings	Fire Clay	Nitric acid solution of pH 0 at 75 °C	83% of total REEs, 86% of LREEs, and 69% of HREEs	[11]
	West Kentucky No. 13	Nitric acid solution of pH 0 at 75 °C	15% of La, 21% of Ce, 31% of Nd, 45% of Y	
	Lower Kittanning	Nitric acid solution of pH 0 at 75 °C	41% of total REEs	
Decarbonized thickener underflow	Fire Clay	Nitric acid solution of pH 0 at 75 °C	31% of La, 26% of Ce, 40% of Nd, 36% of Y	
	West Kentucky No. 13	Nitric acid solution of pH 0 at 75 °C	6% of La, 5% of Ce, 16% of Nd, 34% of Y	
	Lower Kittanning	Nitric acid solution of pH 0 at 75 °C	2% of La, 5% of Ce, 8% of Nd, 25% of Y	

2.2.2. Alkali and Thermal Pretreatment

Alkali and thermal pretreatments of coal and coal refuse have been applied to enhance the acid leaching recoveries of REEs from coal and coal refuse [21,41,44,47,52,53]. A summary of the relevant studies reported in the literature is provided in Table 3. Yang et al. used an 8 M NaOH solution to treat decarbonized fine refuse at 75 °C for two hours prior to acid leaching [41]. It was found that the recovery of REEs was significantly increased from 22% to 75% due to the positive impact on the leachability of the light REEs. Under the same experimental conditions, a small incremental increase in recovery was achieved for the HREEs (38% to 48%). Kuppusamy et al. conducted a study of simultaneous production of clean coal and REEs by alkali-acid leaching of a coal fine refuse material [52]. The ash content of the material was reduced from 46.21% to 14.17% after treatment with a NaOH solution (30 wt.%) at 190 °C for 30 min followed by an HCl solution (7.5 wt.%) at 50 °C for 30 min. Simultaneously, 97% of the LREEs and 76% of the HREEs occurring in the material were extracted.

The enhanced leaching efficiency of REEs resulting from alkali treatment was explained by two mechanisms: (1) Difficult-to-leach rare earth minerals such as monazite were converted to more leachable forms [27,41] and (2) Crystal structures of the dominant minerals such as clays were destroyed resulting in liberation of the encapsulated rare earth minerals [52]. The reactions are as follows:

$$REEPO_{4(s)} + 3NaOH_{(aq)} \rightarrow REE(OH)_{3(s)} + Na_3PO_{4(aq)} \tag{1}$$

$$Al_2O_3 \cdot 2SiO_2 \cdot 2H_2O_{(s)} + 6NaOH_{(aq)} \rightarrow 2Na_2SiO_{3(aq)} + 2NaAlO_{2(aq)} + 5H_2O_{(aq)} \tag{2}$$

A schematic diagram incorporating the use of the alkali-acid leaching process to extract REEs from coal and coal refuse is shown in Figure 1. A negative aspect of the alkali leaching step is the considerable amount of contaminants dissolved into solution along with the REEs, which complicates the downstream purification process. Another negative aspect is the chemical cost, which will likely be prohibitively high given the relatively low content of REEs in coal-based feedstock. Additional studies need to be performed to investigate the possibility of selectively increasing the REE leachability using diluted alkaline solutions and/or weak alkalis.

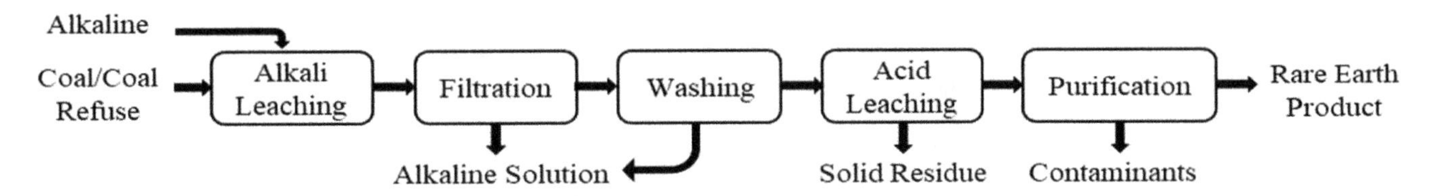

Figure 1. Schematic diagram of the alkaline-acid leaching process to extract REEs from coal and coal refuse [52].

Another scenario to enhance the REE leachability from coal and coal refuse is thermal activation, also known as calcination or roasting [21,41,44,46,47,54]. As reported by Zhang et al., after thermal activation of a coal gangue material at 700 °C for 30 min, 88.6% of the total REEs were leached using 25% HCl at room temperature [46]. The impact of thermal activation on REE leachability has been systematically studied [21,47,54]. As shown in Table 3, for both the clean coal and coal refuse, calcination under 600 °C in static atmosphere without adding any additives significantly improved the REE leaching recovery. For example, the total REE recovery from Pocahontas No. 3 coarse refuse was increased from 14% to 81% by thermal activation using 1.2 M HCl as a lixiviant. In addition, moderate recovery values were also obtained when using mildly acidic conditions. Reducing the acidity by twenty times, i.e., 1.2 M HCl to 0.06 M HCl, resulted in a relatively small drop in REE recovery for thermally activated Pocahontas No. 3 coarse refuse to around 60% [21]. As such, acid consumption is significantly reduced, which is typically the highest cost component of a rare earth extraction process.

Another advantage of thermal activation is that fewer contaminants are dissolved relative to the REEs. During the calcination process, pyrite began to decompose and was converted to iron oxide (primarily hematite) in the temperature range of 400–500 °C. The crystallinity of hematite increased with the elevation in calcination temperature [21,55,56]. When calcined under high temperature, inter-layered structures of clay minerals, especially kaolinite, were destroyed due to dehydration, resulting in disintegration into thin slices and an increase in surface area [21,57,58]. The overall impact of the mineralogical changes was that Al leachability was noticeably increased, while Fe leachability was slightly increased (see Figure 2). However, as shown in Figure 2 the leaching duration can be shortened to less than 15 min, within which the REE leaching reaction is nearly completed while leaching of Al and Fe is minimal. Therefore, a pregnant leaching solution (PLS) containing relatively higher concentrations of REEs can be produced using the thermal activation-mild acid leaching process.

The aforementioned process has been installed and is currently under testing in a rare earth pilot plant funded by the U.S. Department of Energy [59]. A schematic diagram of the process used in the plant is shown in Figure 3. During the testing process, it was found that organic matter associated with the coal refuse helped maintain a constant temperature in the roaster, thus, significantly reducing the energy costs associated with the roaster. In addition to REEs, recent studies [54,60] showed that leaching recovery of other critical metals such as lithium from coal refuse was also positively impacted by calcination. Therefore, given the aforementioned benefits, pre-leach calcination combined with mild acid leaching is one of the most promising approaches for recovering REEs from coal-related materials.

Several mechanisms have been proposed for the positive impacts of thermal activation on REE extraction from coal, i.e.,: (1) Surface area increase resulting from clay dehydration, which liberates some rare earth minerals; (2) Decomposition of the difficult-to-leach rare earth minerals; and (3) Release of a portion of the REEs that were originally associated with the organic matter, which was removed after calcination [21,42,53]. Figure 4 shows the disintegrated kaolinite particles and a REE-enriched particle present in a thermally activated coal refuse sample. Sequential chemical extraction tests have been performed on the calcination products of clean coals of West Kentucky No. 13, Fire Clay, and Illinois No. 6 [47]. It was found that a significant fraction of REEs (50% for the West Kentucky No. 13 material) existed as metal oxides, which originated from the removal of the organic matter present in the clean coals and the decomposition of the difficult-to-leach rare earth minerals.

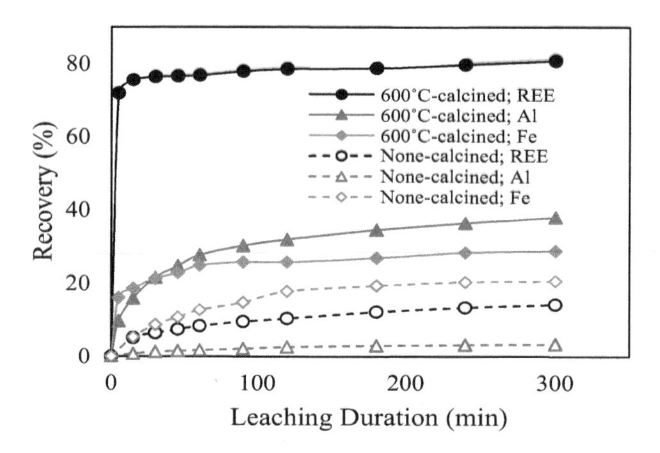

Figure 2. Effects of calcination on the leaching kinetics of REEs, Al, and Fe from Pocahontas No. 3 coarse refuse. (Data were extracted from [21]).

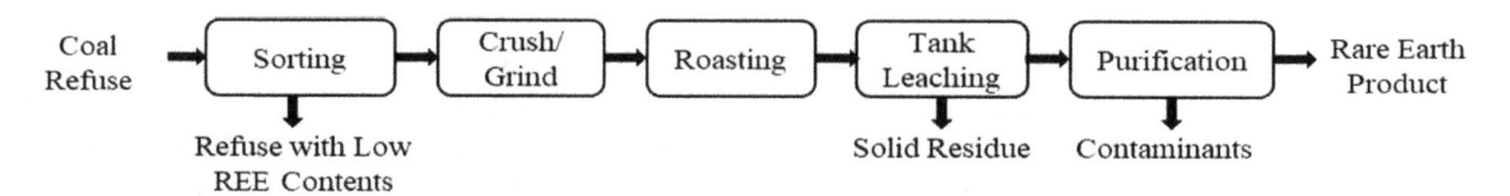

Figure 3. A simplified block diagram of the REE recovery process used in a REE pilot plant located in west Kentucky.

Despite the aforementioned advantages, a few technical and fundamental problems must be resolved to fully validate the approach. As shown in Table 3, the improvements in the HREE recoveries are much smaller than those of the LREEs and elevating the calcination temperature did not enhance the recovery due to the sintering of aluminum silicates [21]. In addition, mineralogical changes of the REEs during calcination is still unclear. Previous studies only listed some possible mechanisms, whereas the direct evidence in supporting the conclusions has not been obtained to date.

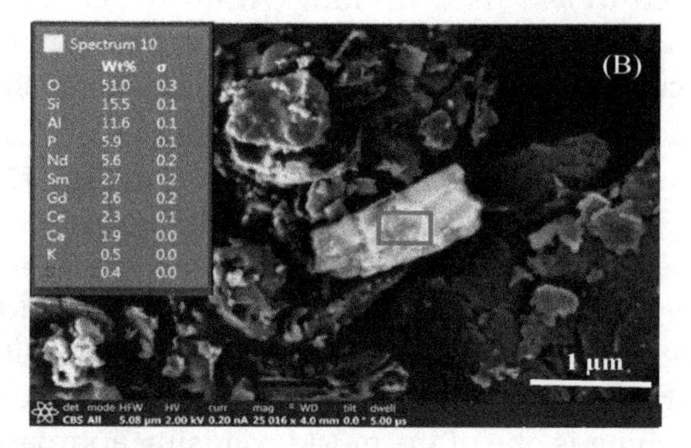

Figure 4. SEM-EDX images of a coal coarse refuse sample after thermal activation: (**A**) Disintegrated kaolinite particles; (**B**) A REE-enriched particle. [21].

Table 3. Summary of the thermal and alkaline pretreatment effects on the leaching recoveries of REEs from coal and coal refuse.

Sample	Coal Seam	Pre-Leach Treatment	Leach Conditions	Recovery			Reference
				TREE	LREE	HREE	
Coarse refuse (2.2 SG float, crushed to below 177 μm)	Pocahontas No. 3	None	1.2 M HCl; 75 °C, 1% (*w/v*) solid concentration, 5 h	14%	12%	23%	[21]
Coarse refuse (2.2 SG float, crushed to below 177 μm)	Pocahontas No. 3	Calcination at 600 °C for 2 h without adding any additives	1.2 M HCl; 75 °C, 1% (*w/v*) solid concentration, 5 h	81%	89%	27%	
Middlings (crushed to below 177 μm)	Pocahontas No. 3	None	1.2 M HCl; 75 °C, 1% (*w/v*) solid concentration, 5 h	28%	31%	19%	
Middlings (crushed to below 177 μm)	Pocahontas No. 3	Calcination at 600 °C for 2 h without adding any additives	1.2 M HCl; 75 °C, 1% (*w/v*) solid concentration, 5 h	76%	80%	57%	
Plant feed (2.2 SG sink, crushed to below 177 μm)	West Kentucky No. 13	None	1.2 M HCl; 75 °C, 1% (*w/v*) solid concentration, 5 h	24%	21%	36%	[54]
Plant feed (2.2 SG sink, crushed to below 177 μm)	West Kentucky No. 13	Calcination at 600 °C for 2 h without adding any additives	1.2 M HCl; 75 °C, 1% (*w/v*) solid concentration, 5 h	79%	87%	41%	
Plant feed (2.2 SG sink, crushed to below 177 μm)	Fire Clay	None	1.2 M HCl; 75 °C, 1% (*w/v*) solid concentration, 5 h	43%	43%	38%	
Plant feed (2.2 SG sink, crushed to below 177 μm)	Fire Clay	Calcination at 600 °C for 2 h without adding any additives	1.2 M HCl; 75 °C, 1% (*w/v*) solid concentration, 5 h	62%	68%	33%	
Plant feed (2.2 SG sink, crushed to below 177 μm)	Illinois No. 6	None	1.2 M HCl; 75 °C, 1% (*w/v*) solid concentration, 5 h	32%	31%	37%	
Plant feed (1.4 SG float, crushed to below 177 μm)	Illinois No. 6	Calcination at 600 °C for 2 h without adding any additives	1.2 M HCl; 75 °C, 1% (*w/v*) solid concentration, 5 h	65%	73%	41%	
Plant feed (1.4 SG float, crushed to below 177 μm)	West Kentucky No. 13	None	1.2 M HCl; 75 °C, 1% (*w/v*) solid concentration, 5 h	25%	30%	15%	[53]
Plant feed (1.4 SG float, crushed to below 177 μm)	West Kentucky No. 13	Calcination at 600 °C for 2 h without adding any additives	1.2 M HCl; 75 °C, 1% (*w/v*) solid concentration, 5 h	86%	88%	82%	
Plant feed (1.4 SG float, crushed to below 177 μm)	Fire Clay	None	1.2 M HCl; 75 °C, 1% (*w/v*) solid concentration, 5 h	41%	47%	20%	
Plant feed (1.4 SG float, crushed to below 177 μm)	Fire Clay	Calcination at 600 °C for 2 h without adding any additives	1.2 M HCl; 75 °C, 1% (*w/v*) solid concentration, 5 h	84%	87%	75%	

Table 3. *Cont.*

Sample	Coal Seam	Pre-Leach Treatment	Leach Conditions	Recovery			Reference
				TREE	LREE	HREE	
Plant feed (1.4 SG float, crushed to below 177 μm)	Illinois No. 6	None	1.2 M HCl; 75 °C, 1% (*w/v*) solid concentration, 5 h	34%	43%	10%	[41] Unpublished data
Plant feed (1.4 SG float, crushed to below 177 μm)	Illinois No. 6	Calcination at 600 °C for 2 h without adding any additives	1.2 M HCl; 75 °C, 1% (*w/v*) solid concentration, 5 h	75%	74%	75%	
Middlings (crushed to below 177 μm)	West Kentucky No. 13	Calcination at 750 °C for 2 h without adding any additives	1.2 M H_2SO_4; 75 °C, 1% (*w/v*) solid concentration, 5 h	41%	81%	40%	
Middlings (crushed to below 177 μm)	West Kentucky No. 13	None	1.2 M H_2SO_4; 75 °C, 1% (*w/v*) solid concentration, 5 h	29%	23%	47%	
Fine refuse	West Kentucky No. 13	Pre-leach using 8 M NaOH solution at a solid/liquid ratio of 1/10 (*w/v*) and 75 °C for 2 h	1.2 M H_2SO_4; 75 °C, 1% (*w/v*) solid concentration, 5 h	75%	82%	48%	
Fine refuse	West Kentucky No. 13	None	1.2 M H_2SO_4; 75 °C, 1% (*w/v*) solid concentration, 5 h	23%	21%	38%	
Flotation Tailings (<500 μm)	East Kootenay	Pre-leach using 30 wt.% NaOH solution at 20% solid concentration and 190 °C for 30 min	7.5 wt.% HCl, 50 °C, 30 min	>85%	97%	76%	[52]
Coal refuse (ground to D_{50} = 3.78 μm)	Junggar coalfield	Calcination at 600 °C for 30 min without adding any additives	25% HCl at 25 °C	88.6%	NA	NA	[46]
Coarse refuse	NA	Calcination at 600 °C for 2 h without adding any additives	6 M HCl, 1/5 solid/liquid ratio, 85–90 °C, 4 h	NA	NA	NA	[44]

3. REE Recovery from Coal Combustion Byproducts

3.1. Modes of Occurrence of REEs in Coal Combustion Ash

Coal combustion fly ash is composed of both amorphous and crystalline phases. The amorphous phases account for 60–90% of bulk fly ash composition, while crystalline material accounts for the remainder [61]. The crystalline phases mainly include quartz, mullite, hematite, magnetite, ferrite spinels, anhydrite, melilite, merwinite, periclase, tricalcium aluminate, and lime [62–64]. Until recently, researchers have found limited success in characterizing the amorphous phases of fly ash due to its disordered nature and heterogeneity [61]. This characteristic along with the low concentration and dispersed nature of REEs have caused difficulties in characterizing the modes of occurrence of REEs in fly ash using traditional approaches. However, many recent studies have addressed this challenge using advanced characterization tools, such as X-ray Absorption Near Edge Structure (XANES), micro-X-Ray Absorption Near Edge Structure (μ-XANES), laser ablation inductively coupled plasma mass spectroscopy (LA-ICP-MS), multimodal image analysis, and sensitive high resolution ion microprobe–reverse geometry SHRIMP-RG [65–70]. In addition, systematic SEM-EDX, TEM-EDX, and sequential chemical extraction (SCE) studies have been performed on coal combustion ashes, which also provided valuable information regarding the REE occurrence modes and potential processing routes [71–78].

Sequential chemical extraction (SCE) tests performed on several class F-type fly ash [79] samples ($SiO_2\% + Al_2O_3\% + Fe_2O_3\% > 70\%$) showed that the majority of REEs were associated with silicates and aluminosilicates (quartz, glass, mullite, zircon, etc.), indicating that REEs are dispersed in the glassy phases and/or associated with the Al–Si–oxide phases [76–78]. Chemical composition analysis of the different size fractions of a class F-type fly ash showed that a strong positive correlation existed between the REEs and the Al plus Si contents [77], which corroborates the above conclusion. In class F fly ashes, REEs associating with carbonates and metal oxides, such as $CaCO_3$ and CaO, accounted for less than 10% of the total REEs, whereas, 50–60% of the total REEs present in class C-type fly ashes produces from Powder River basin coal occurred as carbonates and metal oxides [66,76–78]. Liu et al. combined acid leaching results and solution chemistry modelling findings to predict the percentage of the total REEs that occurred as monazite and hematite in a fly ash sample [66]. It was found that 10–20% of REEs were leached in the pH range of 0–1.5, which corresponds to the range where monazite and hematite dissolve based on solution chemistry modelling. Therefore, 10–20% of the total REEs were reported to exist as monazite and hematite forms. However, many studies have shown that monazite is thermally and chemically stable, and acid cracking or roasting is required to efficiently dissolve monazite [80,81]. Therefore, the solution chemistry modelling findings indicating that monazite and hematite dissolve in solutions having a pH in the range of 0–1.5 is questionable. Furthermore, the conclusion that 10–20% of the total REEs occur as monazite and hematite needs to be re-assessed in further investigations.

The association of REEs with silicates and aluminosilicates has been further proven by SEM-EDX, TEM-EDX, and LA-ICP-MS analyses. Thompson et al. found that, during laser ablation of a REE-enriched fly ash grain, the ion intensities of Al and Si were consistent, whereas the intensities of REEs changed in different ablation periods [70]. This finding indicated that REEs tend to be localized in small grains within fly ash. Using SEM-EDX, some monazite grains within Al–Si cenospheres and aluminosilicate glass particles were found, and particle size of the grains was less than 10 μm [70]. Moreover, several other SEM-EDX studies also showed the dispersion of REE enriched grains within aluminosilicates [66,67,71–73,75,78]. Associations of REEs with the other phases present in fly ash such as iron oxide, zircon, Ca/Fe-rich aluminosilicates, and lime have also been reported [65,66,82]. In addition to associations with the major phases, discrete REE enriched grains such as apatite (Ca, LREE, and P) and monazite (LREE and P) were also found in fly ash [66,70].

Overall, many RE-bearing minerals such as monazite, xenotime, rhabdophane, zircon, ilmenite, lime, and calcite have been found in fly ash, and those minerals have complex association characteristics with the major phases [66,67,70,73,75]. Taggart et al. tested both the bulk and micro speciation of yttrium (Y) in fly ashes using bulk and micro XANES [69]. Bulk XANES analysis indicated that Y coordination states in the fly ashes resembled a combination of Y-oxide, Y-carbonate, and Y-doped glass. However, using micro XANES, some "hotspots" of Y were observed including different mineral forms (e.g., Y-phosphate), which were not observed in the bulk measurements. This result indicated the heterogeneity of REEs in fly ash, and microscale analysis may be unable to represent the REEs in bulk fly ash. Hower et al. reported that the distribution of REEs in a coal ash sample seemed to be in the form of nanoscale crystalline minerals with additional distributions corresponding to overlapping ultra-fine minerals as well as atomic dispersion within the fly ash glass [71]. Therefore, the heterogeneity of REEs in fly ash occurs in nano- and atomic-scales.

3.2. Physical Beneficiation of REEs from Coal Combustion Ash

Coal fly ash particles can be separated into different fractions based on their contrasts in physical characteristics such as density, particle size, magnetism, and surface hydrophobicity [19,24,83–87]. Rather than processing the bulk ash material, REE extraction from certain fractions that are relatively more enriched in REEs will make the overall recovery more economically viable. REEs in fly ash are more concentrated in the finer fractions relative to the entire bulk material [19,24,74,83–86]. Size fractionation analyses of REEs in fly ashes collected from power plants of various countries (China, USA, United Kingdom, Poland) showed that REE contents gradually increased with a decrease in particle size [19,74,83,85]. For example, the minus 500 mesh fraction of a fly ash sample that was collected from the Jungar power plant [74] contained 648 ppm of total REEs, which is more than two times higher than the plus 120 mesh fraction (277 ppm). Several explanations have been suggested for this phenomenon: (1) The finer fractions in fly ash usually contain more glass phase relative to the coarser fractions, while REEs are preferentially associated with the glass phase in fly ash [85]; (2) The organic-bound REEs partially volatize and deposit on the fine particles of fly ash [84,88]; and (3) REEs associated with organic matter may form extremely small particles when the organic matter is combusted and such small particles tend to enrich in the finer fractions of fly ash [84].

Fly ash particles can also be partitioned into different fractions using magnetic separation. Dai et al. analyzed the magnetic, non-magnetic, and glassy fractions of a fly ash and discovered that the magnetic fraction contained less REEs relative to the bulk fly ash (202 ppm versus 261 ppm) [84]. Blissett et al. obtained a magnetic fraction containing 270 ppm of rare earth oxide from a fly ash with 505 ppm of rare earth oxide [83]. Lin et al. separated a fly ash into five fractions by using different magnetic field intensities [24]. It was found that REEs were more enriched in the weak- and non-magnetic fractions. For example, the non-magnetic fraction of a fly ash sample contained more than 600 ppm of REEs, whereas the strongest magnetic fraction only contained around 200 ppm of REEs. Therefore, based on these studies, it can be concluded that REEs are preferentially enriched in the non-magnetic fraction of fly ash.

Fractionation of REEs was also observed in gravity and flotation separations [24,27,83]. In one of our prior studies [19], a float-sink test was performed on a fly ash sample at a density cut point of 1.8 specific gravity (SG). It was found that the 1.8 SG sink fraction contained more REEs than the 1.8 SG float fraction (521 ppm versus 376 ppm). Flotation tests using different collectors and collector dosages also produced a series of products with different REE contents varying from 400 to 650 ppm. Lin et al. performed density fractionation tests on two fly ash samples and it was found that maximum REE contents occurred in the medium density fractions (2.71–2.95 SG and 2.45–2.71 SG) [24].

Given these results, physical separation can be used to pre-concentrate REEs from coal combustion ash, thereby providing a higher-grade feed material to the downstream extraction processes (e.g., acid leaching), resulting in a reduction in the overall recovery cost ($/kg of rare earth oxide). In the patents [89,90], magnetic separation was used to produce a feed material for the subsequent acid leaching process. Two physical beneficiation processes are shown in Figure 5. In both circuits, magnetic separation is performed prior to size fractionation, whereas in another study by Pan et al. [87], the sequence is reversed. Therefore, no fixed strategy has been proposed for the physical beneficiation process. The selection of pre-concentration strategy for a specific ash sample should be based on a comprehensive laboratory evaluation of the REE partitioning characteristics.

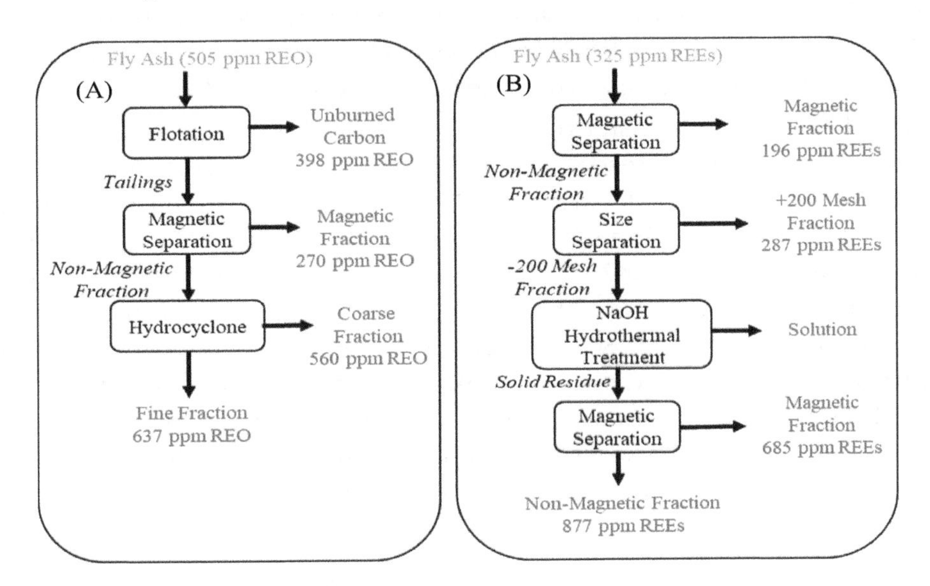

Figure 5. Flowsheet reported in the literature for REE pre-concentration using physical methods: (**A**) reported by [83] and (**B**) reported by [76].

3.3. Chemical Extraction of REEs from Coal Combustion Ash

Many studies have been published in recent years focused on the chemical extraction of REEs from coal combustion ashes [26,53,76,88,89,91–101]. As shown in Table 4, acid leaching has been extensively exploited to extract REEs from coal combustion ash, and often the ash materials must be chemically and/or thermally treated prior to acid leaching to achieve better extraction performance. Chemical and physical characteristics of coal combustion ash change significantly relative to the raw materials after pretreatment [26,96,98]. Therefore, the leaching mechanisms are distinct.

3.3.1. Acid Leaching of REEs from Coal Combustion Ashes

Satisfactory recoveries of REEs from some coal combustion ash produced from conventional boilers has been achieved by using acid leaching alone. For example, Taggart et al. extracted more than 70% of the total REEs from fly ashes of the Powder River basin using 15 M HNO_3 at 85–90 °C [102]. Nearly 100% recovery was obtained from the samples using 12 M HCl at 85 °C, and a considerable amount of REEs (71%) were extracted from a fly ash sample of the same source even under much weaker acidity (1 M HCl). Cao et al. conducted a parametric study to optimize the leaching recovery of REEs from a fly ash sample that was collected from a power plant located in Guizhou, China [91]. It was found that 71.9% of La, 66.0% of Ce, and 61.9% of Nd were leached using 3 M HCl at 60 °C. However, due to the fact that most of the REEs are encapsulated in the amorphous structures of fly ash generated by combusting pulverized coal under high temperature (~1400 °C), relatively low recoveries were usually achieved using acid leaching alone. As shown in Table 4, 35–43% and 40–57% of total REEs were extracted from fly ashes of Illinois and Appalachian basins, respectively, using strong acidity at high temperature (12 M HCl at 85 °C; [96]).

Several studies prepared coal ash samples by combusting coals in muffle furnaces, which were used to simulate fly ash and bottom ash produced from pulverized coal-fired boilers [97,103]. Relatively high leaching recoveries were achieved from artificially prepared coal ashes. However, the temperatures used were much lower than the typical temperatures used in conventional pulverized coal boilers. For example, Kumari et al. burned a coal at 450 °C for 8 h and treated the material as coal bottom ash [97]. Acid leaching optimization showed that 90% of Ce and Nd as well as 35% of Y were extracted using 4 M HCl at 90 °C. It has been realized that mineral matter in coal may oxidize, decompose, fuse, disintegrate, or agglomerate under temperatures as high as 1400 °C, and rapid cooling in the post-combustion zone in boilers results in the formation of spherical, amorphous particles [104,105]. However, when combusting in a muffle furnace at a temperature lower than 900 °C, no glassy phases were detected [21,47,53,57]. Therefore, REEs in the laboratory prepared ash material are more readily leached compared with the REEs in ashes produced in pulverized coal-fired boilers.

Leachabilities of REEs from fly ash produced in fluidized bed combustion (FBC) systems have also been evaluated [53,101]. REEs present in FBC ash are more leachable than ashes produced from pulverized coal-fired boilers. Tuan et al. extracted 62.1% Y, 55.5% Nd, and 65.2% Dy from a FBC bottom ash using 2 M HCl at 80 °C [101]. Honaker et al. achieved around 80% of total REE recovery from an FBC bottom ash using 1.2 M HCl at 75 °C [53]. Relatively low burning temperatures (750–900 °C) are typically utilized in FBC units [106]. Therefore, glassy phases are less likely formed during the combustion and cooling processes in FBC, which contribute to the higher REE leaching efficiency. As shown in Figure 6, the FBC ash samples maintained good crystallization and no noticeable glassy phases were found. Furthermore, similar to the thermal activation of REEs present in coal and coal refuse (Section 3.3.2), combustion in a FBC system may enhance the REE leaching efficiency due to dehydration of the clays, decomposition of the hard-to-leach rare earth minerals, and removal of the organic matters [21,47,53].

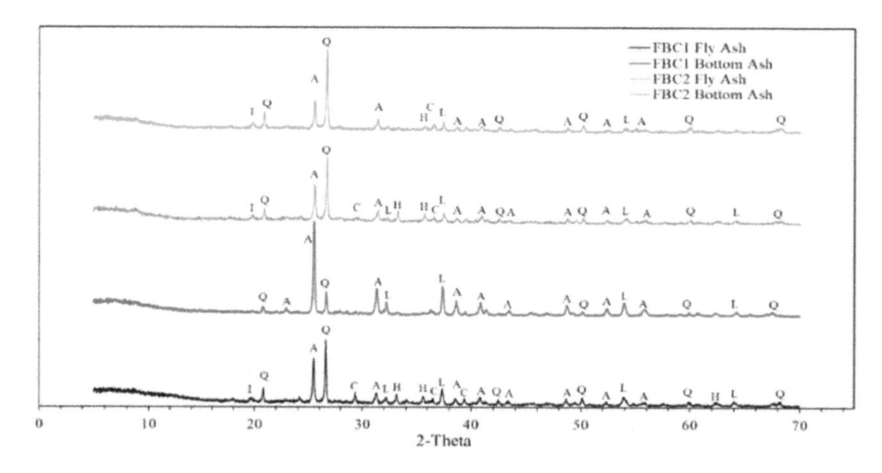

Figure 6. XRD patterns of the fluidized bed combustion (FBC) ash samples: A, anhydrite; C, calcite; H, hematite; I, illite; L, Lime; and Q, Quartz [53].

Parametric studies have been performed to optimize the leaching recovery of REEs from coal combustion ash. Kumari et al. performed leaching tests on a coal ash using three different types of acids, i.e., hydrochloric acid, nitric acid, and sulfuric acid, under the same conditions [97]. It was found that the leaching recoveries follow the order of HCl > HNO_3 > H_2SO_4. However, to the authors' knowledge, sulfuric acid is less volatile and more cost-effective compared with the other two acids. Tuan et al. did not observe a significant difference among the three mineral acids in terms of REE extraction from FBC bottom ashes [101]. Despite the inconsistent findings, hydrochloric acid is the most commonly used lixiviant per literature (see Table 4). Sulfuric acid is expected to provide the worst efficiency since the large amount of calcium present in coal combustion ash will complex with sulfate to form gypsum, and some REEs will be incorporated into the gypsum structure and lost to the precipitate [107]. Leaching recovery of REEs is sensitive to liquid/solid ratio, acid concentration, temperature, and leaching duration [91,97,100]. Leaching recoveries of La, Ce, and Nd from a coal fly ash were nearly doubled when increasing the liquid/solid ratio from 5/1 to 20/1 or prolonging the reaction time from 30 to 180 min [91].

Two flowsheets that were reported in the literature for recovering REEs from coal combustion ash using acid leaching are shown in Figure 7. Solvent extraction was used in both circuits. In the first circuit (Figure 7A), tris-2-ethylhexyl amine (TEHA) was used to complex hydrogen ions in the solvent extraction step, and acid regenerated in the stripping step, which reduced the chemical cost. REEs were recovered from the raffinate of solvent extraction using precipitation. In the second circuit (Figure 7B), fly ash and bed ash produced from FBC combustors are leached using diluted acid (e.g., 1.2 M hydrochloric acid). REEs in the pregnant leach solution are extracted using a three-stage counter-current solvent extraction (SX) system. Finally, a concentrated solution of REEs containing minimal contaminants is produced from the SX-stripping stage. High-purity REE concentrates are produced by selective precipitation using oxalic acid as the precipitant.

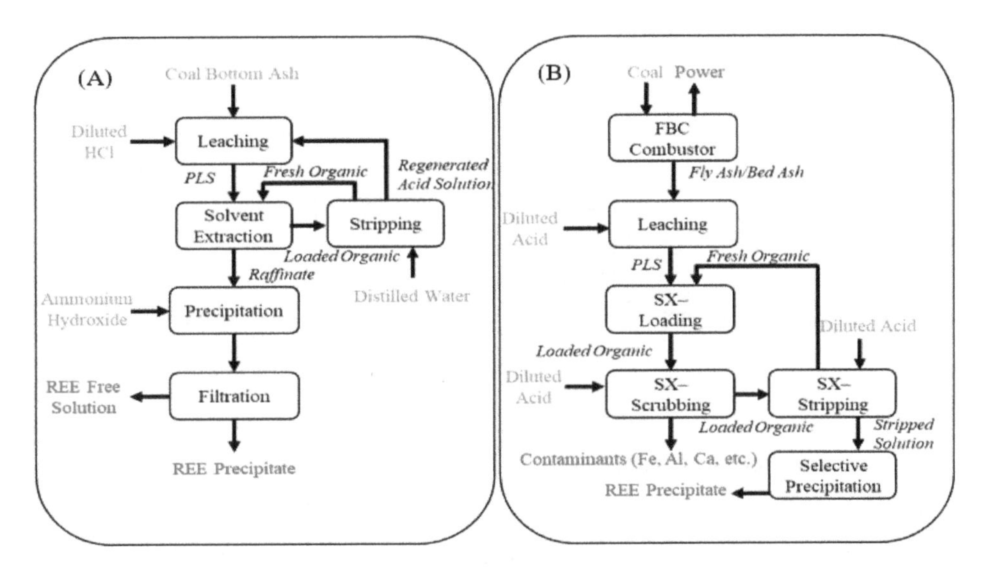

Figure 7. Flowsheets reported in the literature for recovering REEs from coal combustion ash using acid leaching: (**A**) reported by [97] and (**B**) reported by [53].

Overall, REE recovery from coal combustion ashes using direct acid leaching is inefficient. Harsh leaching conditions with concentrated acid solutions at higher temperatures are normally required to achieve satisfactory recovery. In addition to the direct acid leaching, REE extraction from fly ash using bioleaching was also reported [108]. Three microbial strains, *Candida bombicola*, *Phanerochaete chrysosporium*, and *Cryptococcus curvatus*, were tested by Park and Liang in terms of REE extraction from fly ash. *Candida bombicola* provided the optimal results with 63% Sc, 62.2% Y, 67.7% Yb, 64.4% Er, 60% Dy, and 51.9% Gd being extracted at 28 °C for 6 h [108].

3.3.2. Chemical/Thermal Pretreatment

Coal combustion ash was chemically and/or thermally pretreated prior to acid leaching to achieve high REE recoveries [18,26,76,98,100,109]. Lin et al. performed hydrothermal treatment on a coal fly ash and found that 21.3% of the material was dissolved by using 5 M NaOH with a solid/liquid ratio of 1/20 at 100 °C for 120 min [76]. REE content in the solid material was increased from 366 to 803 ppm after hydrothermal treatment, which indicates that REEs present in coal combustion ash remain with the solid residue after the hydrothermal treatment. Wang et al. used an 8 M HCl solution to leach a hydrothermally treated fly ash and achieved a total REE recovery of 88.15% [18]. As shown in Figure 8A, Si, Ga, and Al were also recovered from the fly ash alongside REEs. Ma et al. proposed an alternative NaOH–HCl leaching process, which extracted 55% of REEs, 63% of Si, 72% of Ga, and 78% of Al from the fly ash (see Figure 8B) [98]. Unfortunately, none of the studies produced high-grade final rare earth products.

In the hydrothermal treatment process, NaOH reacts with the major components of fly ash according to the following reactions [110,111]:

$$SiO_2 + 2NaOH \rightarrow Na_2SiO_3 + H_2O \quad (3)$$

$$3Al_2O_3{\cdot}2SiO_2 + 10NaOH \rightarrow 6NaAlO_2 + 2Na_2SiO_3 + 5H_2O \quad (4)$$

$$Al_2O_3 + 2NaOH \rightarrow 2NaAlO_2 + H_2O \quad (5)$$

The above reactions destroy the amorphous glassy structure of fly ash and liberate the RE-bearing particles, which are dissolved in the acid leaching step. In addition, hydrothermal treatment using NaOH is also able to convert hard-to-dissolve rare earth minerals into soluble forms (Equation (1)). All of the above reactions contribute to the extraction of REEs from coal combustion ash.

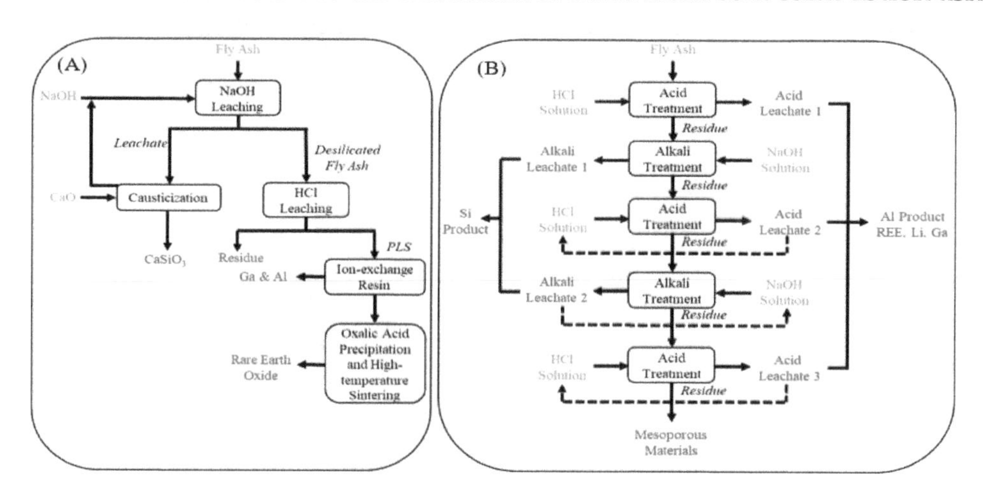

Figure 8. Flowsheets reported in the literature for recovering REEs from coal combustion ash using NaOH hydrothermal treatment followed by acid leaching: (**A**) reported by [18] and (**B**) reported by [98].

Several studies have reported that REE recovery from fly ash was improved by alkali roasting treatment prior to acid leaching [26,100,112]. Taggart et al. compared the performance of various roasting additives, including Na_2O_2, NaOH, CaO, Na_2CO_3, $CaSO_4$, and $(NH_4)_2SO_4$, by performing roasting tests at 450 °C on fly ash samples originating from power stations in the Appalachian, Illinois, and Powder River basins followed by leaching with 1 M HNO_3 [26]. It was found that NaOH roasting often recovered more than 90% of the total REEs, which is more efficient than the other additives. Tang et al. optimized Na_2CO_3 roasting on a coal fly ash collected from Guizhou China [100]. Mass ratio of 1/1 (fly ash/ Na_2CO_3) and roasting temperature of 860 °C provided the maximum total REE recovery (~90%) when leaching with 2 M HCl. Liu et al. proposed a flowsheet to achieve the simultaneous extraction of REEs, Ga, and Nb from a coal fly ash of the Songzao coalfield, which is famous for its significant enrichment in trace elements [84,112,113]. As shown in Figure 9, water leaching was used to extract Ga and Al from the roasting product, and REEs were recovered from the water leaching solid residue by acid leaching. Ion adsorption resin was used to separate Ga and Al. Laboratory test results showed that 68.62% Al, 76.11% Ti, and 80.07% REEs were extracted from the fly ash using the alkali roasting–water leaching–acid leaching method. The reactions between sodium carbonate and fly ash during roasting are as follows:

$$Al_2O_3{\cdot}SiO_2 + 2Na_2CO_3 \rightarrow Na_2SiO_3 + 2NaAlO_2 + 2CO_{2(g)} \quad (6)$$

$$Fe_2O_3 + Na_2CO_3 \rightarrow 2NaFeO_2 + CO_{2(g)} \quad (7)$$

$$TiO_2 + Na_2CO_3 \rightarrow Na_2TiO_3 + CO_{2(g)} \quad (8)$$

$$Ga_2O_3 + Na_2CO_3 \rightarrow 2NaGaO_2 + CO_{2(g)} \quad (9)$$

$$Nb_2O_5 + 3Na_2CO_3 \rightarrow 2Na_3NbO_4 + 3CO_{2(g)} \quad (10)$$

$$REE_2O_3 + Na_2CO_3 \rightarrow 2NaREEO_2 + CO_{2(g)} \quad (11)$$

$$2REEPO_4 + 3Na_2CO_3 \rightarrow REE_2O_3 + 2Na_3PO_4 + 3CO_{2(g)} \quad (12)$$

Table 4. Summary of REE extraction from coal combustion ash using chemical methods.

Sample	Source	Pretreatment	Lixiviant	Leaching Condition	REE Recovery	Reference
Bottom ash	Laboratory prepared	None	4 M HCl	50 g/L pulp density, 90 °C, 120 min	Around 90% for Ce and Nd, and 35% for Y	[97]
Fly ash	Guizhou, China	None	2 M HCl	Liquid/solid ratio 10/1, 120 min	Around 20% for La, 40% for Ce, 5% for Pr, 20% for Nd, and 10% for Y	[100]
Fly ash	Guizhou, China	Na_2CO_3, 1/1 solid/solid ratio, 860 °C	3 M HCl	Liquid/solid ratio 20/1 (*v/w*), 400 rpm stirring speed	72.78% for total REEs	
Fly ash	Upper, Middle, and Lower Kittanning seams, United States	None	1.2 M HCl	1% solid concentration, 75 °C, 5 h	Around 60% for total REEs	[53]
Bottom ash	Illinois No.6 seam, United States	None	1.2 M HCl	1% solid concentration, 75 °C, 5 h	Around 80% for total REEs	
Bottom ash	South Korea	None	2 M HCl	100 g/L pulp density, 80 °C, 12 h	62.1% Y, 55.5% Nd, 65.2% Dy	[101]
Fly ash	Guizhou, China	None	3 M HCl	Liquid/solid ratio 10/1 (*v/w*), 60 °C, 120 min	71.9% La, 66.0% Ce, 61.9% Nd	[91]
Fly ash/Bottom ash	Sichuan, China	None	4% HF	50 g/L, 23–25 °C, 24 h	>90% for total REEs	[18]
Fly ash	Sichuan, China	None	8 M HCl	Liquid/solid ratio 40/1(*v/w*), 80 °C, 6 h	32.36% for total REEs	
Flay ash	Sichuan, China	40% NaOH, 10/1 (*v/w*) solid/liquid ratio, 150 °C, 2 h	8 M HCl	Liquid/solid ratio 30/1 (*v/w*), 60 °C, 2 h	88.15% for total REEs	
Fly ash	Shanxi, China	Acid-alkali based alternate extraction (230 g/L HCl, 200 g/L NaOH, liquid to solid ratio 5/1 (*v/w*), 90 °C)			65% for total REEs	[98]
Fly ash	Powder River Basin, United States	None	12 M HCl	Liquid/solid ratio 100/1, 85 °C, 4 h	Neary 100% for total REEs	[96]
Fly ash	Illinois Basin, United States	None	12 M HCl	Liquid/solid ratio 100/1, 85 °C, 4 h	35–43% recovery for total REEs	
Fly ash	Appalachian Basin, United States	None	12 M HCl	Liquid/solid ratio 100/1, 85 °C, 4 h	40–57% for total REEs	
Fly ash	Appalachian Basin, United States	6.25 M NaOH, liquid/solid mass ratio 10/1, 85 °C, 4 h	20% HCl	NA	48.8–85.9% for total REEs	

Table 4. *Cont.*

Sample	Source	Pretreatment	Lixiviant	Leaching Condition	REE Recovery	Reference
Coal ash	Not Available	None	Super critical CO_2		No experimental tests were performed.	[92]
Fly ash	Ohio, United States	5 M NaOH, solid/liquid ratio 1:20, 100 °C, 2 h	None	None	REE was enriched from 325 to 877 ppm	[76]
Fly ash	Appalachian, Illinois, and Powder River Basins, United States	NaOH roasting (1:1 additive-ash ratio, 450 °C, 30 min)	2 M HNO_3	Room temperature	100% total REE recovery for the Powder River Basin, >70% for the other sources	[26]
Coal ash	Laboratory prepared	None	6 M HNO_3	Liquid/ratio of 33/1 (*v/w*), 60 min	90.5% for total REEs and 90.9% for LREEs	[103]
Fly ash	Powder River Basin, United States	None	15 M HNO_3	10–50 g/L, 85–90 °C, 4 h	69.9% for total REEs	[102]
Coal ash	NA	Physical beneficiation	HNO_3	Approximately 90 °C	NA	[90]
Fly ash	Chongqing, China	Na_2CO_3 roasting (1.5:1 additive-ash ratio, 860 °C, 30 min)	6 M HCl	Liquid/solid ratio 20/1 (*v/w*), 60 °C, 4 h	Around 80% of total REEs were extracted	[112]
Fly ash	Japan	None	9.5% H_2SO_4	Liquid/solid ratio 100/1 (*v/w*), 80 °C, 2 h	Around 10–45% of La was extracted	[88]

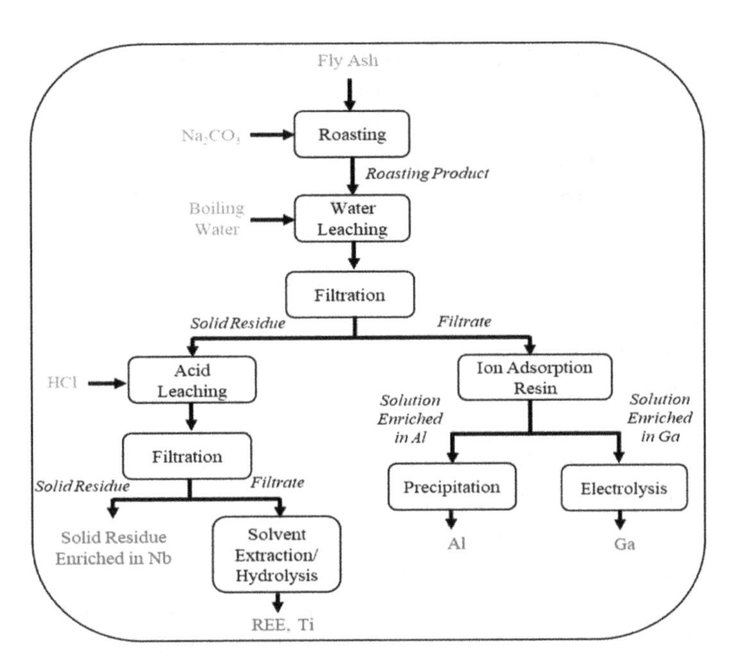

Figure 9. A flowsheet for recovering REEs, Ga, Nb, and Al from a fly ash using alkali roasting [112].

4. REE Recovery from Acid Mine Drainage and Sludge

4.1. REEs in Acid Mine Drainage

The occurrence of REEs in acid coal mine drainage (AMD) is mainly due to the dissolution of RE-bearing species under acidic conditions created by the natural oxidization of sulfide minerals, primarily pyrite. When exposed to the natural environment, pyrite is oxidized as described by the following reactions [114–116]:

$$FeS_{2(s)} + 14Fe^{3+}_{(aq)} + 18H_2O_{(l)} \rightarrow 15Fe^{2+}_{(aq)} + 2SO^{2-}_{4(aq)} + 16H^+_{(aq)} \quad (13)$$

$$FeS_{2(s)} + 7/2O_{2(aq)} + H_2O_{(l)} \rightarrow Fe^{2+}_{(aq)} + 2SO^{2-}_{4(aq)} + 2H^+_{(aq)} \quad (14)$$

$$Fe^{2+}_{(aq)} + 1/4O_{2(aq)} + H^+_{(aq)} \rightarrow Fe^{3+}_{(aq)} + 1/2H_2O_{(l)} \quad (15)$$

As shown in the above equations, both ferric ions and oxygen serve as oxidants for pyrite. Moreover, it has been well realized that Fe^{3+}/S^0-oxidizing microorganisms can significantly accelerate the reaction [116,117].

Many articles have been published focusing on the characterization of REEs in AMD [9,118–123]. Based on the data drawn from 233 samples collected by the United States Geological Survey (USGS) in 1999 and 2011, Ziemkiewicz et al. found that the total REE concentration increases exponentially with a decrease in pH [124]. Other studies also observed the same phenomenon [9,121–123]. For example, Stewart et al. reported that the total REE concentration and solution pH in 18 AMD samples collected from the Appalachian Basin ranged from 0.29 to 1134 μg/L and 2.8–6.6, respectively, with the higher concentration occurring in low pH solutions [9]. Total REE concentration in AMD is normally less than coal and coal refuse, whereas, when reported based on total dissolved solids in AMD, the concentration is similar to or even higher than coal and coal refuse. For example, Honaker et al. obtained the dissolved solid from an AMD sample containing 6.7 mg/L total REEs by completely evaporating the liquid phase [115]. REE content in the dissolved solid was measured to be 380 ppm, which is much higher than the average content of World coals (68 ppm, [4]). Moreover, AMD samples with more than 10 mg/L REEs have been reported in the literature [125]. Therefore, AMD can be used as an alternative resource of REEs. Extremely high concentrations of REEs have been detected in some sludges generated during the passive treatment process [9,22,124]. For example, 3037 mg/kg of total

REEs were estimated from sludge samples from the Saxman Run treatment plant [124]. The same group later conducted a broad survey of 141 treatment sites in the Northern and Central Appalachian coal basins and found that more that 20 of the 623 AMD sludge samples had concentrations exceeding 2000 mg/kg on a dry whole mass basis [22]. Northern Appalachian samples tended to have higher REE concentrations than those from Central Appalachia with statistically significant deviations for all REEs except Ce, Pr, and Nd.

AMD and AMD sludge typically contain more valuable REEs such as Y relative to La and Ce. For example, Y in the coarse refuse collected from the West Kentucky No. 13 and Illinois No. 6 seams represented less than 10% of the total REEs; whereas the REEs in AMD generated from the refuse piles was more than 25% Yttrium. In addition, in the basin comparison study by Vass et al., the ratio of critical REEs (defined as Y, Nd, Eu, Tb, and Dy) to total REEs was nearly 50% for the Northern Appalachian samples [22]. The enrichment of HREEs in AMD, especially the elements located in the middle of lanthanide series (Sm-Dy, middle REEs), has been well recognized by geologists, which may be explained by several mechanisms: (1) the abundance and distribution of mineral phases containing REEs, (2) the stability of RE-bearing mineral phases with respect to the aqueous fluids, (3) the chemistry of the aqueous fluids, and (4) the immobilizing capacity of minerals, precipitates, and colloidal materials to REEs [120,122,123,126,127]. Overall, AMD can be used as a potential resource of REEs due to the relatively high concentration (reported on dissolved solid basis) and the preferential enrichment in HREEs.

4.2. REE Recovery from Acid Mine Drainage

Many studies have been performed to recover valuable components such as Fe, Al, Cu, Zn, Ni, and sulfuric acid from acid mine drainage (AMD). The recovery methods can be classified as precipitation [128–132], adsorption [133,134], diffusion dialysis [135], and ion-exchange [129]. Seo et al. used an oxidation-sequential precipitation method to recover Fe, Al, and Mn from a coal mine drainage [131]. The laboratory test results showed that 99.2–99.3% of Fe, 70.4–82.2% of Al, and 37.8–87.5% of Mn were recovered at pH 4.5, 5.5, and 8.5, respectively. Furthermore, Cu and Zn can be selectively concentrated by collecting the precipitates formed in the pH ranges of 4.49–6.11 and 5.50–7.23, respectively [128]. Instead of artificially adding alkalis, dissolved metals in AMD can also be precipitated and recovered using OH^- produced from electrochemical reactions [136]. Chockalingam and Subramanian found that rice husk is able to uptake 99% Fe^{3+}, 98% Fe^{2+}, 98% Zn^{2+}, and 95% Cu^{2+} from an acid mine water, with a concomitant increase in the pH value by two absolute units [133]. Crane and Sapsford reported that nanoscale zerovalent iron (nZVI) selectively adsorbed Cu, Cd, and Al with more than 99.9% recovery in 1 h [137]. In addition, an acidic pH buffer enabled the formation of copper–bearing nanoparticles from AMD in presence of nZVI. Magnetite nanoparticles have also been successfully prepared from AMD [138–140].

Due to the much higher economic values of REEs relative to the major metals such as iron and aluminum present in AMD, several studies regarding REE recovery from AMD have been reported recently [9,10,12,22,124,141–144]. In one of our prior studies [10], staged precipitation tests were performed on a coal mine drainage (6.14 ppm of REEs), and a REE pre-concentrate containing 1.1% of REEs was produced in the pH range of 4.85–6.11. In addition, the pre-concentrate also contained 17.1% Al, 1.7% Zn, 1.4% Cu, 1.14% Mn, 0.5% Ni, and 0.2% Co, indicating that multiple valuable components can be pre-concentrated simultaneously using staged precipitation. By using selective re-dissolution and oxalic precipitation, a product containing 94% rare earth oxides was finally obtained from the pre-concentrate. A process flowsheet has been developed in a recent study by the authors [145]. The staged precipitation results also explained the observations in the passive treatment systems of coal mine drainage, i.e., >90% REEs were sequestered in the treatment solids when pH was raised above 6.0, and REEs were preferentially retained in the basaluminite ($Al_4(SO_4)(OH)_{10}{\cdot}5H_2O$) [141,142]. Ramasamy et al. (2018) synthesized N- and O- ligand doped mesoporous silica-chitosan hybrid beads for extracting REEs from AMD and the test results showed that more than 90% of REEs were recovered in 5 min [143].

According to Vass et al., REEs in AMD treatment sludge can also be economically recovered by solubilization followed by REE extraction from the solution [22]. Further work by the same group has led to the design and construction of an acid leaching/solvent extraction mini-pilot plant for the recovery of REEs from AMD sludge. Leaching data show that high recoveries of REEs (>80%) can be achieved at a pH value of 1.0 using sulfuric acid. Moreover, the addition of a leaching modifier increased the leaching recovery from 65% to >95% at a pH of 2.0. Together with the downstream solvent extraction operation, the continuous pilot process was able to produce high-grade mixed rare earth oxide products exceeding 80% purity. Techno-economic analysis shows favorable economic outcomes; however, the authors note that the results are very sensitive to consumable costs [146].

Based on the above discussion, a flowsheet for REE recovery and comprehensive utilization of AMD is shown in Figure 10. Cost for the staged precipitation step is minimal given the fact that treatment of AMD is mandated by regulatory agencies. In addition, since only a small quantity of REE-enriched precipitate is obtained from the staged precipitation step, chemical consumptions associated with the downstream processes are low. Therefore, AMD can be considered as a promising source of REEs.

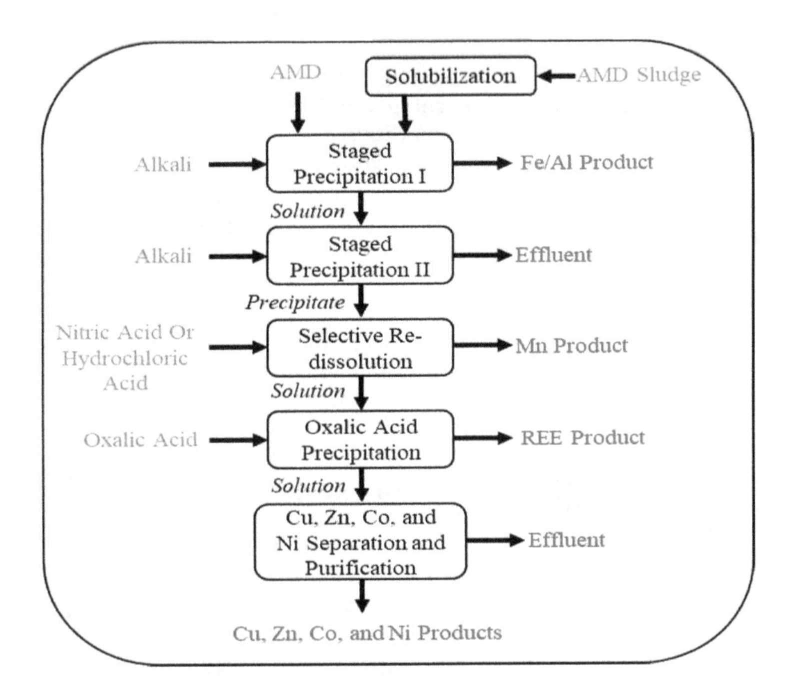

Figure 10. A flowsheet for multiple metals recovery from acid mine drainage (AMD) and acid mine drainage sludge.

5. Summary and Conclusions

Many studies have been conducted for recovering REEs from coal-related materials, primarily coal refuse, coal combustion ash, and acid coal mine drainage. High-purity rare earth concentrates have been successfully produced from coal refuse and acid coal mine drainage. A rare earth pilot plant was constructed and tested, enabling a continuous production of REEs from coal refuse. Reasonable recovery values also have been achieved from coal combustion ash. Altogether, these findings show that significant accomplishments have been made over the past several years in this area.

A summary of the advantages and disadvantages of the methods that have been used for recovering REEs from coal-related materials is shown in Table 5. To achieve optimum recovery performance with minimum cost, process flowsheets integrating various technologies, such as physical beneficiation, acid leaching, pre-leach roasting, and hydrothermal pretreatment, need to be designed and tested. More fundamental studies are required to understand the positive impacts caused by pre-leach calcination on REE recovery from coal refuse. Moreover, this review indicates that not enough attention has been paid to the extraction behavior of other critical elements as well as major elements. This deficiency limits the development of multi-element recovery strategies from coal-related materials. In addition, downstream recovery and purification plans are also ambiguous since the extraction behavior of major elements such as Al, Fe, and Ca were rarely reported.

Table 5. A summary of the advantages and disadvantages of the methods that have been used for REE recovery from coal-related materials.

Material	Method	Advantage	Disadvantage
Coal preparation byproducts	Physical beneficiation	Can be used as a pre-concentration step to generate a higher-grade feed for downstream recovery processes.	Low recovery; Ultrafine grinding is required if a considerable enrichment ratio is expected to be achieved.
	Direct acid leaching	Provide relatively higher recovery compared with physical beneficiation.	Leaching performance depends on the nature of tested samples; Acid consumption is usually high.
	Pre-leach calcination followed by acid leaching	Higher REE recovery; Quick leaching kinetics; Mild leaching conditions; Low chemical consumption.	Recovery of contaminant ions, such as Al^{3+}, is also increased.
Coal combustion ashes	Physical beneficiation	Can be used as a pre-concentration step to generate a higher-grade feed for downstream recovery processes.	Unable to provide a considerable enrichment.
	Direct acid leaching	Able to transfer a portion of REEs from solid into solution, which can be further recovered and purified using other approaches.	Low recovery; High chemical cost; Harsh leaching conditions.
	Alkaline/hydrothermal treatment followed by leaching	High recovery; Quick leaching kinetics.	High alkali consumption; Low selectivity and a lot of contaminants are extracted along with REEs
Acid coal mine drainage	Staged precipitation followed by redissolution and selective precipitation	High recovery; Simple process flowsheet; Low chemical consumption; Can be integrated with existing AMD treatment systems.	AMD containing several ppm of REEs with a large volume may be difficult to find.

Author Contributions: Conceptualization, W.Z.; methodology, A.N. and X.Y.; resources, R.H.; writing—original draft preparation, W.Z. and X.Y.; writing—review and editing, R.H., A.N., and X.Y. All authors have read and agreed to the published version of the manuscript.

References

1. Chu, S. *Critical Materials Strategy*; DIANE Publishing: Darby, PA, USA, 2011.
2. Trump, D. A federal strategy to ensure secure and reliable supplies of critical minerals. *Donald Trump Washington DC Accessed April* **2018**, *3*, 2018.
3. Blengini, G.A.; Nuss, P.; Dewulf, J.; Nita, V.; Talens Peiró, L.; Vidal-Legaz, B.; Latunussa, C.; Mancini, L.; Blagoeva, D.; Pennington, D.; et al. EU methodology for critical raw materials assessment: Policy needs and proposed solutions for incremental improvements. *Resour. Policy* **2017**, *53*, 12–19. [CrossRef]

4. Ketris, M.P.; Yudovich, Y.E. Estimations of clarkes for carbonaceous biolithes: World averages for trace element contents in black shales and coals. *Int. J. Coal Geol.* **2009**, *78*, 135–148. [CrossRef]
5. Seredin, V.V.; Dai, S. Coal deposits as potential alternative sources for lanthanides and yttrium. *Int. J. Coal Geol.* **2012**, *94*, 67–93. [CrossRef]
6. Dai, S.; Xie, P.; Jia, S.; Ward, C.R.; Hower, J.C.; Yan, X.; French, D. Enrichment of U-Re-V-Cr-Se and rare earth elements in the Late Permian coals of the Moxinpo Coalfield, Chongqing, China: Genetic implications from geochemical and mineralogical data. *Ore Geol. Rev.* **2017**, *80*, 1–17. [CrossRef]
7. Seredin, V.V. Rare earth element-bearing coals from the Russian Far East deposits. *Int. J. Coal Geol.* **1996**, *30*, 101–129. [CrossRef]
8. Hower, J.C.; Ruppert, L.F.; Eble, C.F. Lanthanide, yttrium, and zirconium anomalies in the Fire Clay coal bed, Eastern Kentucky. *Int. J. Coal Geol.* **1999**, *39*, 141–153. [CrossRef]
9. Stewart, B.W.; Capo, R.C.; Hedin, B.C.; Hedin, R.S. Rare earth element resources in coal mine drainage and treatment precipitates in the Appalachian Basin, USA. *Int. J. Coal Geol.* **2017**, *169*, 28–39. [CrossRef]
10. Zhang, W.; Honaker, R.Q. Rare earth elements recovery using staged precipitation from a leachate generated from coarse coal refuse. *Int. J. Coal Geol.* **2018**, *195*, 189–199. [CrossRef]
11. Honaker, R.Q.; Groppo, J.; Yoon, R.-H.; Luttrell, G.H.; Noble, A.; Herbst, J. Process evaluation and flowsheet development for the recovery of rare earth elements from coal and associated byproducts. *Miner. Metall. Process.* **2017**, *34*, 107–115. [CrossRef]
12. Vass, C.R.; Noble, A.; Ziemkiewicz, P.F. The occurrence and concentration of rare earth elements in acid mine drainage and treatment by-products: Part 1—Initial survey of the Northern Appalachian Coal Basin. *Min. Metall. Explor.* **2019**, 917–929. [CrossRef]
13. Sarswat, P.K.; Leake, M.; Allen, L.; Free, M.L.; Hu, X.; Kim, D.; Noble, A.; Luttrell, G.H. Efficient recovery of rare earth elements from coal based resources: A bioleaching approach. *Mater. Today Chem.* **2020**, *16*. [CrossRef]
14. Seredin, V.V.; Dai, S.; Sun, Y.; Chekryzhov, I.Y. Coal deposits as promising sources of rare metals for alternative power and energy-efficient technologies. *Appl. Geochem.* **2013**, *31*, 1–11. [CrossRef]
15. Dai, S.; Yan, X.; Ward, C.R.; Hower, J.C.; Zhao, L.; Wang, X.; Zhao, L.; Ren, D.; Finkelman, R.B. Valuable elements in Chinese coals: A review. *Int. Geol. Rev.* **2018**, *60*, 590–620. [CrossRef]
16. Dai, S.; Xie, P.; Ward, C.R.; Yan, X.; Guo, W.; French, D.; Graham, I.T. Anomalies of rare metals in Lopingian super-high-organic-sulfur coals from the Yishan Coalfield, Guangxi, China. *Ore Geol. Rev.* **2017**, *88*, 235–250. [CrossRef]
17. Alvin, M.A. The rarity of rare earth elements (PowerPoint slides). Presentated at National Assocation Regulatorgy Utility Commissioners (NARUC) Winter Policy Summit, Washington, DC, USA, 12 February 2018.
18. Wang, Z.; Dai, S.; Zou, J.; French, D.; Graham, I.T. Rare earth elements and yttrium in coal ash from the Luzhou power plant in Sichuan, Southwest China: Concentration, characterization and optimized extraction. *Int. J. Coal Geol.* **2019**, *203*, 1–14. [CrossRef]
19. Zhang, W.; Groppo, J.; Honaker, R. Ash beneficiation for REE recovery. In Proceedings of the 2015 World Coal Ash Conference, Nashville, TN, USA, 5–7 May 2015.
20. Zhang, W.; Yang, X.; Honaker, R.Q. Association characteristic study and preliminary recovery investigation of rare earth elements from Fire Clay seam coal middlings. *Fuel* **2018**, *215*, 551–560. [CrossRef]
21. Zhang, W.; Honaker, R. Calcination pretreatment effects on acid leaching characteristics of rare earth elements from middlings and coarse refuse material associated with a bituminous coal source. *Fuel* **2019**, *249*, 130–145. [CrossRef]
22. Vass, C.R.; Noble, A.; Ziemkiewicz, P.F. The occurrence and concentration of rare earth elements in acid mine drainage and treatment byproducts. Part 2: Regional survey of Northern and Central Appalachian Coal Basins. *Min. Metall. Explor.* **2019**, *36*, 917–929. [CrossRef]
23. Huang, Q.; Noble, A.; Herbst, J.; Honaker, R. Liberation and release of rare earth minerals from Middle Kittanning, Fire Clay, and West Kentucky No. 13 coal sources. *Powder Technol.* **2018**, *332*, 242–252. [CrossRef]
24. Lin, R.; Howard, B.H.; Roth, E.A.; Bank, T.L.; Granite, E.J.; Soong, Y. Enrichment of rare earth elements from coal and coal by-products by physical separations. *Fuel* **2017**, *200*, 506–520. [CrossRef]
25. Laudal, D.A.; Benson, S.A.; Addleman, R.S.; Palo, D. Leaching behavior of rare earth elements in Fort Union lignite coals of North America. *Int. J. Coal Geol.* **2018**, *191*, 112–124. [CrossRef]

26. Taggart, R.K.; Hower, J.C.; Hsu-Kim, H. Effects of roasting additives and leaching parameters on the extraction of rare earth elements from coal fly ash. *Int. J. Coal Geol.* **2018**, *196*, 106–114. [CrossRef]
27. Zhang, W.; Rezaee, M.; Bhagavatula, A.; Li, Y.; Groppo, J.; Honaker, R. A review of the occurrence and promising recovery methods of rare earth elements from coal and coal by-products. *Int. J. Coal Prep. Util.* **2015**, *35*, 281–294. [CrossRef]
28. Dai, S.; Ren, D.; Chou, C.L.; Finkelman, R.B.; Seredin, V.V.; Zhou, Y. Geochemistry of trace elements in Chinese coals: A review of abundances, genetic types, impacts on human health, and industrial utilization. *Int. J. Coal Geol.* **2012**, *94*, 3–21. [CrossRef]
29. Dai, S.; Finkelman, R.B. Coal as a promising source of critical elements: Progress and future prospects. *Int. J. Coal Geol.* **2018**, *186*, 155–164. [CrossRef]
30. Gupta, T.; Ghosh, T.; Akdogan, G.; Bandopadhyay, S. Maximizing REE enrichment by froth flotation using Box-Behnken design in Alaskan coal. In Proceedings of the 2017 SME Annual Conference & Expo, Denver, CO, USA, 19–22 February 2017; pp. 408–412.
31. Gupta, T.; Ghosh, T.; Akdogan, G.; Srivastava, V.K. Characterizing rare earth elements in Alaskan coal and ash. *Miner. Metall. Process.* **2017**, *34*, 138–145. [CrossRef]
32. Gao, W.; Zhang, X.; Zheng, X.; Lin, X.; Cao, H.; Zhang, Y.; Sun, Z. Lithium carbonate recovery from cathode scrap of spent lithium-ion battery: A closed-loop process. *Environ. Sci. Technol.* **2017**, *51*, 1662–1669. [CrossRef]
33. Zhang, W.; Honaker, R.; Groppo, J. Concentration of rare earth minerals from coal by froth flotation. *Miner. Metall. Process.* **2017**, *34*, 132–137. [CrossRef]
34. Honaker, R.; Groppo, J.; Bhagavatula, A.; Rezaee, M.; Zhang, W. Recovery of rare earth minerals and elements from coal and coal byproducts. In Proceedings of the International Coal Preparation Conference, Lousiville, KY, USA, 25–27 April 2016; pp. 25–27.
35. Honaker, R.; Hower, J.; Eble, C.; Weisenfluh, J.; Groppo, J.; Rezaee, M.; Bhagavatula, A.; Luttrell, G.H.; Bratton, R.C.; Kiser, M.; et al. Laboratory and bench-scale testing for rare earth elements. *Cell* **2014**, *724*, 554–3652.
36. Cheng, T.-W.; Holtham, P.N.; Tran, T. Froth flotation of monazite and xenotime. *Miner. Eng.* **1993**, *6*, 341–351. [CrossRef]
37. Pavez, O.; Peres, A.E.C. Effect of sodium metasilicate and sodium sulphide on the floatability of monazite-zircon-rutile with oleate and hydroxamates. *Miner. Eng.* **1993**, *6*, 69–78. [CrossRef]
38. Gupta, N.; Li, B.; Luttrell, G.; Yoon, R.H.; Bratton, R.; Reyher, J. Hydrophobic-hydrophilic separation (HHS) process for the recovery and dewatering of ultrafine coal. In Proceedings of the 2016 SME Annual Conference and Expo, Phoenix, AZ, USA, 21–24 February 2016; pp. 706–709.
39. Pradip, P.; Fuerstenau, D.W. Design and development of novel flotation reagents for the beneficiation of Mountain Pass rare-earth ore. *Miner. Metall. Process.* **2013**, *30*, 1–9. [CrossRef]
40. Pradip; Fuerstenau, D.W. The adsorption of hydroxamate on semi-soluble minerals. Part I: Adsorption on barite, Calcite and Bastnaesite. *Colloids Surf.* **1983**, *8*, 103–119. [CrossRef]
41. Yang, X.; Werner, J.; Honaker, R.Q. Leaching of rare earth elements from an Illinois basin coal source. *J. Rare Earths* **2019**, *37*, 312–321. [CrossRef]
42. Wei, G.; Bo, F.; Jinxiu, P.; Wenpu, Z.; Xianwen, Z. Depressant behavior of tragacanth gum and its role in the flotation separation of chalcopyrite from talc. *J. Mater. Res. Technol.* **2019**, *8*, 697–702. [CrossRef]
43. Yang, X. Leaching Characteristics of Rare Earth Elements from Bituminous Coal-Based Sources. Ph.D. Thesis, University of Kentucky, Lexington, KY, USA, 2019.
44. Bo, C.; Ya, L.A.I.; Guo, X.I.A.O.; Chang, X.U. Technique for extraction and concentration of rare earth elements in gangue. *Glob. Geol.* **2010**, *28*, 257–260.
45. Kuppusamy, V.K.; Holuszko, M. Rare earth elements in flotation products of coals from East Kootenay coalfields, British Columbia. *J. Rare Earths* 2019. [CrossRef]
46. Zhang, P.; Han, Z.; Jia, J.; Wei, C.; Liu, Q.; Wang, X.; Zhou, J.; Li, F.; Miao, S. Occurrence and Distribution of Gallium, Scandium, and Rare Earth Elements in Coal Gangue Collected from Junggar Basin, China. *Int. J. Coal Prep. Util.* **2017**, *39*, 389–402. [CrossRef]
47. Zhang, W.; Honaker, R. Enhanced leachability of rare earth elements from calcined products of bituminous coals. *Miner. Eng.* **2019**, *142*, 105935. [CrossRef]

48. Rozelle, P.L.; Khadilkar, A.B.; Pulati, N.; Soundarrajan, N.; Klima, M.S.; Mosser, M.M.; Miller, C.E.; Pisupati, S.V. A study on removal of rare earth elements from U.S. coal byproducts by ion exchange. *Metall. Mater. Trans. E* **2016**, *3*, 6–17. [CrossRef]
49. Finkelman, R.B.; Palmer, C.A.; Wang, P. Quantification of the modes of occurrence of 42 elements in coal. *Int. J. Coal Geol.* **2018**, *185*, 138–160. [CrossRef]
50. Honaker, R.; Yang, X.; Chandra, A.; Zhang, W.; Werner, J. *Hydrometallurgical Extraction of Rare Earth Elements from Coal*; Springer International Publishing: Berlin, Germany, 2018. ISBN 978-3-319-95021-1.
51. Zhang, W.; Noble, A. Mineralogy characterization and recovery of rare earth elements from the roof and floor materials of the Guxu coalfield. *Fuel* **2020**, *270*, 117533. [CrossRef]
52. Kuppusamy, V.K.; Kumar, A.; Holuszko, M. Simultaneous extraction of clean coal and rare earth elements from coal tailings using alkali-acid leaching process. *J. Energy Resour. Technol. Trans. ASME* **2019**, *141*, 1–7. [CrossRef]
53. Honaker, R.Q.; Zhang, W.; Werner, J. Acid leaching of rare earth elements from coal and coal ash: Implications for using fluidized bed combustion to assist in the recovery of critical materials. *Energy Fuels* **2019**, *33*, 5971–5980. [CrossRef]
54. Zhang, W.; Honaker, R. Characterization and recovery of rare earth elements and other critical metals (Co, Cr, Li, Mn, Sr, and V) from the calcination products of a coal refuse sample. *Fuel* **2020**, *267*, 117236. [CrossRef]
55. Hu, G.; Dam-Johansen, K.; Wedel, S.; Hansen, J.P. Decomposition and oxidation of pyrite. *Prog. Energy Combust. Sci.* **2006**, *32*, 295–314. [CrossRef]
56. Music, S.; Popović, S.; Ristić, M. Thermal decomposition of pyrite. *J. Radioanal. Nucl. Chem. Artic.* **1992**, *162*, 217–226. [CrossRef]
57. Cao, Z.; Cao, Y.; Dong, H.; Zhang, J.; Sun, C. Effect of calcination condition on the microstructure and pozzolanic activity of calcined coal gangue. *Int. J. Miner. Process.* **2016**, *146*, 23–28. [CrossRef]
58. De la Villa, R.V.; García, R.; Martínez-Ramírez, S.; Frías, M. Effects of calcination temperature and the addition of ZnO on coal waste activation: A mineralogical and morphological evolution. *Appl. Clay Sci.* **2017**, *150*, 1–9. [CrossRef]
59. Honaker, R.Q.; Zhang, W.; Werner, J.; Noble, A.; Luttrell, G.H.; Yoon, R.-H. Enhancement of a process flowsheet for recovering and concentrating critical materials from bituminous coal sources. *Min. Metall. Explor.* **2019**, accepted. [CrossRef]
60. Zhang, W.; Noble, A.; Yang, X.; Honaker, R. Lithium leaching recovery and mechanisms from density fractions of an Illinois Basin bituminous coal. *Fuel* **2020**, *268*, 117319. [CrossRef]
61. Chancey, R.T.; Stutzman, P.; Juenger, M.C.G.; Fowler, D.W. Comprehensive phase characterization of crystalline and amorphous phases of a Class F fly ash. *Cem. Concr. Res.* **2010**, *40*, 146–156. [CrossRef]
62. Goodarzi, F. Characteristics and composition of fly ash from Canadian coal-fired power plants. *Fuel* **2006**, *85*, 1418–1427. [CrossRef]
63. Kukier, U.; Ishak, C.F.; Sumner, M.E.; Miller, W.P. Composition and element solubility of magnetic and non-magnetic fly ash fractions. *Environ. Pollut.* **2003**, *123*, 255–266. [CrossRef]
64. McCarthy, G.J.; Solem, J.K.; Manz, O.E.; Hassett, D.J. Use of a database of chemical, mineralogical and physical properties of North American fly ash to study the nature of fly ashand its utilization as a mineral admixture in concrete. *MRS Online Proc. Libr. Arch.* **1989**, *178*, 3. [CrossRef]
65. Kolker, A.; Scott, C.; Hower, J.C.; Vazquez, J.A.; Lopano, C.L.; Dai, S. Distribution of rare earth elements in coal combustion fly ash, determined by SHRIMP-RG ion microprobe. *Int. J. Coal Geol.* **2017**, *184*, 1–10. [CrossRef]
66. Liu, P.; Huang, R.; Tang, Y. Comprehensive understandings of rare earth element (REE) speciation in coal fly ashes and implication for REE extractability. *Environ. Sci. Technol.* **2019**, *53*, 5369–5377. [CrossRef]
67. Montross, S.N.; Verba, C.A.; Chan, H.L.; Lopano, C. Advanced characterization of rare earth element minerals in coal utilization byproducts using multimodal image analysis. *Int. J. Coal Geol.* **2018**, *195*, 362–372. [CrossRef]
68. Stuckman, M.Y.; Lopano, C.L.; Granite, E.J. Distribution and speciation of rare earth elements in coal combustion by-products via synchrotron microscopy and spectroscopy. *Int. J. Coal Geol.* **2018**, *195*, 125–138. [CrossRef]

69. Taggart, R.K.; Rivera, N.A.; Levard, C.; Ambrosi, J.P.; Borschneck, D.; Hower, J.C.; Hsu-Kim, H. Differences in bulk and microscale yttrium speciation in coal combustion fly ash. *Environ. Sci. Process. Impacts* **2018**, *20*, 1390–1403. [CrossRef] [PubMed]
70. Thompson, R.L.; Bank, T.; Montross, S.; Roth, E.; Howard, B.; Verba, C.; Granite, E. Analysis of rare earth elements in coal fly ash using laser ablation inductively coupled plasma mass spectrometry and scanning electron microscopy. *Spectrochim. Acta-Part B At. Spectrosc.* **2018**, *143*, 1–11. [CrossRef]
71. Hower, J.C.; Cantando, E.; Eble, C.F.; Copley, G.C. Characterization of stoker ash from the combustion of high-lanthanide coal at a Kentucky bourbon distillery. *Int. J. Coal Geol.* **2019**, *213*, 103260. [CrossRef]
72. Hower, J.C.; Qian, D.; Briot, N.J.; Santillan-Jimenez, E.; Hood, M.M.; Taggart, R.K.; Hsu-Kim, H. Nano-scale rare earth distribution in fly ash derived from the combustion of the fire clay coal, Kentucky. *Minerals* **2019**, *9*, 206. [CrossRef]
73. Hower, J.C.; Qian, D.; Briot, N.J.; Henke, K.R.; Hood, M.M.; Taggart, R.K.; Hsu-Kim, H. Rare earth element associations in the Kentucky State University stoker ash. *Int. J. Coal Geol.* **2018**, *189*, 75–82. [CrossRef]
74. Hower, J.; Groppo, J.; Henke, K.; Hood, M.; Eble, C.; Honaker, R.; Zhang, W.; Qian, D. Notes on the Potential for the Concentration of Rare Earth Elements and Yttrium in Coal Combustion Fly Ash. *Minerals* **2015**, *5*, 356–366. [CrossRef]
75. Hower, J.C.; Dai, S.; Seredin, V.V.; Zhao, L.; Kostova, I.J.; Silva, L.F.O.; Mardon, S.M.; Gurdal, G. A note on the occurrence of yttrium and rare earth elements in coal combustion byproducts. *Coal Combust. Gasif. Prod.* **2013**, 39–47. [CrossRef]
76. Lin, R.; Stuckman, M.; Howard, B.H.; Bank, T.L.; Roth, E.A.; Macala, M.K.; Lopano, C.; Soong, Y.; Granite, E.J. Application of sequential extraction and hydrothermal treatment for characterization and enrichment of rare earth elements from coal fly ash. *Fuel* **2018**, *232*, 124–133. [CrossRef]
77. Pan, J.; Zhou, C.; Tang, M.; Cao, S.; Liu, C.; Zhang, N.; Wen, M.; Luo, Y.; Hu, T.; Ji, W. Study on the modes of occurrence of rare earth elements in coal fly ash by statistics and a sequential chemical extraction procedure. *Fuel* **2019**, *237*, 555–565. [CrossRef]
78. Pan, J.; Zhou, C.; Liu, C.; Tang, M.; Cao, S.; Hu, T.; Ji, W.; Luo, Y.; Wen, M.; Zhang, N. Modes of occurrence of rare earth elements in coal fly ash: A case study. *Energy Fuels* **2018**, *32*, 9738–9743. [CrossRef]
79. ASTM standard specification for coal fly ash and raw or calcined natural pozzolan for use. *Annu. B. ASTM Stand.* **2010**, 3–6. [CrossRef]
80. Kumari, A.; Panda, R.; Jha, M.K.; Kumar, J.R.; Lee, J.Y. Process development to recover rare earth metals from monazite mineral: A review. *Miner. Eng.* **2015**, *79*, 102–115. [CrossRef]
81. Cetiner, Z.S.; Wood, S.A.; Gammons, C.H. The aqueous geochemistry of the rare earth elements. Part XIV. The solubility of rare earth element phosphates from 23 to 150 °C. *Chem. Geol.* **2005**, *217*, 147–169. [CrossRef]
82. Hower, J.; Groppo, J.; Joshi, P.; Dai, S.; Moecher, D.; Johnston, M. Location of cerium in coal-combustion fly ashes: Implications for recovery of lanthanides. *Coal Combust. Gasif. Prod.* **2003**, *5*, 73–78. [CrossRef]
83. Blissett, R.S.; Smalley, N.; Rowson, N.A. An investigation into six coal fly ashes from the United Kingdom and Poland to evaluate rare earth element content. *Fuel* **2014**, *119*, 236–239. [CrossRef]
84. Dai, S.; Zhao, L.; Peng, S.; Chou, C.L.; Wang, X.; Zhang, Y.; Li, D.; Sun, Y. Abundances and distribution of minerals and elements in high-alumina coal fly ash from the Jungar Power Plant, Inner Mongolia, China. *Int. J. Coal Geol.* **2010**, *81*, 320–332. [CrossRef]
85. Dai, S.; Zhao, L.; Hower, J.C.; Johnston, M.N.; Song, W.; Wang, P.; Zhang, S. Petrology, mineralogy, and chemistry of size-fractioned fly ash from the Jungar power plant, Inner Mongolia, China, with emphasis on the distribution of rare earth elements. *Energy Fuels* **2014**, *28*, 1502–1514. [CrossRef]
86. Lanzerstorfer, C. Pre-processing of coal combustion fly ash by classification for enrichment of rare earth elements. *Energy Rep.* **2018**, *4*, 660–663. [CrossRef]
87. Pan, J.; Nie, T.; Vaziri Hassas, B.; Rezaee, M.; Wen, Z.; Zhou, C. Recovery of rare earth elements from coal fly ash by integrated physical separation and acid leaching. *Chemosphere* **2020**, *248*, 126112. [CrossRef]
88. Kashiwakura, S.; Kumagai, Y.; Kubo, H.; Wagatsuma, K. Dissolution of rare earth elements from coal fly ash particles in a dilute H2SO4 solvent. *Open J. Phys. Chem.* **2013**, *03*, 69–75. [CrossRef]
89. Joshi, P.B.; Preda, D.V.; Skyler, D.A.; Tsinberg, A.; Green, B.D.; Marinelli, W.J. Recovery of Rare Earth Elements and Compounds from Coal Ash. U.S. Patent 8,968,688, 3 March 2015.
90. Joshi, P.B.; Preda, D.V.; Skyler, D.A.; Scherer, A.; Green, B.D.; Marinelli, W.J. Recovery of rare earth Elements and Compounds from Coal Ash. U.S. Patent 9,394,586, 19 July 2016.

91. Cao, S.; Zhou, C.; Pan, J.; Liu, C.; Tang, M.; Ji, W.; Hu, T.; Zhang, N. Study on influence factors of leaching of rare earth elements from coal fly ash. *Energy Fuels* **2018**, *32*, 8000–8005. [CrossRef]
92. Das, S.; Gaustad, G.; Sekar, A.; Williams, E. Techno-economic analysis of supercritical extraction of rare earth elements from coal ash. *J. Clean. Prod.* **2018**, *189*, 539–551. [CrossRef]
93. Fan, M.; Co, E.M.; Zhao, Y.; Long, Z.; Liu, W. Preprint 17–103. Recovery of valuable elements from Chinese coal by-products. In Proceedings of the Preprint 17–103, 1–6. SME Annual Meeting, Dever, CO, USA, 19–22 February 2017.
94. Huang, Z.; Fan, M.; Tian, H. Rare earth elements of fly ash from Wyoming's Powder River Basin coal. *J. Rare Earths* **2019**. [CrossRef]
95. Huang, C.; Wang, Y.; Huang, B.; Dong, Y.; Sun, X. The recovery of rare earth elements from coal combustion products by ionic liquids. *Miner. Eng.* **2019**, *130*, 142–147. [CrossRef]
96. King, J.F.; Taggart, R.K.; Smith, R.C.; Hower, J.C.; Hsu-Kim, H. Aqueous acid and alkaline extraction of rare earth elements from coal combustion ash. *Int. J. Coal Geol.* **2018**, *195*, 75–83. [CrossRef]
97. Kumari, A.; Parween, R.; Chakravarty, S.; Parmar, K.; Pathak, D.D.; Lee, J.; Jha, M.K. Novel approach to recover rare earth metals (REMs) from Indian coal bottom ash. *Hydrometallurgy* **2019**, *187*, 1–7. [CrossRef]
98. Ma, Z.; Zhang, S.; Zhang, H.; Cheng, F. Novel extraction of valuable metals from circulating fluidized bed-derived high-alumina fly ash by acid–alkali–based alternate method. *J. Clean. Prod.* **2019**, *230*, 302–313. [CrossRef]
99. Shimizu, R.; Sawada, K.; Enokida, Y.; Yamamoto, I. Supercritical fluid extraction of rare earth elements from luminescent material in waste fluorescent lamps. *J. Supercrit. Fluids* **2005**, *33*, 235–241. [CrossRef]
100. Tang, M.; Zhou, C.; Pan, J.; Zhang, N.; Liu, C.; Cao, S.; Hu, T.; Ji, W. Study on extraction of rare earth elements from coal fly ash through alkali fusion–Acid leaching. *Miner. Eng.* **2019**, *136*, 36–42. [CrossRef]
101. Tuan, L.Q.; Thenepalli, T.; Chilakala, R.; Vu, H.H.T.; Ahn, J.W.; Kim, J. Leaching characteristics of low concentration rare earth elements in Korean (Samcheok) CFBC bottom ash samples. *Sustainability* **2019**, *11*, 2562. [CrossRef]
102. Taggart, R.K.; Hower, J.C.; Dwyer, G.S.; Hsu-Kim, H. Trends in the rare earth element content of U.S.-based coal combustion fly ashes. *Environ. Sci. Technol.* **2016**, *50*, 5919–5926. [CrossRef] [PubMed]
103. Peiravi, M.; Ackah, L.; Guru, R.; Mohanty, M.; Liu, J.; Xu, B.; Zhu, X.; Chen, L. Chemical extraction of rare earth elements from coal ash. *Miner. Metall. Process.* **2017**, *34*, 170–177. [CrossRef]
104. Kutchko, B.G.; Kim, A.G. Fly ash characterization by SEM-EDS. *Fuel* **2006**, *85*, 2537–2544. [CrossRef]
105. Ward, C.R.; French, D. Determination of glass content and estimation of glass composition in fly ash using quantitative X-ray diffractometry. *Fuel* **2006**, *85*, 2268–2277. [CrossRef]
106. Mastral, A.M.; Callén, M.S.; García, T. Fluidized bed combustion (FBC) of fossil and nonfossil fuels. A comparative study. *Energy Fuels* **2000**, *14*, 275–281. [CrossRef]
107. Dutrizac, J.E. The behaviour of the rare earth elements during gypsum (CaSO4·2H2O) precipitation. *Hydrometallurgy* **2017**, *174*, 38–46. [CrossRef]
108. Park, S.; Liang, Y. Bioleaching of trace elements and rare earth elements from coal fly ash. *Int. J. Coal Sci. Technol.* **2019**, *6*, 74–83. [CrossRef]
109. Matyas, B.; Gerber, P.; Solymos, A.; Kaszanitzky, F.; Panto, G.; Leffler, J. Process for Recovering Rare Metals from the Combustion Residue of Coal by Digestion. U.S. Patent 4,649,031, 10 March 1987.
110. Ding, J.; Ma, S.; Shen, S.; Xie, Z.; Zheng, S.; Zhang, Y. Research and industrialization progress of recovering alumina from fly ash: A concise review. *Waste Manag.* **2017**, *60*, 375–387. [CrossRef]
111. Yao, Z.T.; Xia, M.S.; Sarker, P.K.; Chen, T. A review of the alumina recovery from coal fly ash, with a focus in China. *Fuel* **2014**, *120*, 74–85. [CrossRef]
112. Liu, H.; Tian, H.; Zou, J. Combined extraction of rare metals Ga-Nb-REE from fly ash. *Sci. Technol. Rev.* **2015**, *33*, 39–43. (In Chinese with English Abstract)
113. Dai, S.; Wang, X.; Zhou, Y.; Hower, J.C.; Li, D.; Chen, W.; Zhu, X.; Zou, J. Chemical and mineralogical compositions of silicic, mafic, and alkali tonsteins in the late Permian coals from the Songzao Coalfield, Chongqing, Southwest China. *Chem. Geol.* **2011**, *282*, 29–44. [CrossRef]
114. Fernando, W.A.M.; Ilankoon, I.M.S.K.; Syed, T.H.; Yellishetty, M. Challenges and opportunities in the removal of sulphate ions in contaminated mine water: A review. *Miner. Eng.* **2018**, *117*, 74–90. [CrossRef]
115. Honaker, R.Q.; Zhang, W.; Yang, X.; Rezaee, M. Conception of an integrated flowsheet for rare earth elements recovery from coal coarse refuse. *Miner. Eng.* **2018**, *122*, 233–240. [CrossRef]

116. Mousavi, S.M.; Jafari, A.; Yaghmaei, S.; Vossoughi, M.; Roostaazad, R. Bioleaching of low-grade sphalerite using a column reactor. *Hydrometallurgy* **2006**, *82*, 75–82. [CrossRef]
117. Casas, J.M.; Martinez, J.; Moreno, L.; Vargas, T. Bioleaching model of a copper-sulfide ore bed in heap and dump configurations. *Metall. Mater. Trans. B Process Metall. Mater. Process. Sci.* **1998**, *29*, 899–909. [CrossRef]
118. Merten, D.; Geletneky, J.; Bergmann, H.; Haferburg, G.; Kothe, E.; Büchel, G. Rare earth element patterns: A tool for understanding processes in remediation of acid mine drainage. *Chemie der Erde* **2005**, *65*, 97–114. [CrossRef]
119. Pérez-López, R.; Delgado, J.; Nieto, J.M.; Márquez-García, B. Rare earth element geochemistry of sulphide weathering in the São Domingos mine area (Iberian Pyrite Belt): A proxy for fluid-rock interaction and ancient mining pollution. *Chem. Geol.* **2010**, *276*, 29–40. [CrossRef]
120. Prudêncio, M.I.; Valente, T.; Marques, R.; Sequeira Braga, M.A.; Pamplona, J. Geochemistry of rare earth elements in a passive treatment system built for acid mine drainage remediation. *Chemosphere* **2015**, *138*, 691–700. [CrossRef]
121. Sahoo, P.K.; Tripathy, S.; Equeenuddin, S.M.; Panigrahi, M.K. Geochemical characteristics of coal mine discharge vis-à-vis behavior of rare earth elements at Jaintia Hills coalfield, northeastern India. *J. Geochem. Explor.* **2012**, *112*, 235–243. [CrossRef]
122. Sun, H.; Zhao, F.; Zhang, M.; Li, J. Behavior of rare earth elements in acid coal mine drainage in Shanxi Province, China. *Environ. Earth Sci.* **2012**, *67*, 205–213. [CrossRef]
123. Zhao, F.; Cong, Z.; Sun, H.; Ren, D. The geochemistry of rare earth elements (REE) in acid mine drainage from the Sitai coal mine, Shanxi Province, North China. *Int. J. Coal Geol.* **2007**, *70*, 184–192. [CrossRef]
124. Ziemkiewicz, P.; He, T.; Noble, A.; Liu, X. *Recovery of Rare Earth Elements (REEs) from Coal Mine Drainage*; West Virginia Mine Drainage Task Force Symposium: Morgantown, WV, USA, 2016.
125. Cravotta, C.A. Dissolved metals and associated constituents in abandoned coal-mine discharges, Pennsylvania, USA. Part 1: Constituent quantities and correlations. *Appl. Geochem.* **2008**, *23*, 166–202. [CrossRef]
126. Grawunder, A.; Merten, D.; Büchel, G. Origin of middle rare earth element enrichment in acid mine drainage-impacted areas. *Environ. Sci. Pollut. Res.* **2014**, *21*, 6812–6823. [CrossRef] [PubMed]
127. Ferreira da Silva, E.; Bobos, I.; Xavier Matos, J.; Patinha, C.; Reis, A.P.; Cardoso Fonseca, E. Mineralogy and geochemistry of trace metals and REE in volcanic massive sulfide host rocks, stream sediments, stream waters and acid mine drainage from the Lousal mine area (Iberian Pyrite Belt, Portugal). *Appl. Geochem.* **2009**, *24*, 383–401. [CrossRef]
128. Balintova, M.; Petrilakova, A. Study of pH influence on selective precipitation of heavy metals from acid mine drainage. *Chem. Eng. Trans.* **2011**, *25*, 345–350. [CrossRef]
129. Feng, D.; Aldrich, C.; Tan, H. Treatment of acid mine water by use of heavy metal precipitation and ion exchange. *Miner. Eng.* **2000**, *13*, 623–642. [CrossRef]
130. Park, S.M.; Yoo, J.C.; Ji, S.W.; Yang, J.S.; Baek, K. Selective recovery of Cu, Zn, and Ni from acid mine drainage. *Environ. Geochem. Health* **2013**, *35*, 735–743. [CrossRef]
131. Seo, E.Y.; Cheong, Y.W.; Yim, G.J.; Min, K.W.; Geroni, J.N. Recovery of Fe, Al and Mn in acid coal mine drainage by sequential selective precipitation with control of pH. *Catena* **2017**, *148*, 11–16. [CrossRef]
132. Wei, X.; Viadero, R.C.; Buzby, K.M. Recovery of iron and aluminum from acid mine drainage by selective precipitation. *Environ. Eng. Sci.* **2005**, *22*, 745–755. [CrossRef]
133. Chockalingam, E.; Subramanian, S. Studies on removal of metal ions and sulphate reduction using rice husk and Desulfotomaculum nigrificans with reference to remediation of acid mine drainage. *Chemosphere* **2006**, *62*, 699–708. [CrossRef]
134. Mohan, D.; Chander, S. Removal and recovery of metal ions from acid mine drainage using lignite-A low cost sorbent. *J. Hazard. Mater.* **2006**, *137*, 1545–1553. [CrossRef] [PubMed]
135. Wei, C.; Li, X.; Deng, Z.; Fan, G.; Li, M.; Li, C. Recovery of H2SO4 from an acid leach solution by diffusion dialysis. *J. Hazard. Mater.* **2010**, *176*, 226–230. [CrossRef] [PubMed]
136. Park, S.M.; Shin, S.Y.; Yang, J.S.; Ji, S.W.; Baek, K. Selective recovery of dissolved metals from mine drainage using electrochemical reactions. *Electrochim. Acta* **2015**, *181*, 248–254. [CrossRef]
137. Crane, R.A.; Sapsford, D.J. Selective formation of copper nanoparticles from acid mine drainage using nanoscale zerovalent iron particles. *J. Hazard. Mater.* **2018**, *347*, 252–265. [CrossRef]

138. Kefeni, K.K.; Msagati, T.M.; Mamba, B.B. Synthesis and characterization of magnetic nanoparticles and study their removal capacity of metals from acid mine drainage. *Chem. Eng. J.* **2015**, *276*, 222–231. [CrossRef]
139. Silva, R.D.A.; Castro, C.D.; Vigânico, E.M.; Petter, C.O.; Schneider, I.A.H. Selective precipitation/UV production of magnetite particles obtained from the iron recovered from acid mine drainage. *Miner. Eng.* **2012**, *29*, 22–27. [CrossRef]
140. Wei, X.; Viadero, R.C. Synthesis of magnetite nanoparticles with ferric iron recovered from acid mine drainage: Implications for environmental engineering. *Colloids Surf. A Physicochem. Eng. Asp.* **2007**, *294*, 280–286. [CrossRef]
141. Ayora, C.; Macías, F.; Torres, E.; Lozano, A.; Carrero, S.; Nieto, J.M.; Pérez-López, R.; Fernández-Martínez, A.; Castillo-Michel, H. Recovery of rare earth elements and yttrium from passive-remediation systems of acid mine drainage. *Environ. Sci. Technol.* **2016**, *50*, 8255–8262. [CrossRef]
142. Hedin, B.C.; Capo, R.C.; Stewart, B.W.; Hedin, R.S.; Lopano, C.L.; Stuckman, M.Y. The evaluation of critical rare earth element (REE) enriched treatment solids from coal mine drainage passive treatment systems. *Int. J. Coal Geol.* **2019**, *208*, 54–64. [CrossRef]
143. Ramasamy, D.L.; Puhakka, V.; Iftekhar, S.; Wojtuś, A.; Repo, E.; Ben Hammouda, S.; Iakovleva, E.; Sillanpää, M. N- and O- ligand doped mesoporous silica-chitosan hybrid beads for the efficient, sustainable and selective recovery of rare earth elements (REE) from acid mine drainage (AMD): Understanding the significance of physical modification and conditioning of th. *J. Hazard. Mater.* **2018**, *348*, 84–91. [CrossRef]
144. Wei, X.; Zhang, S.; Shimko, J.; Dengler, R.W. Mine drainage: Treatment technologies and rare earth elements. *Water Environ. Res.* **2019**, 1–8. [CrossRef] [PubMed]
145. Zhang, W.; Honaker, R. Process development for the recovery of rare earth elements and critical metals from an acid mine drainage. *Miner. Eng.* **2020**, 106382. [CrossRef]
146. Ziemkiewicz, P.; Noble, A. *Recovery of Rare Earth Elements (REEs) from Coal Mine Drainage*; NETL REE Review Meeting: Pittsburgh, PA, USA, 2019.

5

Effect of Mineralogy on the Beneficiation of REE from Heavy Mineral Sands: The Case of Nea Peramos, Kavala, Northern Greece

Christina Stouraiti [1,*], Vassiliki Angelatou [2], Sofia Petushok [1], Konstantinos Soukis [1] and Demetrios Eliopoulos [2]

[1] Faculty of Geology and Geoenvironment, National and Kapodistrian University of Athens, 15784 Athens, Greece; sofpetushok@geol.uoa.gr (S.P.); soukis@geol.uoa.gr (K.S.)

[2] Department of Mineral Processing, Institute of Geology and Mineral Exploration, 13677 Acharnes, Greece; vasaggelatou@igme.gr (V.A.); deliopoulos1000@gmail.com (D.E.)

* Correspondence: chstouraiti@geol.uoa.gr

Abstract: Beneficiation of a rare earth element (REE) ore from heavy mineral (HM) sands by particle size classification in conjunction with high-intensity magnetic separation (HIMS) was investigated. The HM sands of Nea Peramos, Kavala, Northern Greece, contain high concentrations of REE accommodated mainly in silicate minerals, such as allanite. However, the potential of the Northern Greek placer for REE exploitation has not been fully evaluated due to limited on-shore and off-shore exploration drilling data. Characterization of the magnetic separation fractions using XRD and bulk ICP-MS chemical analysis showed that the magnetic products at high intensities were strongly enriched in the light REE (LREE), relative to the non-magnetic fraction. Allanite and titanite are the major host mineral for REE in the magnetic products but mainly allanite controls the REE budget due its high concentration in LREE. SEM/EDS and ICP-MS analysis of the different particle size fractions showed LREE enrichment in the fractions −0.425 + 0.212 mm, and a maximum enrichment in the −0.425 + 0.300 mm. The maximum enrichment is achieved after magnetic separation of the particle size fractions. Mass balance calculations showed that the maximum REE recovery is achieved after magnetic separation of each particle size fraction separately, i.e., 92 wt.% La, 91 wt.% Ce, and 87 wt.% Nd. This new information can contribute to the optimization of beneficiation process to be applied for REE recovery from HM black sands.

Keywords: rare earth elements (REE); heavy mineral sands; EURARE; allanite; monazite; HIMS; mineralogical characterization; geochemical characterization; magnetic separation; particle size fractions

---

1. Introduction

Heavy mineral sands (or black sands) are coastal deposits of resistant dense minerals that locally form economic concentrations of the heavy minerals. They serve as a major source of titanium worldwide with main minerals rutile and ilmenite and, in some cases, show high accumulation in rare earth elements (REE) and Th [1]. The rare earth elements (REE) are a group of 17 chemically similar elements, the lanthanides, scandium (Sc), and yttrium (Y) which behave similarly in most environments in the Earth's crust. REE are considered as "critical metals" for the European Union economy due to the vast application in a variety of sectors, a complicated production process as well as political issues associated with the monopoly in supply from China, especially the supply of heavy rare earths [2–4]. According to the recent EU Joint Research Center (JRC) report among the REE, six are identified as more critical, because their combined importance for strategic sectors of the economy such as high-efficiency electronics and energy technologies with risks of supply shortage. These are Dy, Eu, Tb, Y, Pr, and Nd [5].

Currently REE are not exploited in Europe, however, due to the current situation several exploration projects have been assessed in the course of the recently ended EURARE and ASTER European projects, which showed that some of them are in an advanced stage of exploration and development [6,7]. The most promising cases are the alkaline igneous rock-hosted deposits in South Greenland, the Norra Kärr deposit in Sweden and Fen Complex in Norway (Goodenough et al. [6]) and the alkaline volcanic-derived placers of Aksu Diamas in Turkey [8]. In Greece, the most significant REE concentrations are associated with heavy mineral sands on the coast of Nea Peramos and Strymonikos Gulf. Moreover, the EURARE project highlighted the significance of the secondary REE deposits, such as the bauxite residue (red mud) from the processing of Greek bauxites [4,6,9].

Mudd and Jowitt [1] stretched the economic potential of heavy mineral sands as an important underestimated REE resource especially for monazite and xenotime minerals. HM sands remain excluded from mineral resource considerations mostly due to the environmental problems that are associated with the radioactivity of tailings and the reagents used. However, there has been limited research work on the quantification of these impacts [3].

Geological prospecting by the Institute of Geology and Mineral Exploration (I.G.M.E.) of Greece on the black sands in the broader area of Strymon bay started in 1980's and focused primarily on the natural enrichment in actinides (U-Th) and associated radioactivity in the on-shore and offshore zones of Loutra Eleftheron to Nea Peramos regions [10–13] (Figure 1). In the last decade, there is an increasing number of geochemical and mineralogical studies that have been carried out on the coastal areas of Kavala [14–19], Sithonia Peninsula of Chalkidiki [20], Touzla Cape [21], and the area of Maronia, Samothrace [22]. Most of these studies focused on the characterization of the placers and their natural radioactivity ([23] and references therein). Previous studies have demonstrated that the heavy minerals, monazite, allanite, titanite, uraninite, zircon, and apatite, are traced in the Kavala black sands, derive from the Symvolon/Kavala pluton, a deformed granodioritic complex of Miocene age (Table 1) [6,9,14,24]. Despite the low grade of the Greek placers at Nea Peramos, as emphasized in the EURARE project, there is a good potential of beneficiation due to the coarse particle size and the liberation of REE minerals. The Northern Greece heavy mineral sands potential was not feasible to be fully evaluated as a potential REE resource in the course of EURARE project due to limited exploration data.

Worldwide, the commonly exploited rare earth-bearing minerals in industrial scale are bastnäsite, monazite, and xenotime [25]. Other REE-bearing minerals such as eudialyte, synchysite, samarskite, allanite, zircon, steenstrupine, cheralite, rhabdophane, apatite, florencite, fergusonite, loparite, perovskite, cerianite, and pyrochlore are rarely found in deposits of economic significance [26]. However, there are new deposits being under development containing many new REE minerals that seek further understanding, such as zircon, allanite, and fergusonite [27].

Beneficiation of the three commercially extracted heavy minerals, bastnäsite, monazite, and xenotime involves gravity, magnetic, electrostatic, and flotation separation methods with froth flotation being the most commonly applied REE mineral separation operation ([27–29] and references therein). There are numerous research articles on REE mineralogy and hydrometallurgical processing but there is still a lack of comprehensive descriptions of the beneficiation methods necessary to concentrate REE minerals. A main reason for this lack is the fact that concentrates of monazite and xenotime worldwide are produced from heavy mineral sands, therefore, comminution is scarcely required [26].

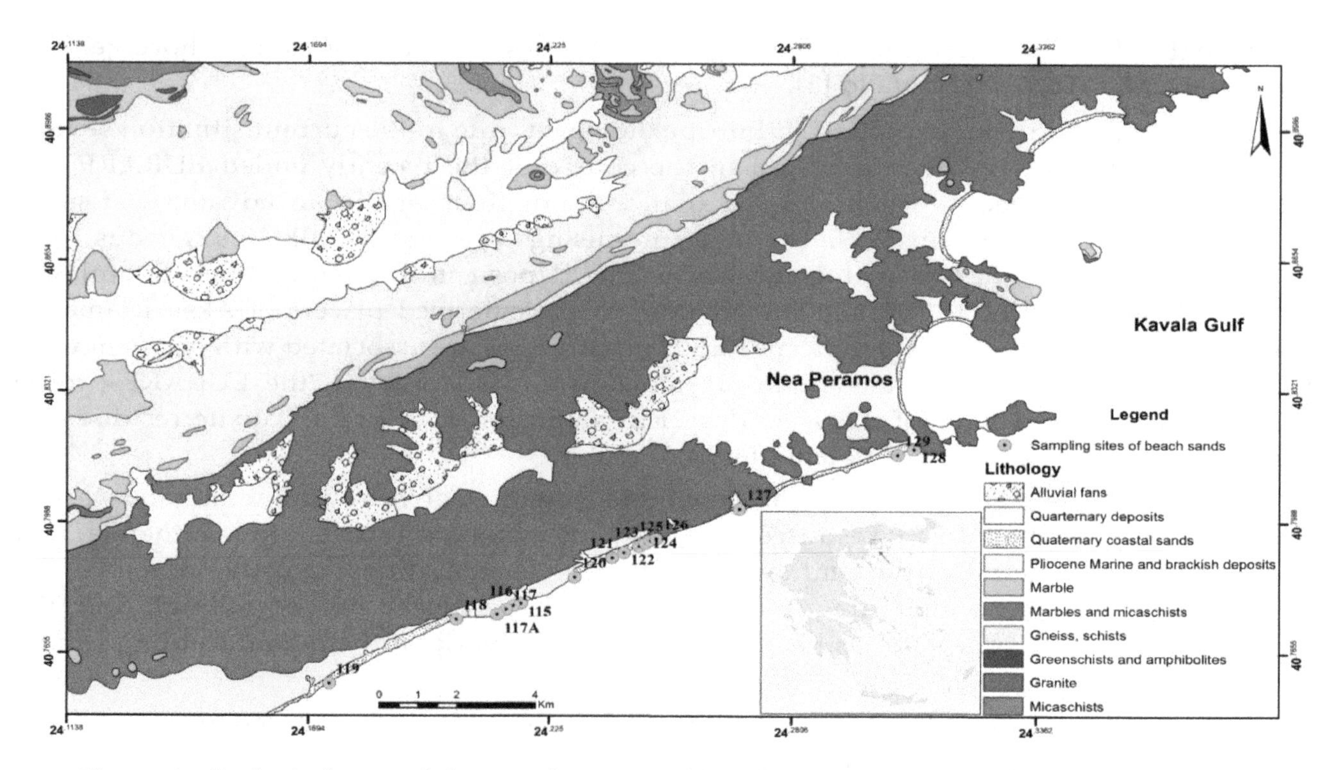

Figure 1. Geological map of the sampling area of Nea Peramos Loutra Eleftheron coast, Kavala region [12,13]. The large igneous body behind the coastline is Symvolon granite.

Table 1. Common silicate, phosphate, and carbonate rare earth element (REE) bearing minerals in heavy mineral (HM) sands (bold, this study) and bauxite residues in Greece [6,9,14,24]. Data for mineral properties from [26] (and references therein) and apatite data from [30].

REE-Mineral	Chemical Formula	Density (g/cm^3)	Magnetic Properties	Weight % REO	ThO_2	UO_2
Silicates						
Allanite (Ce)	$(Ce,Ca,Y)_2(Al,Fe^{2+},Fe^{3+})3(SiO_4)_3(OH)$	3.50–4.20	paramagnetic	3–51	0–3	-
Allanite (Y)	$(Y,Ce,Ca)_2(Al,Fe^{3+})_3(SiO_4)_3(OH)$	n/a	paramagnetic	3–51	0–3	-
Cheralite (Ce)	$(Ca,Ce,Th)(P,Si)O_4$	5.28	n/a	-	<30	-
Sphene (titanite)	$(Ca,REE)TiSiO_5$	3.48–3.60	paramagnetic	<3	-	-
Thorite	$(Th,U)SiO_4$	6.63–7.20	paramagnetic	<3	70–80	10–16
Zircon	$(Zr,REE)SiO_4$	4.60–4.70	diamagnetic	-	0.1–0.8	-
Phosphates						
Apatite	$Ca_5(PO_4)_3(F,Cl,OH)$	3.17	n/a	~19	-	-
Fluorapatite	$(Ca,Ce)_5(PO_4)$	3.10–3.25	n/a	-	-	-
Monazite (Ce)	$(Ce,La,Nd,Th)PO_4$	4.98–5.43	paramagnetic	35–71	0–20	0–16
Monazite (La)	$(La,Ce,Nd,Th)PO_4$	5.17–5.27	paramagnetic	35–71	0–20	0–16
Monazite (Nd)	$(Nd,Ce,La,Th)PO_4$	5.43	paramagnetic	35–71	0–20	0–16
Rhabdophane (Ce)	$(Ce,La)PO_4.H_2O$	3.77–4.01	n/a	-	-	-
Xenotime (Y)	YPO_4	4.40–5.10	paramagnetic	52–67	-	0–5
Carbonates						
Bastnäsite (Ce)	$(Ce,La)(CO_3)F$	4.9–5.2	paramagnetic	70–74	0–0.3	0.09
Bastnäsite (La)	$(La,Ce)(CO_3)F$	n/a	paramagnetic	70–74	0–0.3	0.09
Bastnäsite (Y)	$Y(CO_3)F$	3.90–4.00	paramagnetic	70–74	0–0.3	0.09
Synchysite (Nd)	$Ca(Nd,La)(CO_3)_2F$	4.11 (calc)	n/a	-	-	-

n/a: not available; (-): this information is not known.

This paper presents results of an ongoing beneficiation study of REE-rich HM sands form the Nea Peramos, Kavala (Northern Greece) at laboratory scale. Two process schemes were applied and tested at laboratory scale in this stage in order to improve understanding of mineral separation by different physical processes including, particle size classification through wet sieving as a preconcentration process followed by high-intensity magnetic separation (HIMS) [26,27]. Scanning electron microscopy (SEM), X-ray diffraction analysis, and inductively coupled plasma–mass spectrometry (ICP-MS) were used in combination, for the characterization of feed material as well as beneficiation products.

Geological Setting

The studied area is situated in Northern Greece, in the internal domain of the Aegean arc (Figure 1). It is part of the Rhodope massif, a polymetamorphosed nappe stack, which comprises high-grade Pre-Alpine felsic to intermediate orthogneisses (mostly Variscan age), schists, and marble interlayered or tectonically overlain by basic to ultrabasic rocks, that were exhumed in the Cenozoic [31–38]. The lowermost Pangaion-Pirin or Lower Unit is exposed in several domes in the Rhodope area and as the footwall to the top to SW Strymon detachment [37,39]. The Pangaion-Pirin Unit comprises a thick marble sequence alternating with schists, both underlain by Variscan orthogneiss and schists [37,40,41]. The Pangaion-Pirin Unit is intruded by several Oligocene to early Miocene granitoids [37,42]. The syn-tectonic early Miocene Symvolon (or Kavala) granodiorite, which occupies a large part of the studied area, has intruded and deformed along the southeastern end of the Strymon detachment [39,41]. Marine deposits of Pliocene age, mainly sandstones and marls, are observed along the coast, covering the granodiorite.

2. Samples and Methods

2.1. Sampling

Fifteen samples of coastal sands were collected alongside a coastal line of 10 km from Loutra Eleftheron to Nea Peramos areas, Kavala region, Northern Greece (Figure 1). Sample campaign was carried out in the course of EURARE project during the period of 2013–2015. The sampling technique involved opening holes of 40 cm deep. According to field observations, locally, there were black sand concentrations in layers of few cm thick in alternation with typical light-colored sands. Due to sample inhomogeneity a large amount of ~15 kg of bulk sand from each sampling location was taken for securing a representative sample. A 20 kg sample of placer sand with an initial top size of 1.7 mm was prepared by mixing equal weights of individual collected samples (composite A-mixed sample) and divided by splitting into samples of 2.5 kg.

2.2. Sample Characterization

2.2.1. XRD Analysis

The major mineralogy was determined by powder X-ray diffraction using a Siemens D-5005 diffractometer with Cu K radiation, at the Geology Department of NKUA. Intensities were recorded at 0.02° 2θ step intervals from 3° to 70°, with a 2 s counting time per step. The resultant diffraction patterns were processed using EVA software by Bruker AXS Inc (Madison, WI, USA), in order first to identify peaks and then relate them to selected mineral phases that are present in the Kavala black sands samples (Figure 2). Detection limits are of the order of 1 wt.% approximately, but this is mineral and sample dependent.

The magnetic and non-magnetic fractions and all the particle size fractions were finely ground in the appropriate size, placed in a sample holder, and smeared uniformly onto a glass slide, assuring a flat upper surface.

Semiquantitative analysis was performed using the Reference Intensity Ratio method (RIR) of I/Ic [43]. For the semiquantitative analysis, the Diffract Plus software by Bruker-AXS was used. The minerals that did not have I/Ic ratios, the ratios from similar minerals of the same mineral group were used. For allanite, intensity ratio of epidote (0.9) was used, and for magnesiohornblende, an average ratio of hornblende (0.6) was used.

2.2.2. SEM/EDS Analysis

Sand particles were mounted within a plug of epoxy resin and then polished particle-mount sections were prepared. Before SEM analysis, the thin section was covered with a thin veneer of carbon using a vacuum carbon coater. Textural analysis and semiquantitative elemental analysis of heavy minerals was undertaken using a JEOL JSM-5600 Scanning Electron Microscope (SEM) coupled to an energy dispersive X-ray spectrometer (EDS) of OXFORD LINK ISIS 300 (OXFORD INTRUMENTS), with the use of software for ZAF correction, at the Faculty of Geology and Geoenvironment, NKUA, using secondary electron (SE) and backscatter electron (BSE) modes.

2.2.3. Bulk Chemical Analysis

REE ore sample preparation included initial digestion with Aqua Regia and HF (in order to solve dissolution issues in silicate samples) in Teflon® containers. Subsequently, the residue was chemically attacked with HCl and H_2O_2. Finally, the samples are preserved in 5% concentrated HCl. REE were analyzed by inductively coupled plasma-mass spectrometry (ICP-MS) at I.G.M.E. (Supplementary Table S1). External calibration solutions with matrix correction were used to measure the instrumental sensitivity of ICP-MS.

2.3. Beneficiation Tests

The beneficiation tests were conducted at the Laboratory of Beneficiation and Metallurgy, I.G.M.E.

2.3.1. Particle Size Analysis

From the composite sample A-mixed, a representative sample (of 2327 g initial weight) was prepared properly in order to be submitted to particle size analysis. Samples were submitted to wet sieving in order to provide data on particle size distribution, using sieves with aperture of 1.70, 0.850, 0.500, 0.425, 0.355, 0.300, 0.212, and 0.150 mm. All the samples were placed in a dryer for 24 h at 90 °C.

The information acquired by particle size distribution contributes to the study of the effects of particle size, mineral liberation, and association characteristics aiming at the selection of appropriate process schemes to be tested and used. In the course of wet sieving process, a change in the color was observed as we moved onto smaller particle size fractions.

Preconcentration step helps in rejecting early gangue materials thus achieving benefits such as higher feed grades, lower waste stream production in further processes, and finally, low operating costs.

2.3.2. Magnetic Separation

Magnetic separation was selected as a possible concentration technique as the gangue minerals are known to have lower magnetic susceptibility than value minerals which contain rare earth elements. Based on strong or weak magnetic properties, iron-bearing minerals are characterized as ferromagnetic or paramagnetic, respectively. Ferromagnetic refers to minerals strongly attracted to a magnet, like a piece of iron. Magnetite, maghemite, and pyrrhotite are the most common ferromagnetic minerals [20].

A laboratory-scale High Intensity Magnetic Separator (HIMS; Model 10/1, ERIEZ EUROPE, Caerphilly, UK) with standard intensity was used to recover REE-bearing minerals (Figure 2). More than 20 tests were carried out during magnetic separation process, applying different conditions and parameters (vibration, inclination, and speed). Two process routes were followed:

(i) The whole sample passing through −0.500 mm was driven to magnetic separation.
(ii) Each size fraction separately (down to +0.212 mm) was tested for magnetic separation.

The sample was fed by a vibrating feeder. The moving velocity of the feed carrying conveyor was about 40 str/s, tilt was set initially to 72°. The applied magnetic intensity was 2 T. After recovering magnetic products at 72°, the magnetic sample was fed again to the magnetic separator adjusted to a 74° and repeated to 76° inclination.

Wet high-intensity magnetic separation (WHIMS) was also conducted on the composite A-mixed using an ERIEZ separator by applying different voltages, i.e., at 30 V (0.48 T) and 150 V (2.4 T). The potential difference was selected to correspond to extreme conditions of lower and higher magnetic field strength in order to check the effect of the intensity of the magnetic field on the efficiency of the magnetic separation process. The results after magnetic separation were evaluated on the basis of semiquantitative mineralogical analysis using the Reference Intensity Ratio method (RIR) of I/Ic.

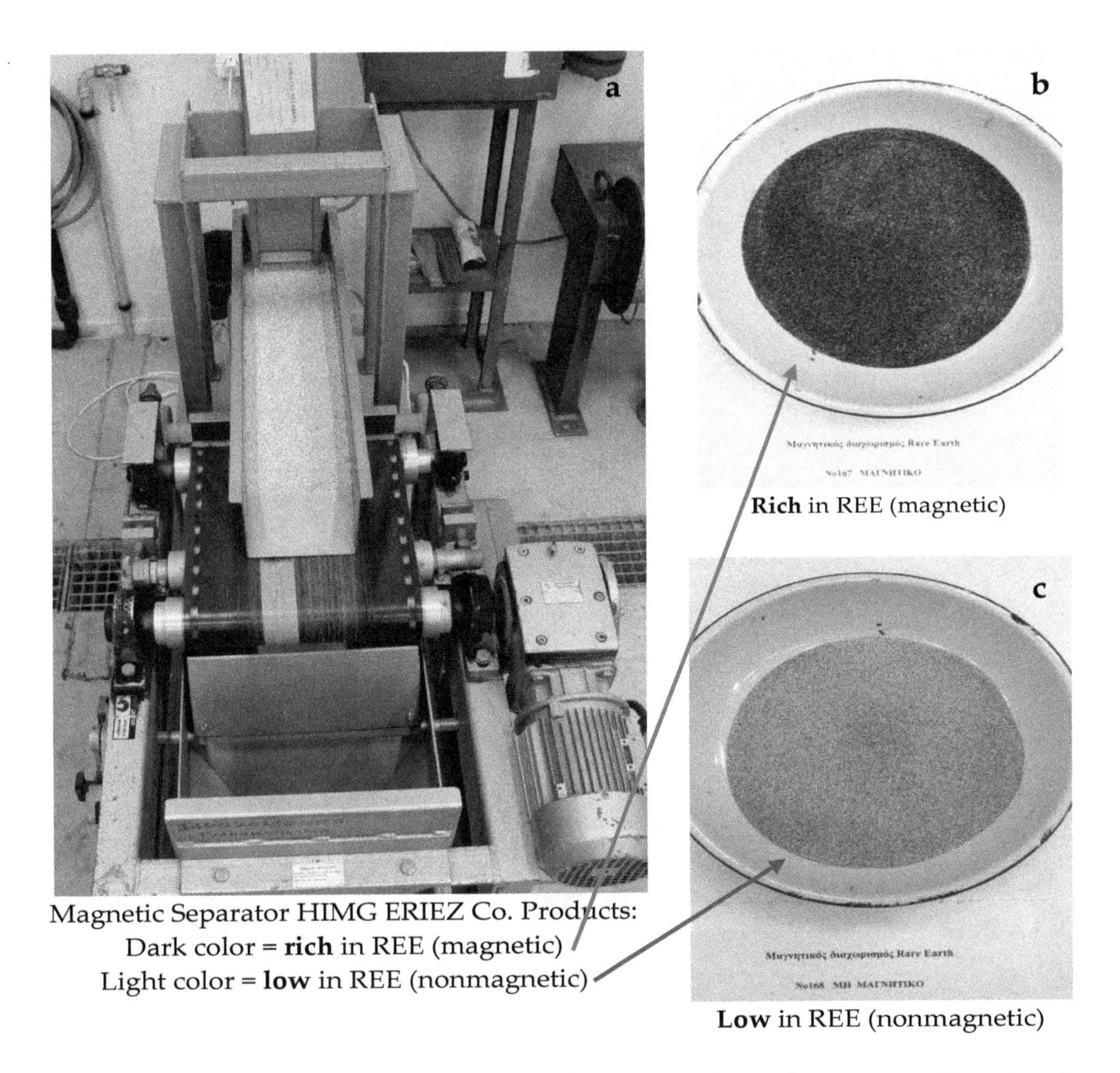

Figure 2. (**a**) Magnetic Separator HIMG ERIEZ Co. and products of separation, (**b**) dark colored magnetic product and, (**c**) non-magnetic product.

2.4. Evaluation of Beneficiation Tests

All the separated fractions from the beneficiation tests were tested for mineralogical and chemical composition, except for the fraction +1.7 mm due to its low quantity and low content in REE. The SEM/EDS analysis of the particle size fractions and the magnetic separates provided a rough evaluation of the test performance. The degree of liberation of the REE-bearing minerals from the gangue at various particle sizes was also established by SEM/EDS analysis. The fractions from particle size analysis and the final magnetic separation of the undersize 0.500 mm fractions were evaluated on the basis of mass balance calculations.

3. Results

3.1. Mineralogy of the Sands

Table 2 and Figure S1 shows the results of mineral identification and semiquantitative modal composition of the composite sample "A-mixed" obtained by XRD analysis and RIR method. The studied HM sands consist mainly of silicate minerals, Fe-oxides (magnetite, hematite), titanite and minor amounts (<3%) of Ti oxides (ilmenite and rutile), and phosphate phases (apatite, monazite, and xenotime) (Table 2). The major heavy minerals contained in the studied sands are amphibole (Mg-hornblende and pargasite), magnetite, titanite, allanite-epidote, and hematite (Table 2). Zircon, monazite, cheralite, ilmenite, rutile, thorite, apatite, xenotime, baryte, and sulfides were identified as minor constituents by SEM/EDS analysis. Allanite is the major host mineral of REE in relative abundance of 3% in the composite sample (Table 2). Based on SEM/EDS analysis and published EPMS data, the studied monazite is classified as a cerian type (monazite-Ce). However, due to the low abundance of the mineral as well as overlapping of monazite peak intensities with those of magnesiohornblende and allanite, this mineral was not clearly identified by the XRD spectra (Figure S1).

Table 2. Semiquantitative mineral contents of the composite sample (A-mixed), based on Reference Intensity Ratio (RIR) method. Specific gravity and magnetic property data from [44–46].

Mineral	Weight %	Nominal Specific Gravity	Magnetic Property
Albite	39	2.68	Diamagnetic
Quartz	31	2.63	Diamagnetic
K-feldspar	11	2.57	Diamagnetic
Titanite	8	3.4–3.6	Paramagnetic
Mg-hornblende	6	3.24	Paramagnetic
Allanite	3	3.75	Paramagnetic
Hematite	2	5.30	Ferromagnetic
Magnetite	1	5.20	Ferromagnetic
Trace minerals	<0.5	-	-
Total	100		

(-): not applicable.

SEM Analysis of REE Minerals

(a) Allanite-(Ce)

SEM/EDS semiquantitative analysis of allanite indicated that total concentration of REE oxides (TREO) range ca. from 10.7 to 16.8 wt.%, which is in agreement with the published data [18]. The stoichiometry of allanite indicates that Ce-allanite is the dominant type (Supplementary Table S2). The backscattered election (BSE) images suggest that allanite is liberated in the particle size −0.500 mm. The surface of the allanite is weathered with several cracks. Backscattered electron images showed that most allanite exhibits pronounced zoning toward the marginal areas of the grain (Figure 3a,b,d). The chemical composition of the peripheral zones indicates replacement of the allanite with epidote. Zoned allanite contains significant Th in the central part of the grain, ca. up to 2.5 wt.%, and they show metamict texture. This is documented by the abundant radiation lines in the structure of the grain displayed by this type of allanite, which results from the destructive effects of its own radiation on the crystal lattice (Figure 3).

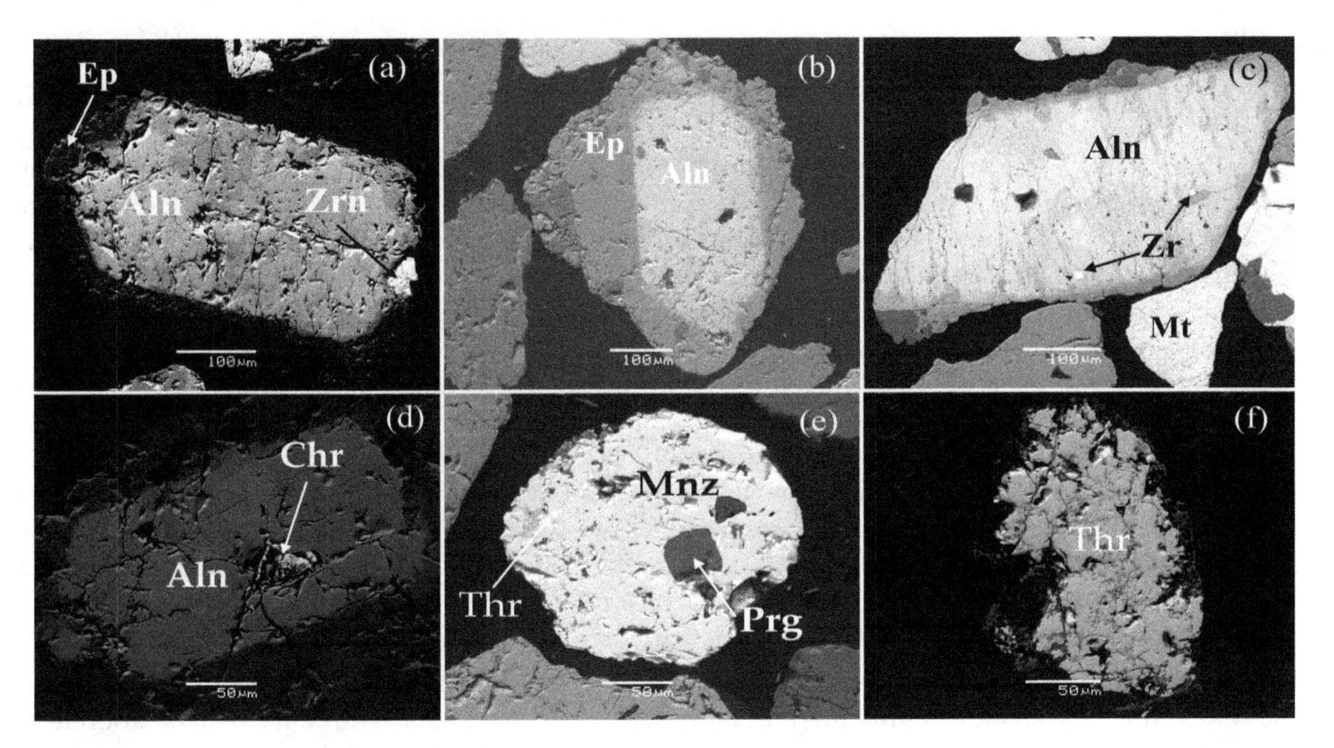

Figure 3. Backscattered electron images from SEM of various rare earth element (REE) minerals and other heavy minerals of different particle size fractions. (**a**,**b**) zoned allanite–(Ce) showing metamict structure and epidote peripheral zone (−0.425 mm to +0.212 mm), (**c**) allanite containing zircon inclusions from the magnetic fraction, (**d**) zoned allanite with weathering features and cracks containing inclusion of Ca-rich monazite (cheralite) (**e**) monazite containing pargasite and thorite inclusions, from the magnetic fraction, (**f**) thorite-weathered grain, darker zones are U depleted); abbreviations: Aln—allanite, Ep—epidote, Zr-zircon, Mnz—monazite, Thr—thorite, Mt—magnetite, Prg—pargasite, and Chr—cheralite.

(b) Monazite (-Ce)

Monazite occurs as very fine grains, i.e., <10–20 µm in diameter, disseminated in allanite, and only rare liberated grains up to 200 µm are found (Figure 3b,e). In most of the cases, monazite is a replacement phase of allanite. Monazite shows higher content in Ce than Nd or La, which agrees for a cerian-type monazite and is consistent with reported data from the same area [18] (Supplementary Table S2). ThO_2 content of the monazite is generally high, i.e., 17 wt.%. The low totals reported in the analyses are probably related to undetected MREE and HREE by SEM/EDS.

(c) Titanite

Titanite is more abundant than allanite (>5%). SEM/EDS analysis of titanites from this study showed that the REE content is below the detection limit of the method (<0.1–0.2 wt.%). The results are consistent with published EPMA analysis for titanites from the same area, which showed that La, Ce, or Nd content is ≤1000 mg/kg [18].

(d) Other Heavy REE Minerals

Zircon is another REE-hosting phase, with relatively higher concentration in heavy REE (HREE) (253–890 mg/kg) relative to light REE (LREE) (8–44 mg/kg) [18,23]. Apatite is a common fine-grained accessory phase disseminated in silicate minerals. Neither apatite nor zircon or thorite was feasible to be analyzed by SEM/EDS for REE due to the low REE content and/or the small particle size of the minerals for this method (see Supplementary Table S2).

3.2. REE Geochemistry

The chondrite-normalized REE contents of the Nea Peramos HM sand samples display similar patterns but a large compositional range (Figure 4). The total REE (TREE) abundances are generally higher than that of the average upper continental crust, i.e., 183 mg/kg [47], except for one sample (#NP120) (Table 3; detailed data in Supplementary Table S1). A comparison of average REE concentrations of the HM sands of Greece based on reported analysis [23] and HM sands from this study shows that the Nea Peramos samples display the highest total REE enrichment. The chondrite-normalized REE patterns of the studied samples show a pronounced enrichment of light REE, e.g., ranging from 70 to 7000 times chondrite values, and flat heavy REE (HREE) with a small negative Eu anomaly (Figure 4). Allanite controls the LREE budget of the studied sands but the flat HREE characteristic is controlled by another major REE mineral, possibly titanite and/or epidote. This characteristic is typically shown by titanite which is the major heavy mineral phase that holds HREE (see Discussion). These observations are in agreement with previous mineralogical studies of the studied sands [18,23]. The average content of Sc is 7.5 mg/kg, which is lower compared to the respective value in upper continental crust, i.e., 14 mg/kg [47]. Average Y concentration in the studied samples is 67 mg/kg and is enriched by a factor of 3 relatively to the upper continental crust, i.e., 21 mg/kg [47]. Yttrium enrichment is associated with xenotime abundance.

Our samples present a large compositional variation in total REE (Figure 4). On the other hand, all samples show a systematic enrichment in light REE. Therefore, the preliminary beneficiation tests were performed on a composite sample (A-mixed) produced by mixing of all the samples.

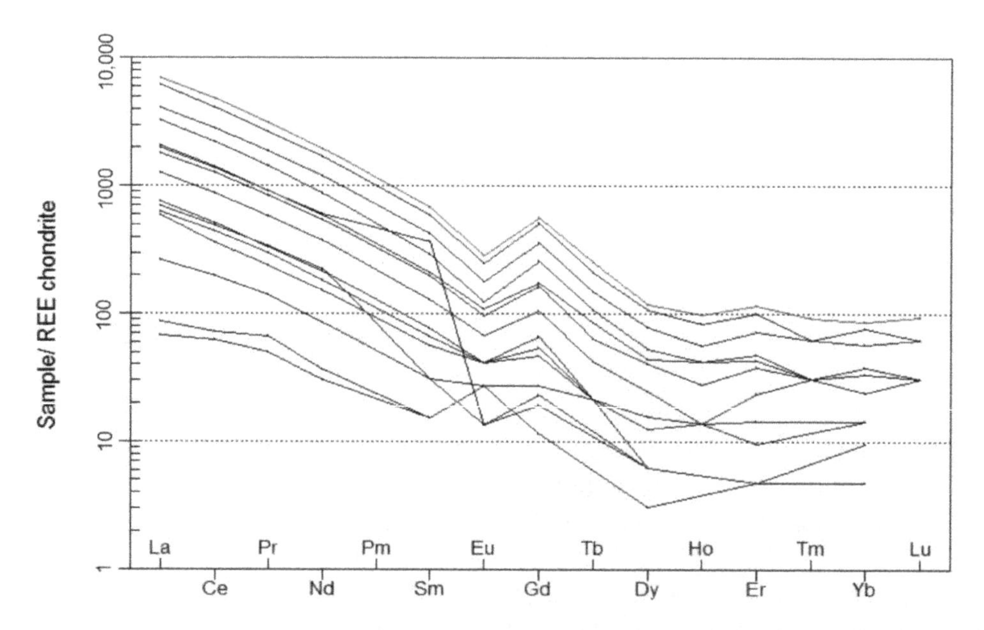

Figure 4. REE patterns of Nea Peramos heavy mineral (HM) sands. Red-colored line highlights the most REE-enriched sample #123. Normalization values of chondrite from [48].

Table 3. REE (Lanthanides, Sc, Y) contents of the Nea Peramos (NP) HM sands (in mg/kg). Light rare earth elements (LREE) group includes the elements from La to Sm, while heavy rare earth elements (HREE) group includes the elements from Eu to Lu; Y and Sc are included in total REE content, according to European Union (EU) definition [2].

Sample	NP 115	NP 116	NP 117	NP 117A	NP 118	NP 119	NP 120	NP 121	NP 122	NP 123	NP 124	NP 125	NP 126	NP 127	NP 128
LREE	600	826	1407	315	707	3504	118	2024	2225	7645	6646	4552	98	2297	787
HREE	47	61	98	27	51	204	18	147	160	465	409	295	15	76	47
TREE	654	894	1509	344	762	3716	143	2181	2393	8124	7068	4857	116	2381	842

3.3. *Particle Size Analysis and REE Distribution*

The distribution of REE and Th in the particle size fractions of the composite sample A-mixed is determined by chemical assays (Table 3).

A combination of SEM/EDS and ICP-MS chemical analysis of the different particle size fractions showed a LREE enrichment in the fractions −0.500 to +0.212 mm and under 0.150 mm (Figure 5a). In Table 4, mass balance calculations show that the −0.150 mm fraction, despite its composition, is insignificant in terms of beneficiation due to its very low mass, i.e., 1.5 wt.% Moreover, the oversize +1.70 mm particle size fraction has a very low REE content and hence is also considered as insignificant for further beneficiation (Figure 5b).

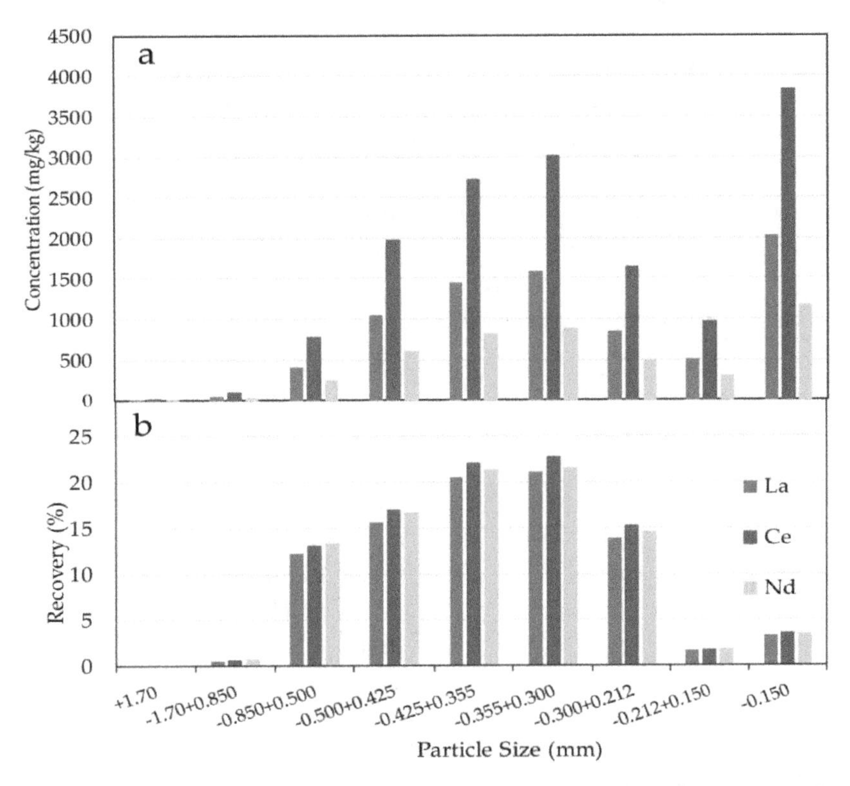

Figure 5. (**a**) Particle size analysis and light rare earth element (LREE) distribution (mg/kg) in sample A-mixed and (**b**) metal recovery in each particle size fraction, based on mass balance calculations (see text).

Table 4. Particle size analysis and REE distribution in each particle size fraction.

Particle Size (mm)	**Mass (g)**	**Weight (%)**	**La**	**Ce**	**Nd**	**Th**	**La**	**Ce**	**Nd**	**Th**
			Concentration (mg/kg)				**Mass (mg)**			
+1.70	55.6	2.42	16	25	10	7	0.89	1.39	0.56	0.39
−1.70 + 0.850	230.7	10.02	54	106	38	25	12.46	24.45	8.77	5.77
−0.850 + 0.500	640.48	27.82	416	777	247	190	266.44	497.65	158.20	121.69
−0.500 + 0.425	324.62	14.10	1044	1983	607	455	338.90	643.72	197.04	147.70
−0.425 + 0.355	307.32	13.35	1445	2719	821	641	444.08	835.60	252.31	196.99
−0.355 + 0.300	286.26	12.43	1598	3018	891	723	457.44	863.93	255.06	206.97
−0.300 + 0.212	353.02	15.33	852	1643	487	390	300.77	580.01	171.92	137.68
−0.212 + 0.150	69.12	3.00	507	968	305	240	35.04	66.91	21.08	16.59
−0.150	35.05	1.52	2031	3840	1162	945	71.19	134.59	40.73	33.12
Total	2302.17						1927.21	3648.27	1105.66	866.90

The maximum effective recovery is achieved in the fractions −0.425 + 0.300 mm; this size range corresponds to the "liberation" size of allanite. A similar trend of enrichment is observed in the sample #123 [17]. Thorium follows the enrichment trend of LREE and its concentration in the undersize

0.500 mm ranges from 240 to 945 mg/kg. Thorium concentration is associated with the abundance of allanite and monazite in the HM sands, which contain 2.3 and 17–35 wt.% ThO_2, respectively, according to EDS/SEM analysis (Supplementary Table S2).

3.4. REE Distribution in the Magnetic Fractions

High-intensity magnetic separation (HIMS) is a common separation step in REE containing beach sands in order to concentrate the targeted paramagnetic REE-bearing part and is usually applied for monazite or xenotime [26,27]. The LREE tend to concentrate in the magnetic fraction because of their magnetic properties. Dry HIMS was applied initially in the most REE-enriched sample #123. Mineralogical evaluation based on XRD spectra shows that from the paramagnetic minerals allanite, titanite and magnesiohornblende tend to concentrate in the magnetic fraction under high intensities. Qualitative evaluation of the mineralogy of magnetic and nonmagnetic separates by XRD spectra is shown in Figure 6. The peak intensities of allanite can be clearly distinguished in the magnetic separate of the sample #123 but are absent in the nonmagnetic part. Following that observation, magnetic separation of particle size fractions from the composite A-mixed sample was carried out. The chemical assays of the magnetic fractions are shown in Table 5 and Figure 7.

The same REE-enrichment trend, as shown in the particle size analysis, is also recorded in the HIMS test (Figure 7). The maximum LREE concentration was achieved in the particle size: −0.355 + 0.212 and a second concentration maximum in the −0.425 + 0.355 mm fraction. Two test of HIMS were conducted: in the first test, all particle sizes (Feed 1 in Figure 8) were processed by magnetic separation, whereas in the second test, only the undersize 0.500 mm were tested after removing the REE-poor +0.500 mm fraction from the initial composite sample A-mixed (preconcentration stage) (Table 5). The results of mass balance calculations and REE distribution in every product are presented in Table 5. Similar recoveries were achieved in the two feeds, i.e., from 92% to 87.6% and 95.9% to 78.87% for the particle size fractions and the undersize 0.500 mm feed, respectively. The maximum LREE enrichment is attained in the magnetic products of the different particle sizes. It is clearly shown in the mass balance calculations in Table 5 that magnetic separation for each particle size separately improves the recovery of REE and reaches recoveries of 75–90% in just the 22% of feed material (fractions −0.500 to +0.212 mm).

Table 5. REE distribution and recovery of REE and Th in the magnetic products of each particle size of the A-mixed sample and the undersize 0.500 mm.

Feed: Sample A (All Fractions)		**La**	**Ce**	**Nd**	**Th**	**La**	**Ce**	**Nd**	**Th**
Mass (g)		**Concentration (mg/kg)**				**Mass (g)**			
	1995	942	1647	512	365	1.87	3.29	1.021	0.728
Particle size		magnetic fractions (mg/kg)				Mass (g)			
−1.70 + 0.850	42.55	261	447	137	93	0.0111	0.0190	0.0058	0.0040
−0.850 + 0.500	162.23	1301	2309	717	392	0.2111	0.3746	0.1163	0.0636
−0.500 + 0.425	101.65	2764	4802	1444	1099	0.2810	0.4881	0.1468	0.1117
−0.425 + 0.355	109.81	3627	6321	1865	1472	0.3983	0.6941	0.2048	0.1616
−0.355 + 0.300	100.6	3838	6649	1968	1595	0.3861	0.6689	0.1980	0.1605
−0.300 + 0.212	142.83	3044	5280	1563	1302	0.4348	0.7541	0.2232	0.1860
Total	659.67					1.7223	2.9989	0.8950	0.6873
Recovery (wt.%)						92.1	91.1	87.6	94.4
Feed: −0.500 mm		**La**	**Ce**	**Nd**	**Th**	**La**	**Ce**	**Nd**	**Th**
Mass (g)		**Concentration (mg/kg)**				**Mass (g)**			
530		1260	2223	670	460	1.9278	3.4012	1.0251	0.7038
Magnetic fraction		490	3774	6089	1650	1300	1.8493	2.9836	0.8085
Recovery (wt.%)						95.9	89.7	78.87	90.5

Notably, Th enrichment follows the LREE enrichment trend, and its recovery is high in the either tested feed. The steps of magnetic separation process and the REE grade of concentrates are summarized in the flowsheet of Figure 8.

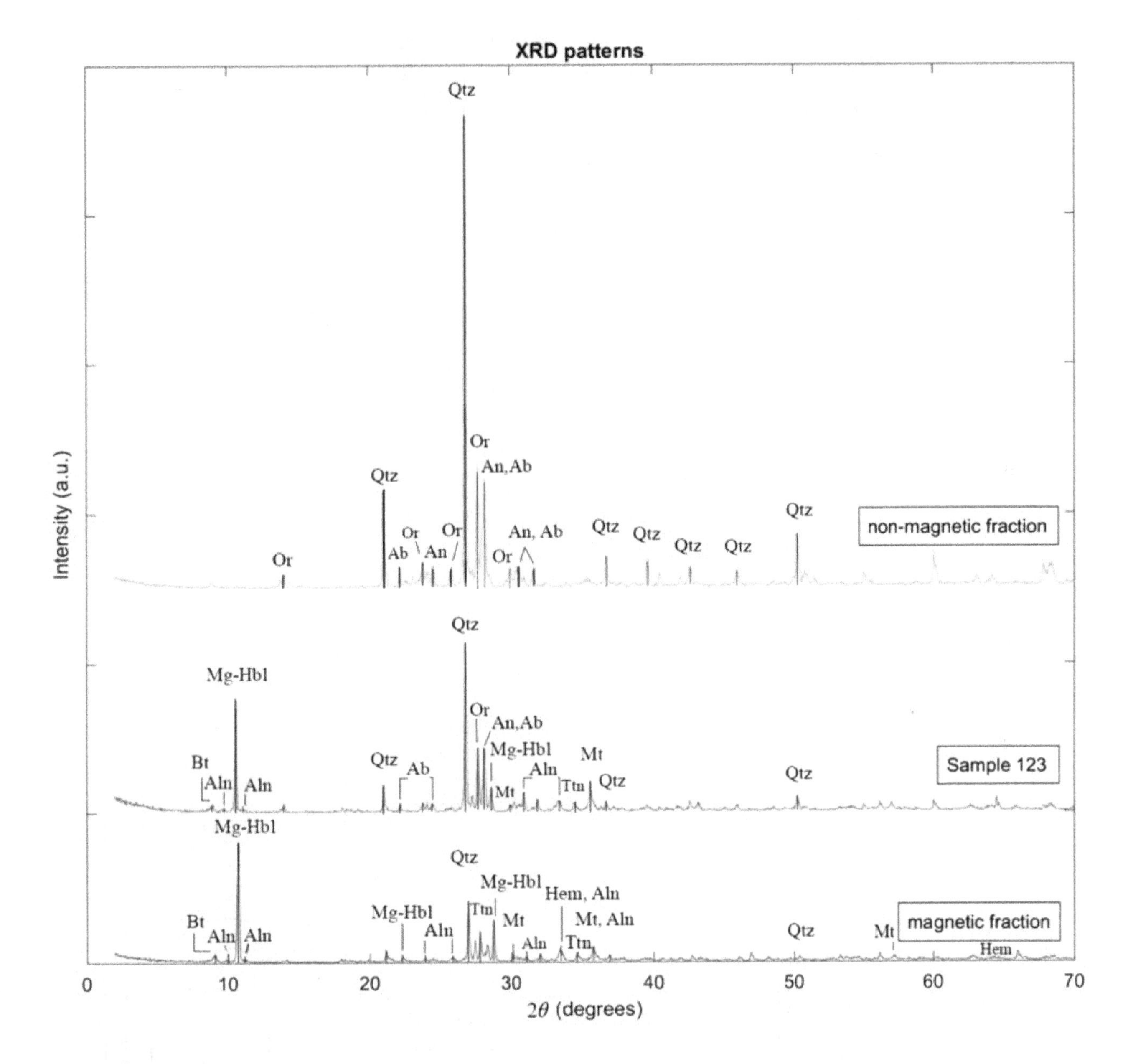

Figure 6. Comparison of XRD patterns of magnetic and nonmagnetic separates after high-intensity magnetic separation (HIMS) from sample #123. Note that the clear peaks of allanite are absent in the nonmagnetic fraction.

Further testing on the magnetic susceptibility of the paramagnetic minerals in the composite sample was conducted under extreme conditions of the magnetic field, i.e., low (0.48 T) and high (2.4 T) strength, by wet HIMS. The results are evaluated by semiquantitative analysis of the XRD spectra of the magnetic and nonmagnetic products. The maximum content of allanite, i.e., 8 wt.%, is achieved in the magnetic fraction at magnetic field strength of 0.48 T (30 V) and remains constant at 2.4 T (150 V) (Supplementary Table S3).

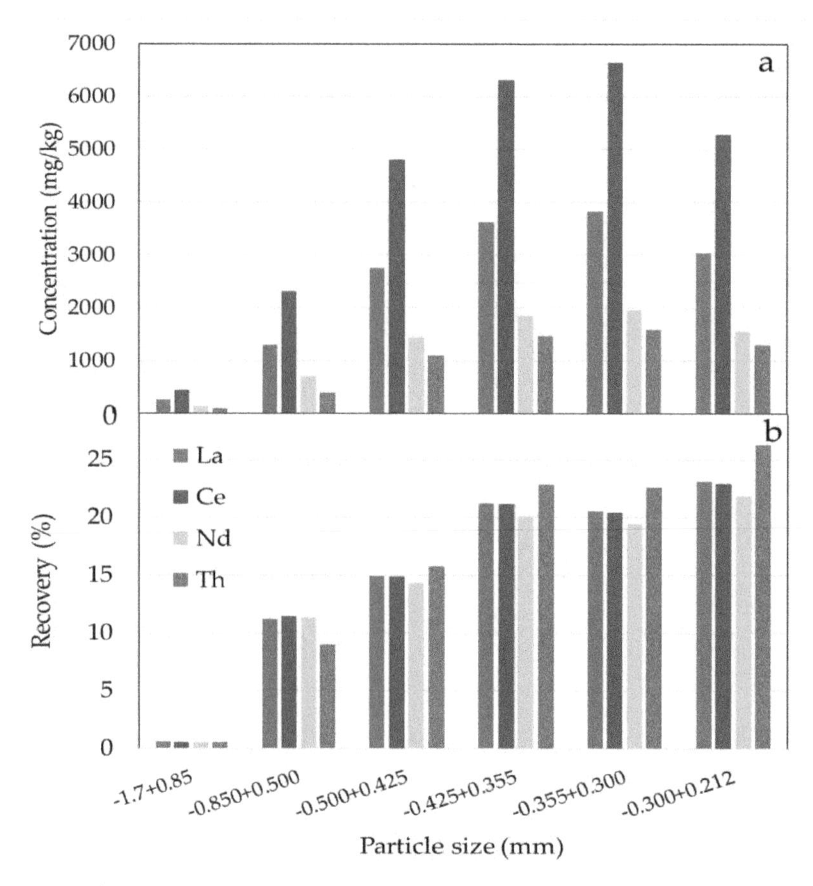

Figure 7. (**a**) LREE and Th distribution in the magnetic products of the particle size fractions and (**b**) metal recovery in each particle size fraction, based on mass balance calculations.

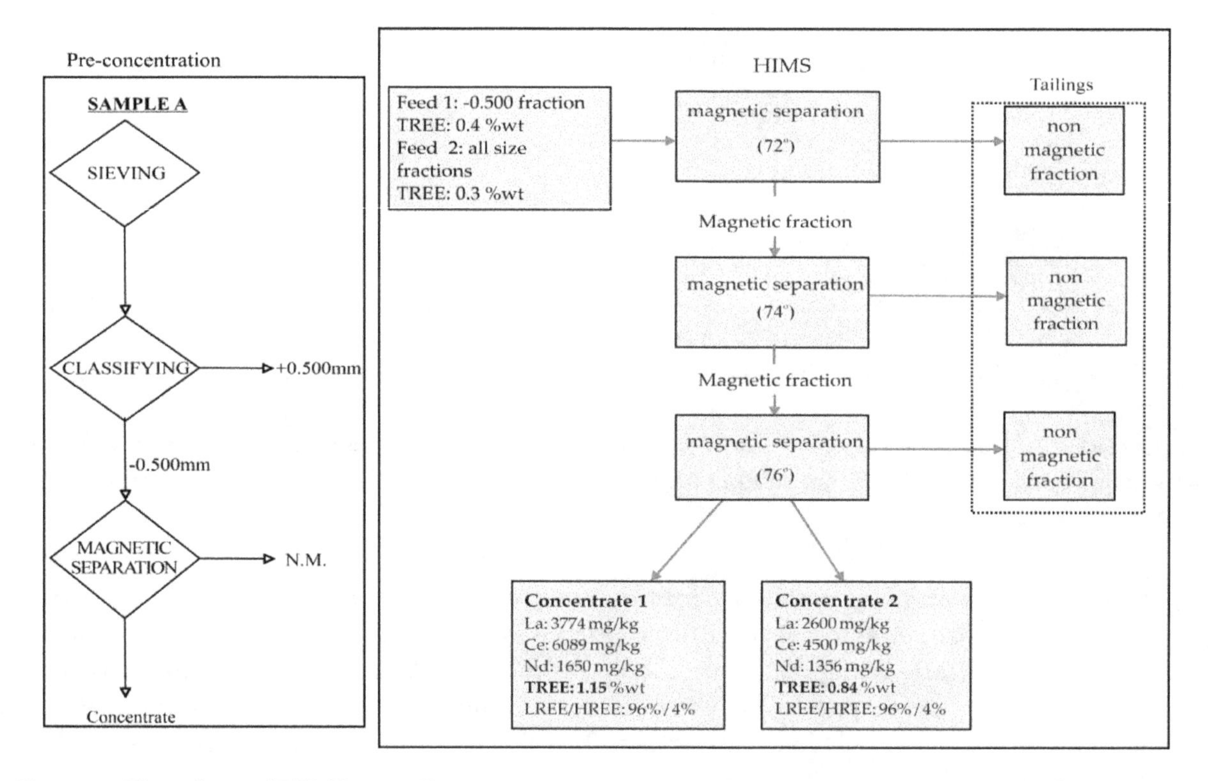

Figure 8. Flowsheet of HIMS tests. Preconcentration step involves sieving and removing of +0.500 mm (Feed 1). Feed 2 corresponds to all the particle size fractions (−1.70 to +0.212 mm) of the initial composite (sample A-mixed).

4. Discussion

The beneficiation of REE ores is challenging from the technological point of view due to complexity in the raw material processing [49–52]. Such processing involves three main stages: an initial beneficiation step, where REE minerals are concentrated from the ore; a second step, where rare earth oxides are extracted from their host minerals producing a mixed rare earth concentrate; and final step, where individual rare earth metals are obtained through metallurgical separation [6].

Mineral characterization is essential for potential process development impacts. Setting the criteria for initial target grind sizes to liberate the economic minerals and determining the possible mineral associations among REE minerals and various gangue phases are both strongly depended on the mineral characterization study [53]. Therefore, the information obtained herein can be used, apart from the prediction of the concentrate quality, in the associated concentration of deleterious elements.

X-ray diffraction (XRD) analysis allows the identification of minerals. However, when mineral modal abundance is less than ~1 wt.% considerable uncertainty and error can be introduced. It is noteworthy that the ore grade in most REE placer deposits is commonly below 0.1% for the REE-host mineral, e.g., monazite [54]. The evaluation of WHIMS magnetic separation products by XRD analysis agrees with the enrichment of allanite in the final cumulative magnetic concentrate where the content of allanite is increased by a factor of 2.5, i.e., from 3 wt.% in the initial composite feed to final 8 wt.% in the magnetic concentrate (Supplementary Table S3). Correspondingly, REE contents of the final magnetic concentrates increased by a factor of 2.8 compared to the initial feed (Table 4, Figure 8). This case study revealed a close association of REE-silicate minerals of allanite with monazite inclusions, as well as a high Th content in the magnetic separate associated with the REE-host minerals, e.g., allanite-monazite/cheralite and thorite. Therefore, the information obtained herein can be used, apart from the prediction of the concentrate quality, in the associated concentration of deleterious elements. One of the most serious issues associated with REE processing is the radioactive wastes produced as by-product of the extraction stage.

The results from the present beneficiation tests of Nea Peramos HM sands in the course of EURARE project also showed the high content of radionuclides, specifically Th. Thorium distribution follows the enrichment trend of LREE in the concentrates and this poses limitations in the concentration process (e.g., [10,11,23]). There are still many questions and knowledge gaps regarding the beneficiation processes of the REE-bearing minerals which require a great deal of investigation [26,28]. Previous research papers described a series of physical separation processes applied to preconcentrated rare earth minerals (REM) and discard iron oxide minerals from the magnetic fractions. These include, gravity, magnetic, electrostatic, and flotation separation techniques ([52] and references therein).

Effects of Mineral Magnetic Susceptibility on the HIMS Treatment

Magnetic separation process of minerals is based on different behavior (magnetic susceptibility) of mineral particles when exposed to an applied magnetic field. The magnetization of a material is a measure of the density of magnetic dipoles induced in the material [44].

The magnetic susceptibility of minerals is mainly controlled by chemical composition. REE minerals such as allanite, monazite, bastnäsite, and xenotime exhibit moderate paramagnetic property [44,48]. Previous studies on the beneficiation applicable for silicate REE minerals, including allanite and cerite from complex ores, have shown that wet high-intensity or high-gradient magnetic separation is more effective than flotation [29,55]. Yang et al. (2015) and Jordens et al. (2014) [55,56] showed that allanite have normally poor floatability using conventional reagents.

The main host mineral for LREE in Nea Peramos HM sands is allanite-(Ce), commonly containing inclusions of other REE-bearing phases such as thorite, zircon, monazite, and cheralite. Therefore, REE concentration in the magnetic fraction is attributed largely to allanite which was efficiently concentrated in the magnetic fraction under high-intensity magnetic field at 0.48 T. The results of magnetic separation from this study are contradictory to a recent study [18]. In the previous study, allanite was recorded to be rather separated in the nonmagnetic part but the results are not comparable since the conditions of the applied magnetic separation test are not described by the authors.

Dry HIMS (DHIMS), applied to each particle size separately, improved significantly the recovery of REE with achieved recoveries of 92–88% (Table 5). Similar REE recoveries were achieved by processing the undersize 0.500 mm, i.e., 96–79%. REE grade of the magnetic concentrate starting from the undersize 0.500 mm is better than the magnetic product of the particle size fractions. In general, the actual TREE concentration in the magnetic products was 2.8 times higher compared to the initial feed contents, in the two feeds tested. The optimum magnetic separation was effected after repetitive passages of the magnetic fraction at increasing belt inclination from 72° to 76°. Moreover, the preliminary results of WHIMS corroborated for an effective recovery of allanite in the magnetic fraction, and the enrichment achieved was in the order of 2.5 times in the magnetic product (Supplementary Table S3).

Additional laboratory tests for REE dissolution and extraction from concentrate was tried by acid treatment, either leaching or acid baking followed by water leaching conducted by I.G.M.E. during EURARE project [57]. The acids used were HCl, H_2SO_4, or both. The results showed that rather low recoveries were achieved. In direct acid leaching, decreasing pulp density leads to increasing rare earth recovery, while the opposite is observed in the recovery procedure with acid baking; an increasing pulp density at acid baking step at about 15% leads to an increasing recovery tendency of about 15–20% (unpublished data, Angelatou pers. comm.). The duration of acid baking test seems to have no effect on the recovery of rare earth elements.

5. Conclusions

HM sands from the shoreline of Nea Peramos, Kavala, Northern Greece was studied in order to be able to determine the best process to concentrate the ore by testing simple screening as a prebenefication method and high-intensity magnetic separation as the main beneficiation method. The results of the test work lead to the following conclusions:

- Allanite-(Ce) is the major host mineral for light REE (LREE), whereas monazite, zircon, and thorite constitute trace amounts. Titanite displays low concentration in LREE <0.1–0.2%. Metamict allanite is common and is thorium-enriched relative to the nonmetamict allanite.
- A simple screening can achieve a satisfactory prebenefication.
- A stepwise magnetic separation improves the recovery of REE.
- Magnetic separation for each particle size fraction separately improves the recovery of REE and reaches recoveries of 75–90% in just the 20% of feed material.
- The grades of magnetic concentrate for the two processes (all particle size fractions and undersize 0.500 mm) were 0.84% and 1.15% TREE, respectively, at the recoveries of 87–92% and 79–96%.
- The increase in REE content is associated with the increase of thorium content in concentrates regarding placer sands from Nea Peramos, Greece.
- The use of gravimetric methods (such as Wilfley shaking table) did not contribute much to the beneficiation of the ores.
- Radioactive wastes from a potential REE processing operation of Nea Peramos HM sands should be carefully encountered for the risk to the environment and humans.

Supplementary Materials
Table S1: REE contents of the Nea Peramos HM sands (in mg/kg), Table S2: SEM/EDS spot analysis of REE-bearing minerals, and Table S3: Semiquantitative mineralogical composition of the wet HIMS-treated samples and the composition of the initial A-mixed composite sample for comparison. Applied voltage values of 30 and 150 V correspond to 0.48 and 150 T magnetic field strength, respectively, Figure S1: X-ray diffraction patterns of composite A-mixed and sample #123, in the range of 20–43° showing in magnification the identified heavy minerals. Quartz appears as the stronger peak. The major peaks of intensities for the heavy minerals only are marked. Monazite (-Ce) peaks shown are indicative (database for RRUFF™ Project [43]). Ttn = titanite, Mz = monazite, Mg-Hbl = magnesiohornblende, Prg = pargasite, Aln = allanite, Mt = magnetite, Hem = hematite).

Author Contributions: V.A. and D.E. collected the samples and provided the field information; V.A. conceived and designed the experiments undertaken by I.G.M.E. in the framework of EURARE project; V.A. and S.P. performed the experiments; C.S. supervised the mineralogical analysis; and C.S., S.P., V.A. and K.S. wrote the paper. All authors have read and agreed to the published version of the manuscript.

Acknowledgments: The authors would like to thank N. Xirokostas for elaborating the chemical analysis of REE ore at the ICP-MS facility of I.G.M.E., V. Skounakis for assistance in performing the SEM/EDS analysis at NKUA, Geology and Geoenvironment, Athens, and I. Marantos from I.G.M.E. for assistance in the quantitative evaluation of the XRD patterns by RIR method. This publication reflects only the authors' view, exempting the Community from any liability.

References

1. Mudd, G.M.; Jowitt, S.M. Rare earth elements from heavy mineral sands: Assessing the potential of a forgotten resource. *Appl. Earth Sci. (Trans. Inst. Min. Metall. B)* **2016**, *125*, 107–113. [CrossRef]
2. EU Commission. *Study on the Review of the List of Critical Raw Materials*; European Commission: Brussels, Belgium, 2017.
3. McLennan, B.; Gorder, G.D.; Ali, S.H. Sustainability of rare earths—An overview of the state of knowledge. *Minerals* **2013**, *3*, 304–317. [CrossRef]
4. Balomenos, E.; Davris, P.; Deady, E.; Yang, J.; Panias, D.; Friedrich, B.; Binnemans, K.; Seisenbaeva, G.; Dittrich, C.; Kalvig, P.; et al. The EURARE Project: Development of a sustainable exploitation scheme for Europe's Rare Earth Ore deposits. *Johns. Matthey Technol. Rev.* **2017**, *61*, 142–153. [CrossRef]
5. Moss, R.; Tzimas, E.; Willis, P.; Arendorf, J.; Thompson, P.; Chapman, A.; Morley, N.; Sims, E.; Bryson, R.; Peason, J.; et al. Critical metals in the path towards the decarbonization of the EU energy sector. In *Assessing Rare Metals as Supply-Chain Bottlenecks in Low-Carbon Energy Technologies*; JRC Report EUR 25994 EN; Publications Office of the European Union: Brussels, Belgium, 2013; p. 242.
6. Goodenough, K.; Schilling, J.; Jonsson, E.; Kalvig, P.; Charles, N.; Tuduri, J.; Deady, E.A.; Sadeghi, M.; Schiellerup, H.; Muller, A.; et al. Europe's rare earth element resource potential: An overview of REE metallogenetic provinces and their geodynamic setting. *Ore Geol. Rev.* **2016**, *72 Pt 1*, 838–856. [CrossRef]
7. Guyonnet, D.; Planchon, M.; Rollat, A.; Escalon, V.; Vaxelaire, S.; Tuduri, J. Primary and secondary sources of rare earths in the EU-28: Results of the ASTER project. In Proceedings of the ERES 2014—1st Conference on European Rare Earth Resources, Milos, Greece, 4–7 September 2014; pp. 66–72.
8. Deady, E.; Lacinska, A.; Goodenough, K.M.; Shaw, R.A.; Roberts, N.M.W. Volcanic-Derived Placers as a Potential Resource of Rare Earth Elements: The Aksu Diamas Case Study, Turkey. *Minerals* **2019**, *9*, 208. [CrossRef]
9. Deady, É.; Mouchos, E.; Goodenough, K.; Williamson, B.; Wall, F. A review of the potential for rare-earth element resources from European red muds: Examples from Seydişehir, Turkey and Parnassus-Giona, Greece. *Mineral. Mag.* **2016**, *80*, 43–61. [CrossRef]
10. Pergamalis, F.; Karageorgiou, D.E.; Koukoulis, A. The location of Tl, REE, Th, U, Au deposits in the seafront zones of Nea Peramos-Loutra Eleftheron area, Kavala (N. Greece) using γ radiation. *Bull. Geol. Soc. Greece* **2001**, *34*, 1023–1029.
11. Pergamalis, F.; Karageorgiou, D.E.; Koukoulis, A.; Katsikis, I. Mineralogical and chemical composition of sand ore deposits in the seashore zone N. Peramos-L. Eleftheron (N. Greece). *Bull. Geol. Soc. Greece* **2001**, *34*, 845–850. [CrossRef]
12. Institute of Geological and Mineral Exploration (IGME). *Geological Map of Greece, Nikisiani-Loutra Eleftheron Sheet, 1:50000*; Kronberg, P., Schenk, P.F., Eds.; IGME: Madrid, Spain, 1974.
13. Institute of Geological and Mineral Exploration (IGME). *Geological Map of Greece, Kavala Sheet, 1:50000*; Kronberg, P., Ed.; IGME: Madrid, Spain, 1974.
14. Eliopoulos, D.; Economou, G.; Tzifas, I.; Papatrechas, C. The potential of Rare Earth elements in Greece. In Proceedings of the ERES2014: First European Rare Earth Resources Conference, Milos, Greece, 4–7 September 2014; pp. 308–316.

15. Papadopoulos, A.; Koroneos, A.; Christofides, G.; Stoulos, S. Natural radioactivity distribution and gamma radiation exposure of beach sands close to Kavala pluton, Greece. *Open Geosci.* **2015**, *7*, 64. [CrossRef]
16. Papadopoulos, A.; Koroneos, A.; Christofides, G.; Papadopoulou, L. Geochemistry of beach sands from Kavala, Northern Greece. *Ital. J. Geosci.* **2016**, *135*, 526–539. [CrossRef]
17. Angelatou, V.; Papamanoli, S.; Stouraiti, C.; Papavasiliou, K. REE distribution in the Black Sands in the Area of Loutra Eleftheron, Kavala, Northern Greece: Mineralogical and Geochemical Characterization of Fractions from Grain Size and Magnetic Separation Analysis (doi:10.3390/IECMS2018-05455). Available online: https://sciforum.net/paper/view/conference/5455 (accessed on 25 April 2020).
18. Tzifas, I.; Papadopoulos, A.; Misaelides, P.; Godelitsas, A.; Göttlicher, J.; Tsikos, H.; Gamaletsos, P.N.; Luvizotto, G.; Karydas, A.G.; Petrelli, M.; et al. New insights into mineralogy and geochemistry of allanite-bearing Mediterranean coastal sands from Northern Greece. *Geochemistry* **2019**, *79*, 247–267. [CrossRef]
19. Papadopoulos, A.; Christofides, G.; Koroneos, A.; Stoulos, S. Natural radioactivity distribution and gamma radiation exposure of beach sands from Sithonia Peninsula. *Cent. Eur. J. Geosci.* **2014**, *6*, 229–242. [CrossRef]
20. Papadopoulos, A.; Christofides, G.; Koroneos, A.; Hauzenberger, C. U Th and REE content of heavy minerals from beach sand samples of Sithonia Peninsula (northern Greece). *J. Mineral. Geochem.* **2015**, *192*, 107–116. [CrossRef]
21. Filippidis, A.; Misaelides, P.; Clouvas, A.; Godelitsas, A.; Barbayiannis, N.; Anousis, I. Mineral, chemical and radiological investigation of a black sand at Touzla Cape, near Thessaloniki, Greece. *Environ. Geochem. Health* **1997**, *19*, 83–88. [CrossRef]
22. Papadopoulos, A.; Koroneos, A.; Christofides, G.; Stoulos, S. Natural Radioactivity Distribution and Gamma Radiation exposure of Beach sands close to Maronia and Samothraki Plutons, NE Greece. *Geol. Balc.* **2015**, *43*, 1–3.
23. Papadopoulos, A.; Tzifas, I.; Tsikos, H. The Potential for REE and Associated Critical Metals in Coastal Sand (Placer) Deposits of Greece: A Review. *Minerals* **2019**, *9*, 469. [CrossRef]
24. Eliopoulos, D.; Aggelatou, V.; Oikonomou, G.; Tzifas, I. REE in black sands: The case of Nea Peramos and Strymonikos gulf. In Proceedings of the ERES, Santorini, Greece, 28–31 May 2017; pp. 49–50.
25. Sengupta, D.; Van Gosen, B.S. Placer-type rare earth element deposits. *Rev. Econ. Geol.* **2016**, *18*, 81–100.
26. Jordens, A.; Cheng, Y.P.; Waters, K.E. A review of the beneficiation of rare earth element bearing minerals. *Miner. Eng.* **2013**, *41*, 97–114. [CrossRef]
27. Jordens, A.; Sheridan, R.S.; Rowson, N.A.; Waters, K.E. Processing a rare earth mineral deposit using gravity and magnetic separation. *Miner. Eng.* **2014**, *62*, 9–18. [CrossRef]
28. Jordens, A.; Marion, C.; Langlois, R.; Grammatikopoulos, T.; Sheridan, R.; Teng, C.; Demers, H.; Gauvin, R.; Rowson, N.; Waters, N. Beneficiation of the Nechalacho rare-earth deposit. Part 2: Characterization of products from gravity and magnetic separation. *Miner. Eng.* **2016**. [CrossRef]
29. Yang, X.; Makkonen, H.T.; Pakkanen, L. Rare Earth Occurrences in Streams of Processing a Phosphate Ore. *Minerals* **2019**, *9*, 262. [CrossRef]
30. Caster, S.B.; Hendrick, J.B. Rare Earth Elements. In *Industrial Minerals and Rocks: Commodities, Markets, and Uses*, 7th ed.; Kogel, J.E., Trivedi, N.C., Barker, J.M., Krudowski, S.T., Eds.; SME: Dearborn, MI, USA, 2006; p. 1568.
31. Burg, J.-P.; Ricou, L.-E.; Ivanov, Z.; Godfriaux, I.; Dimov, D.; Klain, L. Syn-metamorphic nappe complex in the Rhodope Massif. Structure and kinematics. *Terra Nova* **1996**, *8*, 6–15. [CrossRef]
32. Ricou, L.-E.; Burg, J.-P.; Godfriaux, I.; Ivanov, Z. Rhodope and Vardar: The metamorphic and the olistostromic paired belts related to the Cretaceous subduction under Europe. *Geodin. Acta* **1998**, *11*, 285–309. [CrossRef]
33. Mposkos, E.; Kostopoulos, D. Diamond, former coesite and supersilicic garnet in metasedimentary rocks from the Greek Rhodope: A new ultrahigh-pressure metamorphic province established. *Earth Planet. Sci. Lett.* **2001**, *192*, 497–506. [CrossRef]
34. Perraki, M.; Proyer, A.; Mposkos, E.; Kaindl, R.; Hoinkes, G. Raman micro-spectroscopy on diamond, graphite and other carbon polymorphs from the ultrahigh-pressure metamorphic Kimi Complex of the Rhodope Metamorphic Province, NE Greece. *Earth Planet. Sci. Lett.* **2006**, *241*, 672–685. [CrossRef]
35. Brun, J.P.; Sokoutis, D. Kinematics of the Southern Rhodope Core Complex (North Greece). *Int. J. Earth Sci.* **2007**, *96*, 1079–1099. [CrossRef]

36. Liati, A.; Gebauer, D.; Fanning, C.M. Geochronology of the Alpine UHP Rhodope Zone: A review of isotopic ages and constraints on the geodynamic evolution. In *Ultrahigh–Pressure Metamorphism: 25 Years after the Discovery of Coesite and Diamond*; Dobrzhinetskaya, L., Faryad, S.W., Wallis, S., Cuthbert, S., Eds.; Elsevier: Amsterdam, The Netherlands, 2011; pp. 295–324.
37. Burg, J.P. Rhodope: From Mesozoic convergence to Cenozoic extension. Review of petro-structural data in the geochronological frame. *J. Virtual Explor.* **2012**, *42*, 1–44. [CrossRef]
38. Tranos, M.D. Slip preference analysis of faulting driven by strike-slip Andersonian stress regimes: An alternative explanation of the Rhodope metamorphic core complex (northern Greece). *J. Geol. Soc.* **2017**, *174*, 129–141. [CrossRef]
39. Dinter, D.A.; Royden, L. Late Cenozoic extension in northeastern Greece: Strymon Valley detachment system and Rhodope metamorphic core complex. *Geology* **1993**, *21*, 45–48. [CrossRef]
40. Papanikolaou, D.; Panagopoulos, A. On the structural style of southern Rhodope, Greece. *Geol. Balc.* **1981**, *11*, 13–22.
41. Dinter, D.A.; MacFarlane, A.; Hames, W.; Isachsen, C.; Bowring, S.; Royden, L. U-Pb and 40Ar/39Ar geochronology of the Symvolon granodiorite: Implications for the thermal and structural evolution of the Rhodope metamorphic core complex, northeastern Greece. *Tectonics* **1995**, *14*, 886–908. [CrossRef]
42. Pe-Piper, G.; Piper, D.J.; Lentz, D.R. *The Igneous Rocks of Greece: The Anatomy of an Orogeny*; Gebruder Borntraeger: Berlin, Germany, 2002; 573p.
43. Hubbard, C.; Snyder, R. RIR—Measurement and Use in Quantitative XRD. *Powder Diffr.* **1988**, *3*, 74–77. [CrossRef]
44. Anthony, J.W.; Bideaux, R.A.; Bladh, K.W.; Nichols, M.C. Handbook of mineralogy. In *Mineralogical Society of America*; Mineral Data Publishing: Chantilly, VA, USA, 2001.
45. Rosenblum, S.; Brownfield, I.K. *Magnetic Susceptibilities of Minerals—Report for US Geological Survey*; U.S. Geological Survey: Reston, VA, USA, 1999; pp. 1–33.
46. Gupta, C.K.; Krishnamurthy, N. *Extractive Metallurgy of Rare Earths*; CRC Press: Boca Raton, FL, USA, 2005; p. 484.
47. Rudnick, R.L.; Gao, S. Composition of the continental crust. *Crust* **2003**, *3*, 1–64.
48. McDonough, W.F.; Sun, S.S. The Composition of the Earth. *Chem. Geol.* **1995**, *120*, 223–253. [CrossRef]
49. Reisman, D.; Weber, R.; McKernan, J.; Northeim, C. *Rare Earth Elements: A Review of Production, Processing, Recycling, and Associated Environmental Issues*; EPA Report EPA/600/R-12/572; U.S. Environmental Protection Agency (EPA): Washington, DC, USA, 2013. Available online: https://nepis.epa.gov/Adobe/PDF/P100EUBC.pdf (accessed on 30 August 2019).
50. Weng, Z.H.; Jowitt, S.M.; Mudd, G.M.; Haque, N. Assessing rare earth element mineral deposit types and links to environmental impacts. *Appl. Earth Sci.* **2013**, *122*, 83–96. [CrossRef]
51. Dutta, T.; Kim, K.-H.; Uchimiya, M.; Kwonc, E.E.; Jeon, B.-H.; Deep, A.; Yun, S.-T. Global demand for rare-earth resources and strategies for green mining. *Environ. Res.* **2016**, *150*, 182–190. [CrossRef]
52. Wall, F.; Rollat, A.; Pell, R.S. Responsible sourcing of critical metals. *Elements* **2017**, *13*, 313–318. [CrossRef]
53. Grammatikopoulos, T.; Mercer, W.; Gunning, C. Mineralogical characterisation using QEMSCAN of the Nechalacho heavy rare earth metal deposit, Northwest Territories, Canada. *Can. Metall. Q.* **2013**, *52*, 265–277. [CrossRef]
54. British Geological Survey. Rare Earth elements profile. In *Mineral Profile Series*; BGS NERC: Keyworth, UK, 2011; p. 53.
55. Yang, X.; Satur, J.V.; Sanematsu, K.; Laukkanen, J.; Saastamoinen, T. Beneficiation studies of a complex REE ore. *Miner. Eng.* **2015**, *71*, 55–64. [CrossRef]
56. Jordens, A.; Marion, C.; Kuzmina, O.; Waters, K.E. Physicochemical aspects of allanite flotation. *J. Rare Earths* **2014**, *32*, 476–486. [CrossRef]
57. Angelatou, V.; Drossos, E. Beneficiation of green black sands for REE recovery. In Proceedings of the ERES, Santorini, Greece, 28 June–1 July 2017; pp. 49–50.

6

Leaching of Rare Earth Elements from Central Appalachian Coal Seam Underclays

Scott N. Montross [1,2,*], Jonathan Yang [1,3], James Britton [4], Mark McKoy [5] and Circe Verba [1]

1 National Energy Technology Laboratory, Albany, OR 97321, USA; Jonathan.yang@netl.doe.gov (J.Y.); Circe.Verba@netl.doe.gov (C.V.)
2 Leidos Research Support Team, Albany, OR 97321, USA
3 Oak Ridge Institute for Science and Education, Oak Ridge, TN 37830, USA
4 West Virginia Geological and Economic Survey, Morgantown, WV 26507, USA; britton@geosrv.wvnet.edu
5 National Energy Technology Laboratory, Morgantown, WV 26507, USA; Mark.McKoy@netl.doe.gov
* Correspondence: Scott.Montross@netl.doe.gov

Abstract: Rare earth elements (REE) are necessary for advanced technological and energy applications. To support the emerging need, it is necessary to identify new domestic sources of REE and technologies to separate and recover saleable REE product in a safe and economical manner. Underclay rock associated with Central Appalachian coal seams and prevalent in coal utilization waste products is an alternative source of REE to hard rock ores that are mainly composed of highly refractory REE-bearing minerals. This study utilizes a suite of analytical techniques and benchtop leaching tests to characterize the properties and leachability of the coal seam underclays sampled. Laboratory bench-top and flow-through reactor leaching experiments were conducted on underclay rock powders to produce a pregnant leach solution (PLS) that has relatively low concentrations of gangue elements Al, Si, Fe, and Th and is amenable to further processing steps to recover and produce purified REE product. The leaching method described here uses a chelating agent, the citrate anion, to solubilize elements that are adsorbed, or weakly bonded to the surface of clay minerals or other mineral solid phases in the rock. The citrate PLS produced from leaching specific underclay powders contains relatively higher concentrations of REE and lower concentrations of gangue elements compared to PLS produced from sequential digestion using ammonium sulfate and mineral acids. Citrate solution leaching of underclay produces a PLS with lower concentrations of gangue elements and higher concentrations of REE than achieved with hydrochloric acid or sulfuric acid. The results provide a preliminary assessment of the types of REE-bearing minerals and potential leachability of coal seam underclays from the Central Appalachian basin.

Keywords: rare earth elements; coal utilization byproducts; pregnant leach solution; underclay; organic acid

---

1. Introduction

Rare earth elements (REE) are essential for the development of low-carbon, renewable energy technologies. In the United States (U.S.), a lack of domestic REE production is forcing end-users in energy, high-end technology, and manufacturing sectors to seek overseas sources. Exploration and production of new domestic sources of REE and critical minerals (CM) is essential to meet future demands. The U.S. Department of Energy report—2017 Report to Congress on Rare Earth Elements from Coal and Coal Utilization Byproducts—on rare earth elements from coal and coal byproducts outlines the strategic plan for expanding the U.S. REE reserve base [1]. The plan calls for identification of coal and coal byproducts with the highest known concentration of REE and the development of cost-effective separation technologies to recover the resource. A diversified REE product slate that includes recovery of REE from domestic coal byproducts and various types of sedimentary geologic materials can contribute to supply security and help to limit risks to market disruptions [1,2].

Production of REE in the United States, primarily sourced from bastnaesite and other accessory minerals, has increased between 2018–2019 from 18,000 to 26,000 tons, but is low compared to Chinese production (132,000 tons) [2]. The United States' domestic coal and coal utilization byproducts (CUB) are nevertheless promising sources of recoverable REE [1–6]. The term CUB includes a range of materials that are produced during coal utilization [1,3–6]. Coal mining waste rock and coal preparation plant refuse are two types of byproduct that contain underclay, a clay-rich sedimentary rock that is found adjacent to a coal seam. Underclay is commonly categorized as roof or floor rock and it is exposed and sometimes excavated during the mining of a coal seam.

Clay-rich horizons in Central Appalachian (CentApp) coal seams, for example, commonly contain higher concentrations of REE than the coal or other non-clay bearing rock adjacent to the coal seam [3,4]. In the CentApp region, there are approximately 840 coal refuse piles that overlie nearly 40 square kilometers of abandoned mines and coal fields. The amount of coal refuse in Pennsylvania alone is estimated to be 1.5×10^9 cubic meters [7]. Waste refuse piles, which plausibly contain a high percentage of clay minerals, may be heap leached or processed with limited beneficiation techniques (e.g., crushing and grinding, calcining, roasting, and floatation), compared to mineral bound ore. REE and CM can be leached from produced and stockpiled waste materials. With the availability of potential resource material, numerous studies have investigated REE recovery from coal-related materials, including coal fly ash [8,9], coal middlings [5,10], and underclays [3,4,10]. Underclays have an increased resource potential [4] as the rock is often subjected to previous diagenetic events and natural processes that transport and concentrate REE and CM in forms that may be easier to extract, compared to minerals bound in crystalline rock. Ease of extraction makes this type of material a more promising geologic source of REE and CM.

Organic acids and their degradation products provide ligands and chelating agents for heavy metals [11]. The citric acid-citrate system forms a relatively stable complex with alkaline earth metals [12] as well as heavy metals and lanthanides. The citric acid molecule is composed of one alpha position hydroxyl and carboxyl group and two beta position carboxylic acid groups, together the molecule contains at least seven potential O-donor sites that are capable of coordinating metal ions [12,13]. Carboxyl groups of citric acid have been shown to complex with both bivalent and trivalent metal ions in biological systems [13,14] and during interaction with alkaline earth metal ions [12].

Effects of organic acid on the leaching process of REE from ion-adsorbed clays was investigated by Wang et al. [15]. The leaching experiments by Wang et al. [15] showed REE recovery using citric acid was highest (10.4 mmol/kg) at pH range 3.5–4.0. The experiments were conducted at varying pH with the same carboxylic group concentration of 10^{-4} mol/L. Rare earth element concentrations decreased in the solutions with increasing pH (from 2–6). Increased pH should lead to greater acid dissociation because of pKa shifts increasing the number of complexation sites for REE on the organic ligands. The results confirm that organic anions, including anions of citric acid, can act as assistant leaching agents both through the complexation of REE in solution and the interaction with the clay surface to promote changes in the zeta potential of the clay. This process can lead to greater leaching of sorbed cations from the clay surface or for better dispersion of individual clay grains [15].

Citric acid anion recovery of REE from coal seam underclay is a promising method that may liberate higher concentrations of ion exchangeable REE from the clay compared to traditional lixiviants such as ammonium sulfate or sodium chloride. We chose to investigate the influence of organic acids on the recovery of exchangeable, or weakly bonded, REE ions from the surfaces of clay grains, as well as liberation of the nonexchangeable (stronger bonded) REE ions. Our selective approach of using pH buffered organic acid-based solutions amended with ionic constituents (e.g., $(NH_4)_2SO_4$ or NaCl) is designed to isolate and recover the exchangeable REE fraction type through partial dissolution of the clay matrix. The approach is based on previous studies which have shown that optimum recovery of REE adsorbed on clay minerals using ionic lixiviants occurs at the 3–4 pH range [15–17].

A mildly acidic organic acid-based ionic recovery solution with the presence of a monovalent salt likely liberates clay surface adsorbed REE and some inorganic REE mineral phases embedded in the clay matrix. The leaching solution may be pH buffered for optimum recovery from different clay mineral assemblages. Clay grains have a high surface area typically with a negative net charge. Exchangeable and nonexchangeable ions are present on the surface of clay grains. The amount of cation exchange capacity (CEC) or anion exchange capacity (AEC) is dependent on the clay mineral type or presence of organic matter. Exchangeable ions are weakly held in contact with the clay solution and are readily replaced by ions in solution. Positively charged ions, such as Al^{3+}, Ca^{2+}, K^{+}, and REE^{3+}, may be present on clay surfaces as exchangeable ions. Nonexchangeable ions are typically adsorbed by strong bonds or held in inaccessible places within the clay matrix (e.g., K^{+} between layers of illite). The use of organic acids can potentially increase the recovery of REE from clay-rich rocks by: (a) Maintaining a balanced charge on clay surfaces and increasing cation exchange capacity (CEC); (b) selectively dissolving matrix rock and increasing pore space connectivity and transmissivity of fluids; and (c) solubilizing phosphate bound REE [18,19].

Rozelle et al. [3] identified clay-rich ore deposits that contain up to 90% of the total REE in the rock, bound as ion-adsorbed REE, with the balance existing as colloids (e.g., Fe, Mn-oxides) and crystalline minerals (e.g., REE-phosphates). In China, about 10,000 tons of rare earth oxide (REO) concentrate are produced annually from weathered elution-deposits derived from lateritic weathering of granitic rock and in situ aqueous mining yields ~200 tons of REE annually [17]. This study characterizes and tests the leaching behavior of underclay rock from geologic formations associated with coal production in West Virginia, Pennsylvania, and Ohio.

2. Materials and Methods

2.1. Sample Preparation

The National Energy Technology Laboratory Research and Innovation Center (NETL-RIC) obtained underclay core samples from the West Virginia Geological and Economic Survey (WVGES) (Figure 1). The core samples were taken from strata associated with production coal seams—Lower Freeport, Middle Kittanning, and Pittsburgh No.5—in West Virginia.

Figure 1. Photographs of as-received underclay core samples examined in this study. Information and descriptions of rock from the core samples are shown in Table 1.

A sample of approximately 200 g was either cut from the core or collected as chips and pieces. Each sample was pulverized first using a small jaw crusher then Zr-Ti lined shatter box to reduce the grain size to 149 μm or less. A mortar and pestle was used to hand-grind the material until it all passed through a 100 mesh (<149 μm) sieve. After grinding and sieving the powdered sample was homogenized using the cone and quarter technique [20]. The homogenized sample was split into specific subsamples for X-ray diffraction (30 g), elemental analysis (3 g), sequential digest (5 g), and leaching tests (0.5 g and ~40 g). Sample preparation for X-ray diffraction analysis included an additional step. The homogenized sample was ground for 60 s in a micronizing mill to achieve a grain size of <65 μm. The subsamples for the different analyses and tests were stored in chemical-free paper envelopes in a nitrogen purged desiccator until needed. The goal for sampling and sample preparation was to create a bulk homogenized underclay powder that could be used to test the recovery of REE from clay-rich material using different leaching solutions composed of organic acid anions. Bulk analysis and characterization of each sample was conducted to determine basic mineralogical and physical properties of the material. Rock chips or slices of core (3 cm × 2 cm) were used for electron microscopy imaging and X-ray microanalysis. Contextual information about the samples and core descriptions of the material are shown in Table 1.

Table 1. Sample information and core descriptions for underclay samples.

Sample ID	Material Type Formation	Depth (ft)	Core Description
UC-01	Underclay, shale Lower Freeport	1352.0–1352.5	Dark black to gray; fine grain matrix with visible pyrite and calcite cement; irregular sharp wavy contact. Directly underlies base of lower Freeport coal.
UC-02	Flint-clay, underclay Middle Kittanning	1463.0–1463.5	Medium to light gray/olive green rock fragments. Sub angular fine to medium clasts; few fine root traces and plant debris; few fine to medium black shale and coal streaks; clear lower contact.
UC-03	Underclay Pittsburgh	740.5–741.0	Medium gray and olive yellow-brown; common fine distinct olive mottles; few fine faint black mottles; few fine faint red mottles; clear lower contact.
UC-06	Paleosol, seat earth Lower Freeport	712.6–713.0	Pistachio green; extremely brecciated; paleosol; spiderweb calcite cement; siderite banding; soft sediment deformation structures present.

2.2. Scanning Electron Microscopy and X-Ray Microanalysis

Thin section and epoxy-mounted samples were evaporatively coated with carbon and imaged with a field emission scanning electron microscope (FE-SEM, FEI Inspect F) equipped with an energy-dispersive X-ray spectrometer (EDS, Oxford Instruments, Abingdon, UK). SEM imaging and EDS analysis was done at 20 kV, ~100 nA; a working distance of 10.4 mm, beam aperture 3, and spot size 5.0–5.5 nm. The entire area of the sample—thin section or epoxy mounted rock slice—was viewed frame by frame in x and y directions at low magnification (300×) in backscattered electron mode (BSE). Electron microscope images and EDS data were collected from single spots and full fields of view at multiple locations within the sample. Large area images (4 mm^2) with corresponding EDS maps were collected and constructed using the automate function in the Oxford INCA SEM-EDS software package (Version 5.05, Oxford Instruments, Abingdon, UK). Standards-based quantitative EDS was accomplished using REE-phosphate standards REEP25-15+ FC (Astimex Standards Ltd., Toronto, ON, Canada) and REE-oxide standard #489 (Gellar Analytical, Topsfield, MA, USA) for all analyses. Standard block #489 is certified to ISO 9001 and 17025 standards. Putative mineral phase identifications were made using images and elemental data from SEM-EDS analysis.

2.3. X-Ray Diffraction

Bulk mineralogy of rock samples was determined by X-ray diffraction (XRD) of randomly oriented powder mounts. Each sample was powdered to <63 μm using a micronizing mill. The powdered samples were spiked with 10 wt. % ZnO and mounted on an automatic 6-position sample changer equipped with a sample spinner. XRD patterns were collected using a Rigaku III Ultima diffractometer with Cu K-alpha radiation at 40 kV and 44 mA from 3.0–65.0 degrees-two-theta with a step size of 0.02° at 2.4 s. Initial peak alignments and identifications, and mineral IDs were made via comparison of the diffraction peaks against the ICDD-4 database using HighScore Plus XRD software (Version 3.0, Malvern PANalytical Ltd., Malvern, UK). Basic Rietveld fitting was performed using the software to quantify mineral percentages and estimate amorphous content (wt. %). Semi-quantitative analysis of crystalline components and mineral phase identifications were done by diffraction pattern analysis using the RockJock 7.0 computer program (U.S. Geological Survey, Boulder, CO, USA) [21].

Oriented mounts were prepared for clay identification by XRD analysis following the methods outlined in the U.S. Geological Survey (USGS) Open file report 01-041 [22]. The prepared mounts of each sample were scanned with a Rigaku III Ultima diffractometer (Rigaku, Tokyo, Japan) with Cu K-alpha radiation at 40 kV and 44 mA from 2.0–30.0 degrees-two-theta with a step size of 0.02° at 0.5° per minute. After the initial scan, the samples were treated sequentially with ethylene glycol and two separate heat treatments (400 and 550 °C). The samples were scanned after each treatment. Phase IDs, peak alignments, and mineral identifications of clay mineral peaks were made via comparison of the diffraction peaks against the ICDD-4 database using MDI Jade 6.0 XRD software (MDI Jade, Livermore, CA, USA). Diffraction patterns for untreated and treated samples were compared using Jade and basic Rietveld fitting was performed using the software to quantify mineral percentages. Presence of specific clay mineral phases was determined by changes in diffraction peak patterns across treatments following the identification flow chart in the USGS report [22].

2.4. Particle Size

Particle size analyses were completed on unreacted and reacted solid samples using a Malvern Mastersizer2000 (Malvern PANalytical Ltd., Malvern, UK) following procedures outlined in Sperazza et al. [23]. Briefly, ~5.5 g/L sodium hexametaphosphate was added to the solid samples as a dispersant and vortexed. The resulting slurry was then added to the sample introduction unit and the laser obscuration value adjusted to fall between 10–20% by adding tap water or additional sample. For unreacted samples, 60 s of ultrasonication was applied in the pre-measurement routine. For reacted samples, ultrasonication was turned off to preserve the particle size distribution from the reaction. Standardization and accuracy of measurements was monitored with QA standard QAS3002 (15–150 μm) from Malvern. Analytical procedure and results are shown in the Supplementary Materials.

2.5. Sequential Acid Digestion

Underclay powders were reacted sequentially using ammonium sulfate $(NH_4)_2SO_4$, hydrochloric acid (HCl), sulfuric acid (H_2SO_4) and the residual solids were subjected to lithium borate ($LiBO_2$) fusion and digestion. The procedure was designed to operationally evaluate common lixiviants used in commercial leaching and extraction of REE from geologic materials [24] and to provide a first-order comparison to leaching with organic acid-based reagents. The reagents and conditions for each step of the sequential digest are shown in Table 2. Dry, powdered sample was combined with different reagents and mixed in polypropylene tubes on a rotator at 25 rpm or stirred and heated in 100-mL Teflon beakers on a magnetic hot plate. At the end of each step the extraction solutions were separated by centrifugation 3500× *g* for 20 min. The extraction solution was collected from the centrifuge tube with a syringe and the liquid passed through a 0.45 μm nylon filter and collected for analysis. Major, trace, and rare earth element concentrations in the liquids collected were determined by inductively coupled plasma optical emission spectroscopy (ICP-OES) and mass spectroscopy (ICP-MS) following the methods in Bank et al. [25]. Post extraction solids were collected by centrifugation and washed by rinsing the solids with ~30 mL of MilliQ, mixing solids and water on the rotator for 5 min and then separated by centrifugation (3500× *g*, 20 min). The wash step was repeated three times. The washed solids were dried at 60 °C overnight and weighed for dry weight. Weight loss for each step was always less than 5%. The steps were repeated for each solution. The remaining solid material was collected for $LiBO_2$ fusion and digestion [25] and reported as residual.

Table 2. Extraction reagents and conditions used in sequential extraction. Concentration of reagents shown in mol/L (M). Mass of starting.

Step	Reagent/Target Fraction	Solids (%)	Temperature (°C)	Time (h)	pH
1	0.5 M $(NH_4)_2SO_4$ Exchangeable	1	22	4	5.0
2	1 M HCl Colloid	1	22	24	1.0
3	1.2 M H_2SO_4 Colloid + Mineral	1	70	1	0.86
4	$LiBO_2$-Digestion Mineral + Residual	200 mg	-	-	-

2.6. Citrate Leaching of REE

Benchtop leaching experiments were conducted on subsamples of underclay powders using various formulations of a water based leaching solution. The composition of each solution tested and a list of samples and conditions for each leach are shown in Table 3. The solutions were composed of combination of citric acid, sodium chloride, and conjugate buffer salt sodium citrate tribasic dihydrate ($HOC(COONa)(CH_2COONa)_2{\cdot}2H_2O$). Initial benchtop tests were conducted using rock powders from sample UC-01, 02, 03, and 06 and liquid–solid ratio of 1% solids (e.g., 50 mL of leaching solution—0.5 g of powder). The powders and solution were combinedin 50 mL polypropylene tubes and mixed using a tube mixer. Subsequent testing was performed on sample UC-02 using 20 g of powder and 200 mL of citrate leaching solution (10% solids)The test was conducted to (a) to verify the 10% slurry was properly mixed during the leaching steps, and (b) produce a PLS from a larger sample of material. Element concentrations in pregnant leach solutions (PLS) were determined by ICP-OES and ICP-MS. All leaching solutions contained a citrate concentration [citric acid + Na-citrate] of 0.1 mol/L (M) (See Table 3). In some cases, $(NH_4)_2SO_4$, or NaCl was added to the organic acid solution to provide an additional source of ions for ion exchange. Solutions were buffered to pH 3, 4, 5, or 6 using a mixture of citric acid and sodium citrate at a final concentration of 0.1 M citrate. The leaching solutions were mixed with a 0.5 g subsample of rock powder that was originally collected from the bulk homogenized sample (see above section) dried and powdered (<150 μm, 100 mesh) underclay sample in a 50 mL polypropylene centrifuge tube. The tubes were mixed on a rotational mixer for a range of times from 4–24 h at room temperature. The solution was separated from the slurry at the end of the leaching time by centrifugation (3500 rpm for 25 min.). The liquid was recovered from the tube with a syringe then passed through a 0.45-micron nylon filter. The solution pH was measured, and the remaining liquid sample was analyzed by ICP-MS at the NETL Pittsburgh Analytical Laboratory following the methods in Bank et al. [25].

Table 3. Leaching solutions and conditions used in benchtop experiments.

Solution ID	Composition	Samples Tested	Solid (%)	Temp (°C)/Time (h)	pH
RS-1	0.1 M Citrate * ($C_6H_5O_7^{3-}$ + 0.5 M Ammonium Sulfate ($(NH_4)_2SO_4$)	UC-02, UC-06	1	22/4	5.1
RS-2	0.1 M Citrate * + 0.5 M Sodium Chloride (NaCl)	UC-01, UC-02, UC-06	1, 10	22/24	3.5
RS-3	0.1 M Citrate *	UC-02, UC-06	1	22/24	2.0
RS-4	0.5 M Sodium Chloride (NaCl)	UC-01, UC-02, UC-06	1	22/24	5.0
RS-7	0.1 M Citric acid + Na-Citrate ($NaC_6H_5O_7^{3-}$) + 0.5 M NaCl	UC-01, UC-02,UC-03, UC-06	1	21/24	3.0–6.0

* Citrate from citric acid.

2.7. Flow Through

Powdered samples and fractured core samples were flooded with leaching solution RS-2 (see Table 3) a citrate buffered solution amended with NaCl. Fluid flow was established for time periods of 1–24 h. Hold-in times, referring to the length of time the solution is in contact with the sample (either powder or core) without flow, were varied from 20 min to 5 days. For the powdered clay samples, saturated flow was initially established and maintained for 5–6 h, after which flow was discontinued and sampling proceeded in a stepwise function at discrete time points of 6 h, 7 h, 24 h, and 5 days from the initial contact of fluid with the sample. Detailed descriptions of equipment, experimental setup, and the experimental parameters are found in the Supplementary Materials section on flow-through experiments.

3. Results and Discussion

3.1. Characterization of Underclay

The concentration of elements in the underclay samples, expressed as oxides, are shown in Table 4. Silicon (as SiO_2) and Al (as Al_2O_3) are the dominant cations in all samples. The range of Al_2O^3 concentrations (see Table 4) indicate the material comes from highly weathered crustal materials [26]. Additionally, low Ca, Na, and K values are also indicative of highly weathered horizons or zones of intense leaching. The exception is the Ca concentration in UC-06, withcalcite and siderite cement present throughout the rock matrix and visible in hand specimens (See Figure 1).

Table 4. Concentration of major cations as oxides (wt. %) in underclay samples.

Sample	Al	Ca	Fe	K	Mg	Na	P	Si	Ti	Zr
UC-01	18.5	0.4	9.0	3.1	1.6	0.3	0.1	62.8	0.9	<0.04
UC-02	28.0	0.8	1.7	3.2	0.5	0.4	0.1	63.3	1.4	<0.04
UC-03	18.9	0.7	4.5	4.0	1.2	0.6	0.0	68.4	0.9	<0.04
UC-06	24.9	5.4	2.4	3.9	0.4	0.4	0.4	65.3	1.2	<0.04

The extensive network of carbonate cement throughout the clay matrix is due to late diagenesis. The presence of diagenetic calcite in sample UC-06 results in a high concentration of Ca (5.4%) compared to the other underclay samples analyzed <1.0%. Rare earth element concentrations in each sample are shown in Table 5. The four samples have REE concentrations ranging from 262–353 mg/kg (See Table 5). The results reported in Table 5 are from the powdered and homogenized samples used for characterization, sequential digest, and 1% solids leaching tests. The results from a subsequent analysis of powder samples used in the leaching tests under different pH conditions are shown in Table S1 of the Supplementary Materials. Sample UC-02, a flint clay/underclay from the Middle Kittanning formation, had the highest REE concentration.

Table 5. Rare earth element concentrations (mg/kg) in West Virginia coal underclay samples. Values are reported on a whole sample basis.

Sample	La	Ce	Pr	Nd	Sm	Eu	Gd	Tb	Dy	Y	Ho	Er	Tm	Yb	Lu	ΣREE
UC-01	51.0	103.1	12.6	49.5	10.0	2.1	10.9	1.2	8.9	43.8	1.7	4.6	0.6	4.1	0.6	305
UC-02	82.9	119.3	13.2	45.4	9.4	2.8	9.7	1.0	8.0	50.3	1.5	4.3	0.6	4.2	0.6	353
UC-03	58.3	74.5	8.7	32.9	7.3	2.1	7.2	0.7	6.7	54.0	1.2	3.7	0.6	3.8	0.5	262
UC-06	71.3	102.1	9.7	30.9	7.1	1.9	6.0	0.4	5.1	35.4	0.9	2.9	0.4	3.0	0.5	278

Scanning electron microscope images of polished samples (see Figure 2) and various powdered samples (data not shown) were used to evaluate the composition and texture of the rock samples. SEM images and EDS results were used to make putative identifications of primary and secondary minerals in the samples, including the REE-bearing phases present. Rare earth element phosphate minerals were observed in all samples using SEM backscatter mode. The predominant REE-bearing mineral phases observed were rhabdophane, apatite, churchite, monazite, xenotime, and crandallite sp. Examples of REE-bearing minerals observed in the underclay samples analyzed are shown in Figure 2. Yang et al. [27] provides a comprehensive characterization of all REE-bearing minerals in the samples discussed here, as well as other underclay samples from West Virginia coal seam strata.

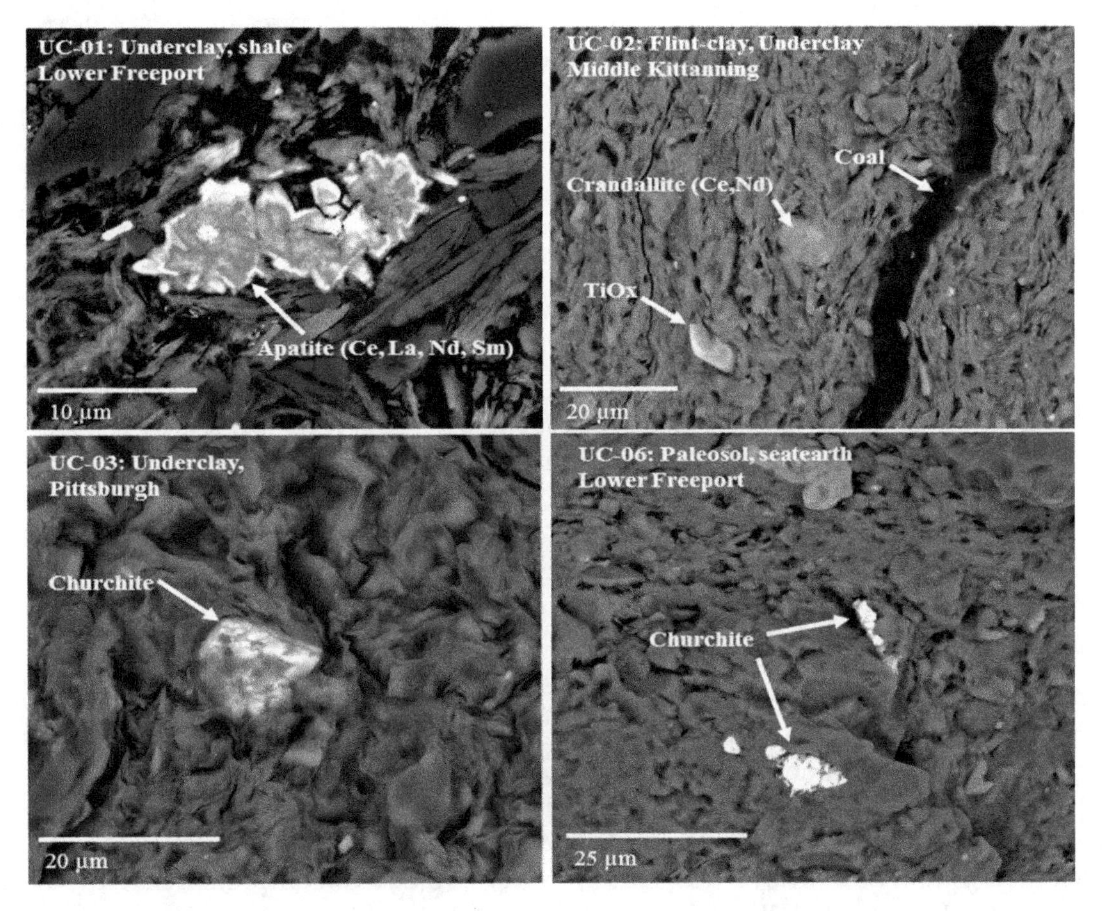

Figure 2. SEM micrographs of rock matrix and examples of rare earth elements (REE)-bearing minerals present in West Virginia underclay samples.

In sample UC-04 trace concentrations (1–4 wt. %) of Ce, La, and Nd were detected in grain-coating clay associated with pore filling framboidal pyrite and pyrite cement. Sample UC-02 contained Ce, La, Nd associated with aluminum phosphate ($AlPO_4$) mineral grains that were dispersed throughout the rock matrix and have a similar chemical composition and morphology to the mineral crandallite, a hydrous aluminum phosphate. Crandallite contains LREE, Ba, and Sr and is present throughout the clay matrix as 5–50-μm sized crystals that oval to round in grain mount samples and appear as spongy, porous in thin section (Figure 2, upper right). A summary of the REE-bearing minerals identified in the samples and general observations of texture and other mineral phases present in the samples are shown in Table 6.

Table 6. General characterization results from SEM-EDS analysis.

Sample ID	REE-Bearing Minerals Identified	General Observations
UC-01	Apatite in pore space contains Ce, La, Nd REE phosphate grains with Ce, La, Nd, Sm, and Eu up to 20 μm long, also contain U/Th. Ce, La, and Nd detected in clay coating on framboidal pyrite. Ytterbium detected in pore filling pyrite cement.	Abundant pyrite in bands and isolated matrix grains Large euhedral pyrite grains (up to 50 μm) in matrix and as pore filling cements. Apatite grains contain ~1–3 wt. % U and Th.
UC-02	Ce, La, Nd phosphate (rhabdophane) and monazite) in clay pore space. Y, Gd, Dy, Er, Yb phosphate (xenotime) grains present in clay pore space. Range of size 1–10 μm long. Xenotime grains bound in massive iron oxide. Ce, La, Nd associated with aluminum phosphate ($AlPO_4$) dispersed throughout the rock matrix. Aluminum phosphate grains with similar chemical composition and morphology of the mineral crandallite. Crandallite present throughout the matrix as 5–50 μm size grains.	Abundant Ti Oxide with Hf (0.5 wt. %) and Sc (0.25 to 0.5 wt. %). Aluminum phosphate grains present. $AlPO_4$ contains equimolar concentrations of S, Sr, Ba, and REE (Ce, La, Nd, Sm). Stoichiometrically constant with the mineral crandallite an illite conversion product. Massive iron oxide. Mixed Cu, Se, Pb sulfides Plant root fossils
UC-03	Ce, La, Nd phosphate (rhabdophane or monazite) mineral grains in clay. Y phosphate (churchite or xenotime) grains in matrix.	Sample contains abundant Fe-oxide and Ti oxide. Fe oxide band collocated within coal layer. Zircon present. Clay matrix composed of illite and smectite. Abundant quartz
UC-06	Monazite present as large crystals 10–25 microns. Monazite contains up to 5.0 wt. % U/Th. Xenotime and monazite present as embedded grains within siderite or calcite. Cementation by Ca/P mineral phase with LREE detectable using EDS.	Abundant pore filling and pore lining clay. Matrix clay composed of fibrous, tubular morphology typical of Halloysite and platy particles resembling kaolinite. Diagenetic spider web calcite, banded siderite, and pore filling clay, Calcium phosphate mineralization Pyrite, barite, zircon, rutile, and galena present in the matrix. Light grey zones have high quartz content, lack extensive carbonate cementation. Dark grey zones contain massive siderite and calcite.

3.2. XRD Results

Analysis of diffraction patterns collected from randomly oriented and oriented powders showed that predominant crystalline non-clay components are quartz, calcite, and ilmenite. Illite and smectite are the most abundant clay minerals in the samples. Halloysite is present in minor (5–7%) abundance in all samples except UC-03. Clay minerals make up more than 55% of the bulk material in each sample. Semi-quantitative results for all non-clay and clay minerals identified are shown in the Table S2 in the Supplementary Materials. The samples are all composed of two-component mixed clays from the groups Kaolin (kaolinite and Halloysite) and Mica (illite). Kaolinite and halloysite were identified by evaluating changes to the 7 Å XRD peak present in the scans. In all samples the peak remained unchanged or there was a small increase in d-spacing following glycol treatment, both of which are attributed to the presence of kaolinite or halloysite. The distinguishing treatment was the destruction of the 7 Å XRD peak after heating to 550 °C. Further confirmation of the presence and classification of kaolinite and halloysite were made using electron micrographs. Kaolinite displays a hexagonal morphology whereas halloysite has a tubular morphology. Analysis of SEM images confirms the presence of grains with hexagonal and tubular morphology. Illite was confirmed through the presence of a 10 Å XRD peak that remains unchanged by glycol and heat treatments.

3.3. Leaching of Rare Earth Elements from Underclay

The results of the sequential extraction of underclay powders and other leaching tests conducted on the samples in this study show that REEs are distributed across different mineral phases in the underclay. The results provided a basic screening of the distribution of REE in the bulk samples, specifically we were interested in the fraction of REE that was in the exchangeable phase. Other authors have presented results on ion-exchangeable clay from CentApp coal seam strata [2,18] and the results are not consistent across units in the basin [2,18,25]. Heterogeneity likely exists between individual clay units at the formation scale and plausibly within core samples collected from specific units. The samples may not be representative of an entire formation or basin, however, the results presented here are meaningful to evaluating different leaching solutions to recover REE from the material. The results of basic characterization efforts and leaching tests provide valuable information on not only the mineralogy and nature of REE-bearing minerals in the clay, but how the material responds to leaching with citrate solutions. In order to mature this technology and raise to a higher readiness level (TRL), currently at TRL 4-5, both the type and scale of sampling must be reconsidered, and the amount of material tested (e.g., scale) will have to be increased several fold. This is necessary due to the prevalence of heterogeneity and spatial variability of elements in geologic materials, as well as the presence of hot spots in such materials and within specific areas [28].

The samples discussed here and in Yang et al. [27] were taken from existing core samples. The characterization and analysis results reported here are from subsamples taken from the bulk, homogenized material prepared for the leaching tests. The results of basic sequential digest of the material, which can be considered a step-wise leaching test, are compared to the results of leaching using citrate. The results of the leaching tests are significant to evaluating alternative lixiviants (e.g., citric acid-citrate) and to compare the results to standard conventional mineral acid or salt leaches. The characteristics of the new solution, at minimum, should leach the ion-exchangeable fraction of REE from the samples, not produce excess liquid or solid hazardous waste, and be amenable to downstream processing to separate and recover REE in its pure form. Our leaching tests and demonstration of sorbent capture of REE from the citrate leachate demonstrate the potential of this technology.

We trialed a suite of different organic acids (e.g., acetic, indole-3-acetic, citrate) and formulations during our initial testing and development of an organic acid based lixiviant for recovering REE from clay [29]. Based on initial results we chose to pursue further testing and optimization of a citrate-based solution. A majority of the REE in the underclay samples analyzed here are bound in the residual phase (Table 7) and not extractable by exchange, or dissolution using 1 M HCl, or warm (70 °C) 1.2 M H_2SO_4. The complete set of analytical results for the sequential digests are shown in the Table S3 in the Supplementary Materials.

Table 7. Fraction of REE in pregnant leach solution (PLS) from sequential digestion of underclay powders. Values reported as percent of total REE leached from the solid sample.

Sample	$(NH_4)_2SO_4$ Exchangeable	HCl Extractable	H_2SO_4 Extractable	Residual
UC-01	0.3	12.1	8.2	79
UC-02	7.5	19.0	6.8	67
UC-03	4.9	31.1	20.7	43
UC-06	1.3	10.1	3.5	85

The $(NH_4)_2SO_4$ exchangeable fraction of REE in the samples ranged from 0.3–7.5% and are low but generally fall within a range of other published values for ion-exchangeable REE in CentApp coal and coal byproducts [6,29,30], with the exception of the values reported by Rozelle et al. [2].

Underclay from the Middle Kittanning formation (MKT)—sample ID UC-02—and from the Pittsburgh coal seam—sample ID UC-03—contained the highest concentrations of $(NH_4)_2SO_4$ exchangeable and HCl extractable REE of the samples tested. Both samples contain ~25–35% of the total REE in the exchangeable and HCl leached fraction (Figure 3). This fraction is likely comprised of REE bound to the surface of clay minerals, carbonates, and Fe-oxides. While the remaining REE is bound to phosphatic minerals or within the structure of more recalcitrant phases and may only be recovered using hot sulfuric acid or more destructive dissolution techniques such as microwave digestion or treatment with hydrofluoric acid. The concentration of REE in the HCl extracts may be higher in these samples due to the abundance of metal oxides that may bind colloidal ions or carbonate that may have ion adsorbed/exchangeable REE (See Table 8). The concentration of REE and gangue elements in leachates recovered from sequential extraction were compared to bench-top leaching experiments with citrate, and the results are shown in Figure 3 and Table 8. Ammonium sulfate used for recovery of ion exchangeable leached ~10 μg of REE or approximately 7% of the total REE present in UC-02 (Figure 3 and Table 8), other samples tested had low recovery (<10% of the total REE) when using ammonium sulfate. Citrate leaching using a cycle of citrate solution amended with NaCl (pH 5) leached ~45 μg of REE into the PLS. This amount correlates to nearly 33% of the total REE in the clay sample. Two different slurry concentrations using sample UC-02, 1 and 10%, were leached with citrate + NaCl. The results of the comparison are shown in Table S5.

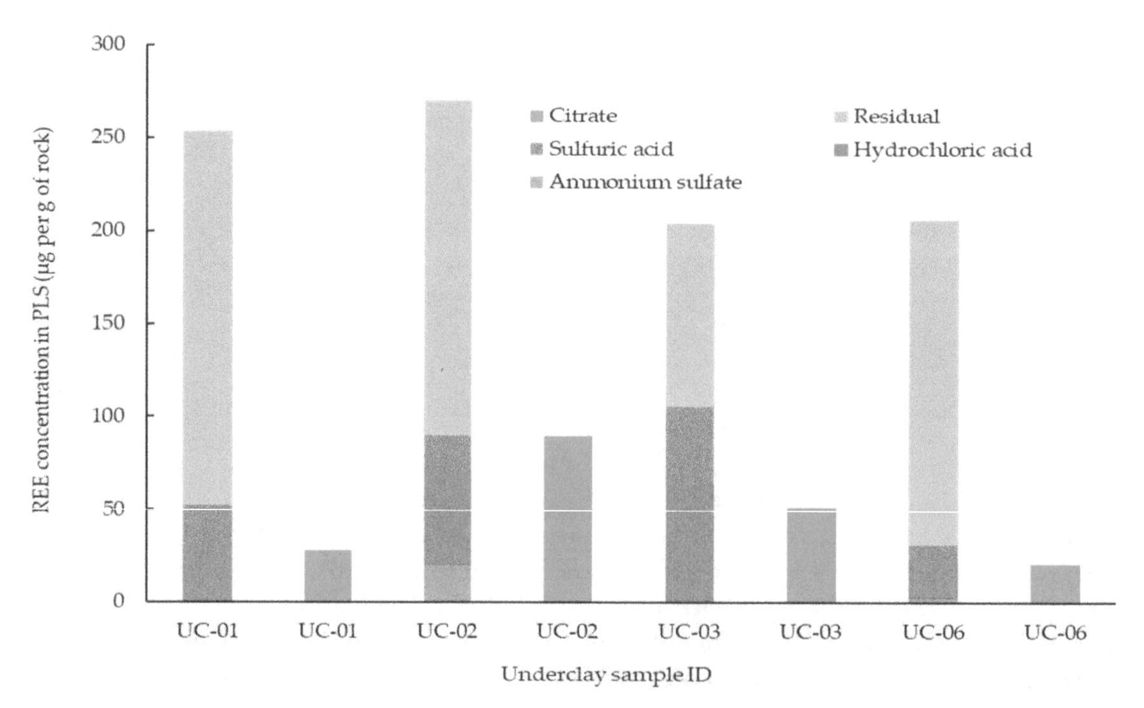

Figure 3. Concentration of REE in PLS after leaching with citrate solution and sequential digestion solutions. Concentrations leached into PLS for each step are denoted by the different bar colors. Citrate recovery values are shown as green bars. Values are reported as concentration of element in leached into PLS per gram of material (μg/g).

The results from the benchtop leaching experiments indicates that the citrate anion solution is more effective at recovering potentially ion exchangeable, or easily liberated REE that is present in the clay-rich sample. Additional REE, excess of ion adsorbed, is likely leached via chelation or complexation of the element from the clay or other mineral surface and solubilized into the leachate, presumably as an REE-citrate complex. In samples with less than < 1% ion exchangeable REE (e.g., UC-01, 03, and 06), citrate leached 10× more REE from the underclay than ammonium sulfate or NaCl, both of which are used commonly to recover exchangeable ions from soil and rock (Table S6, Supplementary Materials). One plausible explanation for the greater recovery of REE from citrate + sodium chloride treatment compared to ammonium sulfate is the effect of solution electrolyte concentration on dispersion of clay grains [31,32].

Table 8. Concentration (in μg/L) of Al, Si, Fe, Th, and REE in mineral acid and organic acid PLS generated from leaching of Middle Kittanning underclay (sample UC-02). Values are adjusted to mass of material leached.

PLS Composition	Al	Si	Fe	Th	REE
Mineral acids					
$(NH_4)_2(SO_4)$	5.0	1320	<DL	0.1	202
HCl	27,610	35,006	56,860	16	508
H_2SO_4	21,696	19,766	39,230	6	184
Σ Mineral acids + $(NH_4)_2(SO_4)$	76,907	56,092	57,252	24	894
Organic acids					
Citrate + NaCl (step 1 of 2)	21,230	13,491	3246	7.0	976
NaCl only (step 2 of 2)	1060	bd	843	0.5	44
Σ Citrate + NaCl	24,520	13,491	4099	7.5	1020
Citrate + $(NH_4)_2(SO_4)$	22,860	16,740	3380	18	146

Sodium chloride amended citrate may balance surface charge and increase total dissolved solids (TDS) conductivity in the leachate that may enhance colloid and particle dispersion [32]. We hypothesize that an increase in clay particle dispersion would lead to increased solution-clay grain interaction and higher recovery. However, the citrate anion leaches a greater fraction of REE than simply the exchangeable fraction, determined by $(NH_4)_2SO_4$. It is plausible that exchangeable RE concentrations in these samples are low and do not exceed more than 10% and that citrate liberates an additional fraction of the RE that is not exchangeable and not liberated by ammonium sulfate. Future work will be aimed at optimizing the solution in order to determine if additional mechanisms are at work where clay grains or RE mineral bearing phases are more susceptible to solubilization in the presence of citrate, compared to mineral acids or inorganic salts such as ammonium sulfate.

The concentration of REE and gangue elements in the different leachates from tests conducted on the Middle Kittanning underclay (UC-02) are shown in Table 8. The highest concentration of REE was released (during leaching using one single solution) by leaching with 0.1 M citrate solution amended with NaCl (solution RS-2). The high concentration of Al, Si, and Fe released from samples UC-01 and 06 during treatment with RS-2 leach solution indicates some dissolution of the mineral phases in these samples, as Si is not typically present as an exchangeable ion. The presence of weakly crystalline, amorphous phases and water-soluble species may also contribute to the higher release of these elements. Whether or not the solution can leach REE from refractory or phosphatic minerals such as monazite was not clear. Notably, the citrate PLS from leaching of UC-02 contained measurable P. Further testing is required to evaluate the leachability of these phases using a citrate or other organic acid-based solution. The recovery results for the different solutions tested indicate that the organic acid solution Na-citrate buffered citric acid (solution RS-2) leaches a greater mass of REE compared to ammonium sulfate (Table 8 and Figure 4), in some cases citrate leaching exceeds leaching with inorganic mineral acids (see Figure 4 and Tables S4 and S5).

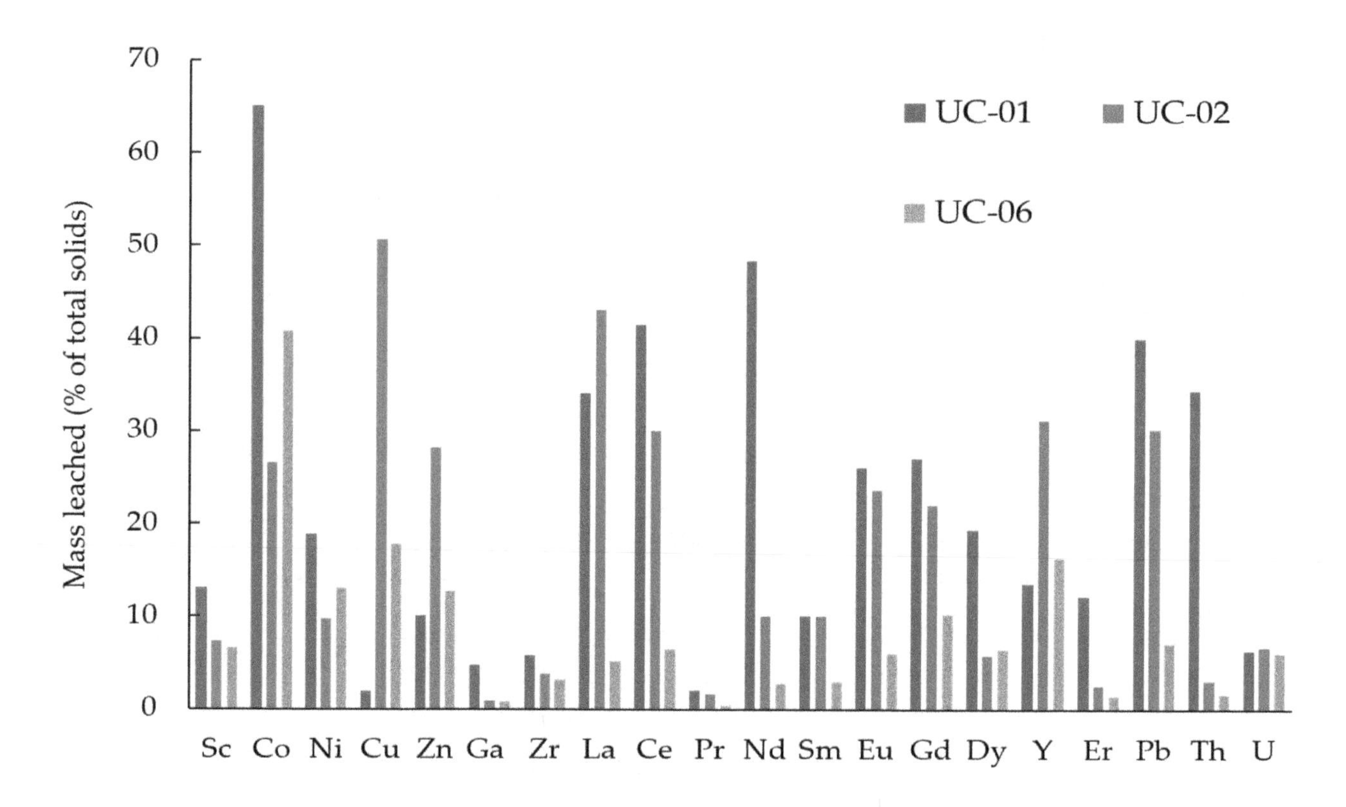

Figure 4. Concentration of REE and other trace metals leached from underclay with 0.1 M citrate + NaCl. Values are reported as the percent of total solids leached from the sample.

Citrate + sodium chloride leaching solution (RS-2) and the other citrate-based solutions tested released a higher concentration of base cations—Al, Si, and Fe—than ammonium sulfate (SEQ-1) but far less than the concentrations of cations released by low-pH inorganic mineral acids (Table 8, Table S3). The ratio of REE to base cations in the leachate accounts for the proportion of REE recovered to the proportion of "contaminants" such as cations Si, Al, and Fe. In recovering 45.9 μg of REE, RS-2, a citrate solution at pH 5, had a REE to base cation ratio of 2.5, compared to a significantly lower value of 0.3–0.2 for the acid and heat-based recovery solutions tested during the sequential extraction. The citrate solution (RS-2) performed best, based on the quantity of REE recovered and the reduction in the concentration of base cations and radioactive Th in the leachate. The citrate + sodium chloride solution, had a significantly higher recovery of REE than $(NH_4)_2SO_4$ from all three different underclay materials tested (Table 8, Tables S2 and S3).

The citrate solution leached greater than 30% of the total Ce, La, Nd, Eu, Gd, and Dy from samples 02 and 03. Leaching from UC-06 was significantly less than the other samples, apart from cobalt. For sample UC-02, both Cu, Y, and Pb are abundant in the leachate. Leaching of the radioactive element Th is low in PLS from samples UC-02 and 06, less than 3% of the total Th in the solid. Th leaching from sample UC-01 is nearly 8× that from the other samples. The concentration of Th in UC-01 starting solid material was lowest of the samples tested, 14 μg/g, compared to the concentration of Th in UC-02 and 06 which was 22 and 24 μg/g respectively (see Figure 5 and Table S3). The difference in Th leaching between the samples may be due to the mineral associations of Th in the different clay strata or formations. Notably, there is evidence that Th leaching is minimum, with less than 3% of the total leached from the clay when using citrate for leaching. Reducing Th in leachates, as well as other elements such as U, can help to reduce costs associated with waste disposal and remediation.

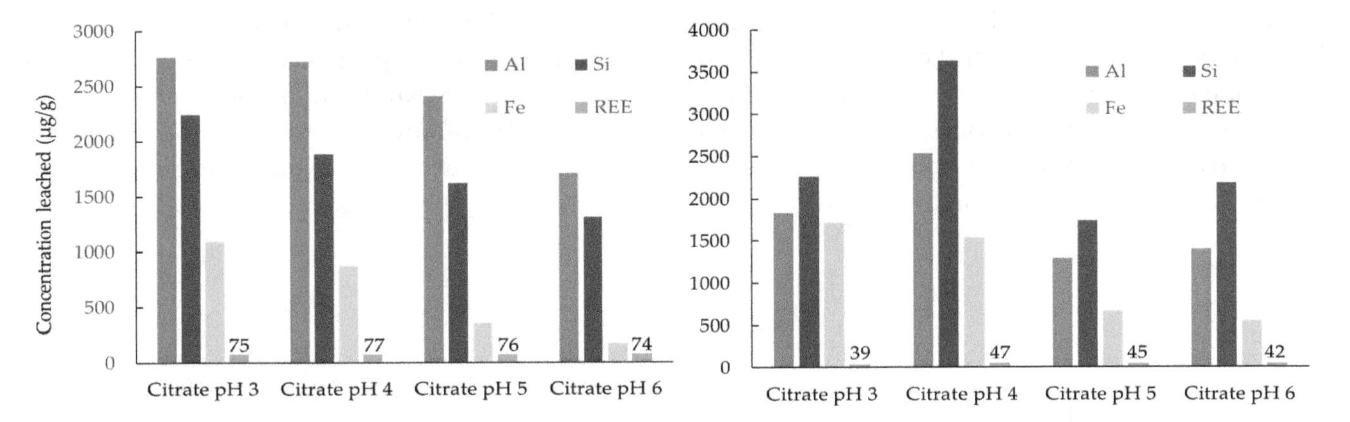

Figure 5. Concentration of Al, Si, Fe, and REE in PLS generated by citrate solution at different pH values for underclay sample UC-02 (**left**) and UC-03 (**right**). Values are converted to micrograms of element leached into PLS per gram of underclay (μg/g).

We evaluated how the pH of the leaching solution impacts the release of REE and gangue elements from samples UC-02 and UC-03 at a range of pH values from 3 to 6. Figure 5 shows the results from leaching of samples UC-02 and UC-03. For both samples tested, there was no significant change in the concentration of REE leached between pH 3–6. Decrease in the concentration of Al, Si, and Fe in the leachate was observed as pH increased (Figure 5, Table S7). Conversely, calcium in the leachate increased with increasing pH. Sequential acid digestions indicate that a majority of the REE are bound in mineral phases in the residual pool and are not extractable by $(NH_4)_2SO_4$, 1 M HCl, or 1.2 M H_2SO_4. The exceptions are for samples UC-02 and UC-03 which have a greater proportion of REE bound as exchangeable or extractable using hydrochloric and sulfuric acid.

At different pH values, range 3–6, there is a reduction in the concentration of gangue in the leachates (Figure 6). The solution can be pH buffered across a range of pH vales from 3–6 which allows for selective leaching of elements from the material. Leaching of non-REE base cations decreases with increasing pH and provides a system to selectively leach REE and minimize leaching of Al, Si, and Fe. Increasing pH should yield greater chelation recovery due to the increase in the number of deprotonated anion sites associated with pKa values for citric acid [12,13]. This is likely due to the monodentate bonding of the REE-citrate complex which is typical of the citrate-metal ion complex [12–14]. This bonding regime is the strongest of the bonds associated with chelation/complexation to carboxylic acid, the functional group on the citrate molecule [14]. There is little change in the concentration of REE in the citrate leachate across pH range 3–6 (see Figure 5, Table S7), which crosses two pKa boundaries for citric acid (e.g., 3.1, 4.7). At pH 4.7, a second carboxylic acid group becomes deprotonated. If this functional group could attract an additional RE ion, there should be a commensurate increase in REE concentration observed between pH 4 and 5 leachates. However, since there is little to no change in REE concentration across the range of pH it is likely the complexation of REE-citrate is more influenced by other factors such as saturation of the clay surface with the solution, dispersion of the clay grains, or amount of carboxylic acid group present (e.g., initial molarity of citrate leaching solution) for chelation and recovery of the metal ions.

The preliminary results from the bench-top experiments indicate that citrate is a chemically effective lixiviant that can be used to generate a PLS from clay-rich sedimentary rock. Presumably, REE and other ionically bound base cations that are present on clay surfaces are solubilized via complexation/chelation with the carboxylic functional group on the citrate anion. Simple chelation/complexation using the citrate anion in solution at pH 5 reduces the potential for dissolution of crystalline mineral phases and concurrent release of elements gangue elements. Such is the case with refractory minerals such as monazite and xenotime that are not only sources of REE but also radioactive elements Th and U. The process presented here is not aimed at leaching of crystalline bound, phosphatic REE. Rather, REE is sorbed to the surface of mineral grains that can be recovered with basic hydrometallurgical leaching.

The citrate PLS generated in the experiments contains a relatively low concentration of gangue elements because of the limited mineral dissolution. The overall economics of the process cannot be assessed at this stage of the research or at the scale of this work. However, future experiments and analysis will be aimed at evaluating the economics of the process including subsequent steps to purify REE from the PLS. Currently there are few processes that have been developed at a higher Technology Readiness Level (TRL) on the extraction and recovery of REE from secondary products and wastes.

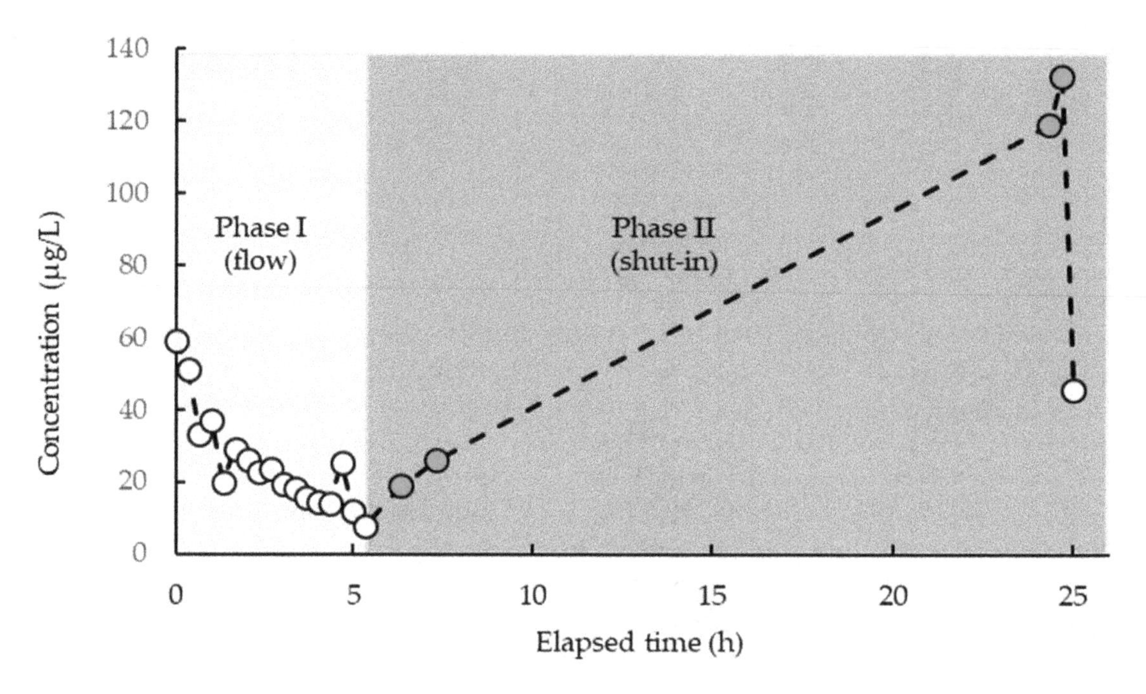

Figure 6. REE concentration (in µg/L) in effluent PLS collected during flow-through leaching of Middle Kittanning underclay (UC-02) powder. Black dots denote sampling points taken during continuous flow. Green dots indicate samples taken after various shut-in times.

3.4. Leaching of REE from Powdered Underclay Using Core Flow-Through

An additional set of leaching tests were conducted on the remaining unreacted powders from the benchtop test tube leaching tests. Flow-through leaching reactors (See Supplementary Materials) packed with rock powder were flooded and subjected to pressurized flow at ~1600–1800 psi to compensate for low porosity and correspondingly low permeability of clay. The highest concentration of REE in effluent leachates occurred during the initial 20–40 min of fluid flow through the powder. Powder reactor runs exhibited lower concentrations of REE released to the fluid and relative to amount of material in reaction vessel. Shut-in periods up to 5 days in some cases recovered equal amounts as continuous flow for 7 h (based on cumulative concentrations). The results indicate that shut-in time may be necessary to fully saturate the material for increased fluid interaction with the surfaces of the clay grains. Peak REE concentrations were also noted to occur within the first 20–40 min of fluid flow, denoted as phase I in Figure 6. Though absolute concentrations increased with increasing fluid shut-in times, the peak REE concentrations were consistently noted to occur within 20–40 min of initializing fluid flow (Figure 6).

Increased recoveries of the REE in the effluent solutions of the underclay powders were evident after increased shut-in times. Peak concentrations increased from the initial start of fluid flow to the samples flowed at 24 h (after ~17-h shut-in period) to the samples flowed at 5 days. An example of the effect shut-in time has on the liberation of REE into solution is shown in Figures 6 and 7.

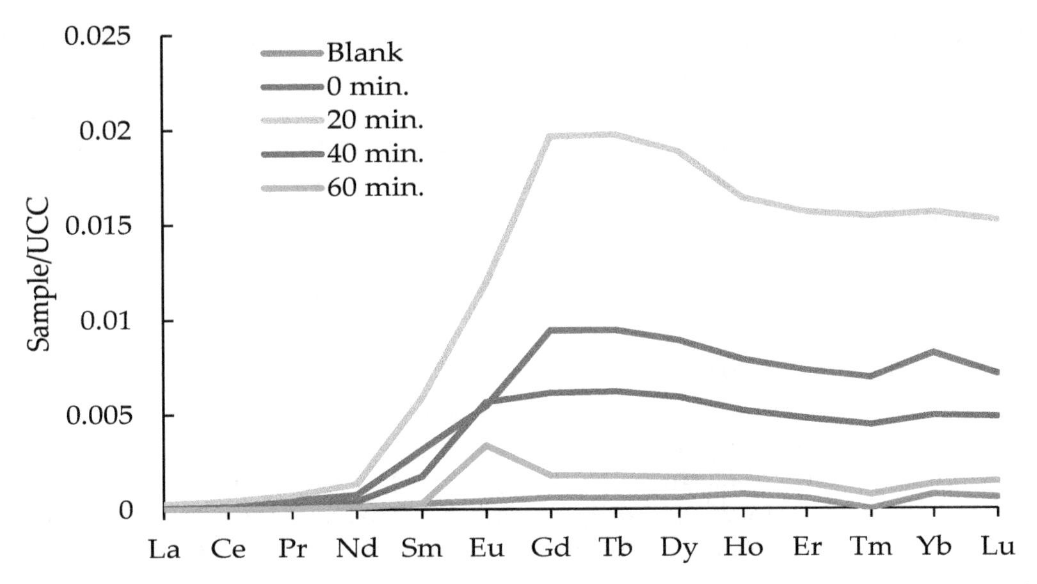

Figure 7. Concentration of REE, normalized to upper continental crust values, in flow-through PLS analyzed at 0, 20, 40, and 60 min.

The highest concentrations of REE were observed in effluents after a 5-day shut-in period. In Figure 6, the initial opening of the reactor, after 5 days, occurs at 0 min. Subsequent measurements were taken at 20, 40, and 60 min after flow was established. The highest concentration of REE in the leachate occurred at 20 min after reestablishing flow (Figure 7). Additionally, the pregnant leach solutions exhibit middle—to heavy—REE enrichments.

Results from these initial flow-through experiments highlight the considerations and parameters that need to be optimized to maximize the extraction of REE from underclay feedstocks using an organic lixiviant such as a sodium citrate solution. The flow-through experiments demonstrate that the initial 20–40 min of flow is the most critical for recovering the highest concentrations of the REE. Additional flow beyond that initial 20–40 min recovers significantly lower concentrations of the REE and may not be economical. The relatively quick release of REE in the first 20–40 min of fluid flow is consistent with our hypotheses that the sodium citrate solution targets the sorbed/colloidal components of the underclays. Increased shut-in times of the fluid with the underclay sample at pressure also led to increased concentrations in the PLS. The highest concentrations of REE in PLS produced came after a 5-day shut-in period. Notably, after these extended shut-in periods, the highest concentrations of the REE were still observed as occurring in the first 20–40 min of fluid flow. We suggest that the mechanism of REE extraction remains the same as in the powder benchtop tests, i.e., desorption from clay surfaces and/or complexation of ions from colloidal phases, but that the additional shut-in time is needed for wetting of micropores and packed grains. These observations are consistent between the powdered samples and the fractured core experiments. However, extraction efficiencies for the flow-through experiment were low, <1% of the total REE content, and likely due to the lower liquid–solid ratio and decreased contact between solution and material due to low transmissivity of the fluid through the material. Increasing the extraction efficiencies for flow-through applications remains an area of active research and further development and testing of a flow-through method that could eventually be employed at the field scale for in situ solution mining is necessary.

3.5. A Method for Recovery of REE from Citrate PLS

In this study a citrate-based al, the subsequent processing, and quantitative recovery of REE from the pregnant leach solution (PLS) is an essential consideration for mining operations and technology economic evaluation of the application. Sorbent capture is a promising technology for the recovery of a high purity REE fraction from PLS, including citrate-based solutions. In conventional mining operations of the REE, the PLS is subjected to a series of purification steps to remove impurities such as Fe, Al, P, and Th (i.e., gangue elements) before the REE can be recovered [33,34]. While these downstream processes are still an area of active research for the citrate solutions described herein, there are a number of potential pathways that include the use of a novel amine-based sorbent to selectively recover the REE. In order to process the pregnant leachate solution for the recovery of a high purity REE fraction, the leachate is passed over a bed of sorbent material [34]. After leaching, the PLS is passed over a bed of sorbent material [35]. During capture, alkali and alkaline earth elements, as well as the lixiviant, pass through the bed while gangue elements (e.g., Al, Si, Fe, and Th) and REE bind to the bed. Once bound to the sorbent bed, the REE is selectively eluted, away from the gangue elements to produce high purity REE fraction [33,35,36].

In a separate series of laboratory experiments we produced and tested the ability to capture REE from a citrate-based PLS and simultaneously remove gangue elements from the leachate. A citrate PLS—from the leaching of coal prep plant fines with the citrate leaching solution RS-2—was used. The initial, unoptimized sorbent capture test using the citrate PLS showed approximately 60–70% uptake of REE from the feed solution, up through holmium, with a bias toward the light and middle rare earth elements in the unoptimized recovery of bound REE from the sorbent. The overall recovery for the light and middle REE was found to be 80–100% for elements lanthanum to dysprosium, while the recovery ranged from 50–70% for MREE and HREE, with the greatest recovery in this subset occurring with holmium. These results show great promise toward the concentration of REE from a citrate-based PLS generated by leaching of clay-rich geologic material. Future work will be aimed at using the solid sorbent to remove all base metals and other elements from the citrate lixiviant in order to recycle the solution and reuse multiple cycles of leaching. Reuse of the lixiviant during sequential cycles will add cost savings to the process. When coupled with the use of solid polymer sorbents to recover and concentrate REE, citrate leaching may be a promising method for the leaching and concentration of REE from clay-rich coal mine waste rock and coal preparation plant refuse.

4. Conclusions

Underclay associated with coal seams in the Lower Freeport, Middle Kittanning, and Pittsburgh formations contain REE concentrations ranging from 250–353 ppm. Clay minerals such as illite, halloysite, and kaolinite are the predominant clay minerals that make up >55% of the total bulk mineralogy of the rock. The introduction of leaching solutions into underclay rock powder initiates chemical reactions such as ion exchange, hydrolysis, and mineral dissolution that result in the release of ion constituents into the citrate PLS. Bench-top leaching and flow-through experimental results indicate that citrate is a chemically effective molecule for leaching weakly bound REE or other elements of interest from clay-rich sedimentary rock. Rare earths and other metal elements may exist as water soluble, ion exchangeable, or ion/colloidal where the REE is adsorbed to clay or other minerals such as metal oxyhydroxides that are present in the rock. The properties of the citrate molecule provide an added benefit of pH-controlled selectivity against leaching gangue elements from the rock matrix. The process of using organic acid anions for chelation and complexation of target elements is a promising method to leach REE and other critical metals from a variety of different sedimentary lithologies (e.g., underclay, sandstone, and shale) and produced materials such as coal mining waste and coal preparation plant refuse. A chief advantage of the citrate leaching technology is the demonstrated ability to recover REE from feedstocks with minimal release of gangue and radioactive elements, using an environmentally benign and relatively cost-effective leaching solution.

Changes to the physical and/or chemical properties of the clay rock may have both a positive and negative outcome pertaining to the leaching of specific elements from the rock matrix. Bench top powder leach and the flow-through experiments conducted here provide first-order results on the efficacy of organic acid anions for leaching of REE and other elements. Additionally, the results provide observational data on the minimal impact the leaching has on the physical structure of the rock because of the likelihood that there is minimal dissolution of the rock matrix when leaching with citrate at a pH range of 3–6. Continuing work will be aimed toward maximizing the extraction efficiency of the organic acid-based citrate leaching solution. As described in the methods, a total accounting by mass of the extracted REE from flow-through experiments was not possible but estimates of the extraction efficiency of the system remain low, <1% of the total REE content for the underclay samples. Comparatively, benchtop experiments where powdered underclay samples were reacted with sodium citrate solution at 1% and 10% solids (e.g., Table S5) demonstrated leaching up to ~30% of the total REE. Future work using flow-through experiments will explore parameters such as increasing the leach solution ratio of fluids reacting with the underclay samples, the concentration of the sodium citrate solution, and sampling schemes such as a step-wise shut-in reaction. Information gained from these results and the guided work of future studies will aid in developing a technology economic evaluation of the process to determine costs associated with upscaling this technology for use in larger scale operations.

Supplementary Materials
Figure S1: Photograph of underclay reaction vessel used in flow-through leaching experiments, Figure S2: Schematic of flow-through apparatus used in powder leaching experiments, Table S1: Concentration of elements in bulk underclay powders used in pH tests, Table S2: Semiquantitative XRD results from random and oriented mounts, Table S3: Results of sequential digest of powdered underclay samples, Table S4: Concentration of trace elements in sequential digest PLS, Table S5: Concentration of elements in PLS from leaching of UC-02 (1 and 10% solids) with 0.1 mol/L citrate and 0.5 mol/L NaCl, Table S6: Concentration of elements (μg/g) leached into PLS from underclay samples using citrate solution RS-2, Table S7: Concentration of elements leached from underclay samples UC-02 and UC-03 using citrate solutions buffered to pH 3, 4, 5, and 6, Table S8: Tabulated parameters for particle size distributions of unreacted (initial) underclay samples UC-01, UC-02, UC-03, and UC-06.

Author Contributions: Conceptualization, S.N.M., M.M., and C.V.; methodology, S.N.M.; formal analysis, S.N.M., J.Y. and J.B.; investigation, S.N.M. and J.Y.; resources, J.B., M.M. and C.V.; data curation, S.N.M., and C.V.; writing—original draft preparation, S.N.M., J.Y., J.B., M.M. and C.V.; writing—review and editing, S.N.M., J.Y., J.B., and M.M.; supervision, M.M. and C.V.; project administration, C.V.; funding acquisition, M.M. and C.V. All authors have read and agreed to the published version of the manuscript.

Acknowledgments: We thank Mary Anne Alvin (DOE Rare Earths Technology Manager) and Thomas Tarka (REE FWP Technical Portfolio Lead) for their support. We thank MacMahan Gray (USDOE-NETL) and Brian Kail (LRST) for their time and efforts with sorbent capture tests and analysis, Ward Burgess and Randal Thomas (LRST) for reviewing the manuscript prior to submission.

References

1. DOE Report. *Report on Rare Earth Elements from Coal and Coal Utilization Byproducts, Report to Congress*; United States Department of Energy: Washington, DC, USA, 2017. Available online: https://www.energy.gov/sites/prod/files/2018/01/f47/EXEC-2014-000442%20-%20for%20Conrad%20Regis%202.2.17.pdf (accessed on 29 May 2020).
2. U.S. Geological Survey. *Mineral Commodity Summaries 2020*; U.S. Geological Survey: Reston, VA, USA, 2020; 200p. [CrossRef]
3. Rozelle, P.; Khadikar, A.; Pulati, N.; Soundarrajan, N.; Klima, M.; Mosser, M.; Miller, C.; Pisupati, S. A study on removal of rare earth elements from U.S. coal byproducts by ion exchange. *Metall. Mater. Trans.* **2016**. [CrossRef]
4. Montross, S.N.; Verba, C.A.; Chan, H.L.; Lopano, C. Advanced characterization of rare earth element minerals in coal utilization byproducts using multimodal image analysis. *Int. J. Coal Geol.* **2018**, *195*, 362–372. [CrossRef]
5. Zhang, W.; Yang, X.; Honaker, R.Q. Association characteristic study and preliminary recovery investigation of rare earth elements from fire clay seam coal middlings. *Fuel* **2018**, *215*, 551–560. [CrossRef]
6. Honaker, R.Q.; Zhang, W.; Werner, J. Acid leaching of rare earth elements from coal and coal ash: Implications for using fluidized bed combustion to assist in the recovery of critical materials. *Energy Fuels* **2019**, *33*, 5971–5980. [CrossRef]
7. Appalachian Region Independent Power Producers Association (ARIPPA). 2018 Coal Refuse Whitepaper. Available online: https://arippa.org/wp-content/uploads/2018/12/ARIPPA-Coal-Refuse-Whitepaper-with-Photos-10_05_15.pdf (accessed on 2 April 2019).

8. Lin, R.; Stuckman, M.; Howard, B.; Bank, T.; Roth, E.; Macala, M.; Lopano, C.; Soong, Y.; Granite, E. Application of sequential extraction and hydrothermal treatment for characterization and enrichment of rare earth elements from coal fly ash. *Fuel* **2018**. [CrossRef]
9. Taggart, R.; Hower, J.; Hsu-Kim, H. Effects of roasting additives and leaching parameters on the extraction of rare earth elements from coal fly ash. *Int. J. Coal Geo.* **2018**. [CrossRef]
10. Huang, Q.; Noble, A.; Herbst, J.; Honaker, R. Liberation and release of rare earth minerals from Middle Kittanning, fire clay, and west kentucky No. 13 coal sources. *Powder Technol.* **2018**. [CrossRef]
11. Burckhard, S.R.; Schwab, A.P.; Banks, M.K. The effects of organic acids on the leaching of heavy metals from mine tailings. *J. Hazard. Mater.* **1994**, *41*, 135–145. [CrossRef]
12. Kondoh, A.; Oi, T. Interaction of alkaline earth metal ions with carboxylic acids in aqueous solutions studied by 13CNMR spectroscopy. *Z. Für Nat. A* **1998**, *53*, 77–91.
13. Wyrzykowski, D.; Chmurzyński, L. Thermodynamics of citrate complexation with Mn^{2+}, Co^{2+}, Ni^{2+}, and Zn^{2+} ions. *J. Term. Anal. Calorim.* **2010**, *102*, 61–64. [CrossRef]
14. Zabiszak, M.; Nowak, M.; Taras-Goslinksa, K.; Kaczmarek, M.T.; Hnatejko, Z.; Jastrzab, R. Carboxyl groups of citric acid in the process of complex formation with bivalent and trivalent metal ions in biological systems. *J. Inorg. Biochem.* **2018**, *182*, 37–47. [CrossRef] [PubMed]
15. Wang, L.; Lioa, C.; Yang, Y.; Xu, H.; Xiao, Y.; Yan, C. Effects of organic acids on the leaching process of ion-adsorption type rare earth ore. *J. Rare Earths* **2017**, *35*, 1233–1238. [CrossRef]
16. Moldovean, G.A.; Papangelakis, V.G. Recovery of rare earth elements adsorbed on clay minerals: II. Leaching with ammonium sulfate. *Hydrometallurgy* **2013**, *131–132*, 158–166. [CrossRef]
17. Zhi Li, L.; Yang, X. China's rare earth ore deposits and beneficiation techniques ERES2014. In Proceedings of the 1st European Rare Earth Resources Conference, Milos, Greece, 4–7 September 2014; pp. 26–36.
18. Brisson, V.L.; Zhuang, W.; Alvarez, L. Bioleaching of rare earth elements from monazite sand. *Biotechnol. Bioeng.* **2015**, *113*, 339–348. [CrossRef]
19. Shan, X.Q.; Wen, J.J.B. Effect of organic acids on adsorption and desorption of rare earth elements. *Chemosphere* **2002**. [CrossRef]
20. Schumacher, B.; Shines, K.; Burton, J.; Papp, M. *Comparison of Soil Sample Homogenization Techniques*; Lewis Publishers: Chelsea, MI, USA, 1990.
21. Eberl, D.D. *User's Guide to RockJock—A Program for Determining Quantitative Mineralogy from Powder X-ray Diffraction Data, Open-File Report 03-78*; U.S. Geological Survey: Boulder, CO, USA, 2003.
22. Poppe, L.J.; Paskevich, V.F.; Hathaway, J.C.; Blackwood, D.S. *A Laboratory Manual for X-Ray Powder Diffraction, Open-File Report 01-041*; U.S. Geological Survey: Woods Hole, MA, USA, 2010.
23. Sperazza, M.; Moore, J.N.; Hendrix, M.S. High-resolution particle size analysis of naturally occurring very fine-grained sediment through laser diffractometry. *J. Sed. Res.* **2004**. [CrossRef]
24. Peelman, S.; Sun, Z.H.; Siestma, J.; Yang, Y. Leaching of rare earth elements: Past and present. In Proceedings of the 1st European Rare Earth Resources Conference, ERES2014, Milos, Greece, 4–7 September 2014.
25. Bank, T.; Roth, E.; Tinker, P.; Granite, E. *Analysis of Rare Earth Elements in Geologic Samples using Inductively Coupled Plasma Mass Spectrometry*; US DOE Topical Report-DOE/NETL-2016/1794[R]; National Energy Technology Lab. (NETL): Pittsburgh, PA, USA, 2016.
26. Taylor, A.; Blum, J. Relation between soil age and silicate weathering rates determined from chemical evolution of a glacial chronosequence. *Geology* **1995**, *23*, 979–982. [CrossRef]
27. Yang, J.; Montross, S.N.; Britton, J.; Stuckman, M.; Lopano, C.; Verba, C. Microanalytical approaches to characterizing REE Appalachian basin underclays. *Minerals* **2020**, *10*. [CrossRef]
28. Komnitsas, K.; Modis, K. Geostatistical risk estimation at waste disposal sites in the presence of hot spots. *J. Haz. Mat.* **2009**, *164*, 1185–1190. [CrossRef]
29. Montross, S.; Verba, C.; Falcon, A.; Poston, J.; McKoy, M. Characterization of rare earth element minerals in coal utilization byproducts and associated clay deposits from Appalachian basin coal resources. In Proceedings of the 34th International Pittsburgh Coal Conference, Pittsburgh, PA, USA, 3–9 September 2019.
30. Hower, J.C.; Groppo, J.G.; Joshi, P.; Dai, S.; Moecher, D.P.; Johnston, M. Location of cerium in coal-combustion fly ashes: Implications for recovery of lanthanides. *Coal Combust. Gasificat. Prod.* **2013**, *5*, 73–78. [CrossRef]
31. Allen, J.R.L. *Principles of Physical Sedimentology*; Springer: New York, NY, USA, 2012; p. 212.

32. Norrström, A.-C.; Bergstedt, E. The impact of road de-icing salts (NaCl) on colloid dispersion and base cation pools in roadside soils. *Water Air Soil Pollut.* **2001**, *127*, 281–299. [CrossRef]
33. Gray, M.L.; Kail, B.W.; Wang, Q.; Wilfong, W.C. Stable immobilized amine sorbents for REE and heavy metal recovery from liquid sources. *Environ. Sci. Water Res. Technol.* **2018**. [CrossRef]
34. Wilfong, C.W.; Kail, B.W.; Wang, Q.; Shi, F.; Shipley, G.; Tarka, T.J.; Gray, M.J. Stable immobilized amine sorbents for heavy metal and REE removal from industrial wastewater. *Environ. Sci. Water Res. Technol.* **2020**, *6*, 1286–1299. [CrossRef]
35. Wang, Q.; Kail, B.W.; Wilfong, W.C.; Shi, F.; Tarka, T.J.; Gray, M.L. Amine sorbents for selective recovery of heavy rare-earth elements (Dysprosium, Ytterbium) from aqueous solution. *ChemPlusChem* **2020**, *85*, 130–136. [CrossRef]
36. Wang, Q.; Wilfong, W.C.; Kail, B.W.; Yu, Y.; Gray, M.L. Novel polyethylenimine–acrylamide/SiO_2 hybrid hydrogel sorbent for rare-earth-element recycling from aqueous sources. *ACS Sustain. Chem. Eng.* **2017**, *5*, 10947–10958. [CrossRef]

7

Factors Controlling the Gallium Preference in High-Al Chromitites

Ioannis-Porfyrios D. Eliopoulos * and George D. Eliopoulos

Department of Chemistry, University of Crete, Heraklion GR-70013, Crete, Greece; giorgoshliop@yahoo.gr
* Correspondence: disaca007@hotmail.com

Abstract: Gallium (Ga) belongs to the group of critical metals and is of noticeable research interest. Although Ga^{3+} is highly compatible in high-Al spinels a convincing explanation of the positive Ga^{3+}–Al^{3+} correlation has not yet been proposed. In the present study, spinel-chemistry and geochemical data of high-Al and high-Cr chromitites from Greece, Bulgaria and the Kempirsai Massif (Urals) reveals a strong negative correlation (R ranges from −0.95 to −0.98) between Cr/(Cr + Al) ratio and Ga in large chromite deposits, suggesting that Ga hasn't been affected by re-equilibration processes. In contrast, chromite occurrences of Pindos and Rhodope massifs show depletion in Ga and Al and elevated Mn, Co, Zn and Fe contents, resulting in changes (sub-solidus reactions), during the evolution of ophiolites. Application of literature experimental data shows an abrupt increase of the inversion parameter (x) of spinels at high temperature, in which the highest values correspond to low-Cr^{3+} samples. Therefore, key factors controlling the preference of Ga^{3+} in high-Al chromitites may be the composition of the parent magma, temperature, redox conditions, the disorder degree of spinels and the ability of Al^{3+} to occupy both octahedral and tetrahedral sites. In contrast, the competing Cr^{3+} can occupy only octahedral sites (due to its electronic configuration) and the Ga^{3+} shows a strong preference on tetrahedral sites.

Keywords: chromite; gallium; spinel; structure; composition; correlation; ophiolites; disorder

---

1. Introduction

Gallium (Ga) is a vital metal for the economy, due to its use in high-technology applications, such as electronics industry, electric cars, solar panels. Although bauxite deposits are traditionally mined for their Al content and are important sources of Ga as a byproduct commodity [1], the distribution of Ga in chromite ores may be of particular research interest, due to its relationship with the major element composition of chromite.

Chromite belongs to the subgroup of spinels, which accommodate a wide variety of cations in their structure with the general formula AB_2O_4. Many authors emphasized that despite their simple structure, many spinels exhibit complex disordering phenomena involving the two cation sites, which play an important role both in their thermochemical and their physical properties [2–6]. The movements of cations between tetrahedral and octahedral sites, as a result of cation substitution, have been discussed under the aspect of structural parameters, such as tetrahedral and octahedral bond lengths, cation-cation and cation-anion distances, bond angles and hopping lengths, which were calculated by experimental lattice constants and oxygen parameters [2–4,7–12]. The ability of Ga^{3+} (r = 0.62 Å) to replace Al^{3+} (r = 0.54 Å) in aluminum minerals is related to their geochemistry (Group III of the periodic table), while Ga might be expected to behave in a similar way in chromite and magnetite as they share similar ionic radii (Cr^{3+}, r = 0.62 Å; Fe^{3+} = 0.64 Å) for octahedral and tetrahedral coordination (Al^{3+}, r = 0.39; Ga^{3+} = 0.47; Fe^{3+} = 0.49) [3]. Gallium levels reported in Cr-spinel grains from ophiolites, varying from 10 to 50 ppm [13,14] are consistent with experimental mineral-melt data on the partition coefficient (D_{Ga} = 0.9–11.2) [15]. Although mineral–melt partition coefficients are not constants, depending on a number of factors (pressure, temperature, oxygen fugacity or mineral and melt composition) on the basis of experimental data it has been suggested that Ga is volatile and there is no significant effect of temperature, magma composition and at very low oxygen fugacity conditions [15].

A tectonic discrimination of peridotites, using the oxygen fugacity ($f\,O_2$)–Cr#[Cr/(Cr + Al)] diagram, and the Ga–Ti–Fe^{3+}# [Fe^{3+}/(Fe^{3+} + Cr + Al)] systematics in chrome-spinels, has been proposed [16]. A negative correlation between Ga and Cr in chromitites has been established, that may be related to the composition of parental magmas [17–24] or to the outer electronic structure of Ga that is similar to that of Al [25]. Also, the investigation of spinels in lithospheric mantle xenoliths from distinct tectonic settings has demonstrated that trace elements contribute in discriminating between spinels hosted in peridotites and those crystallized from magmas [11]. However, a convincing explanation of the positive correlation between Ga^{3+} and Al^{3+} has not yet been offered.

In the present study we characterize the spinel chemistry, bulk ore composition, including Ga, from chromitite samples of selected ophiolite complexes in Greece (Pindos, Central Vourinos and Skyros), the Rhodope–Serbo–Macedonian zone (SMZ) massifs (all of Mesozoic age) and the Kempirsai Massif (Kazakhstan) in the Urals (Palaeozoic age), all hosting both high-Cr and high-Al chromitites. The investigated samples are representative of large chromite deposits and small occurrences, in order to define potential relationships between major, minor or trace elements and Ga and the effect of re-equilibration processes, during a long evolutionary time of the ophiolites. We apply available platinum-group element (PGE) data to define potential correlations between the Ga content and fractional crystallization, and experimental literature data for the structure of spinels, aiming to investigate the role of intra-crystalline cation exchange, and contribute to still uncertain factors controlling the positive Al-Ga correlation of chromitites.

2. Materials and Methods

2.1. Mineral Analysis

Polished sections of all chromitite samples were examined using a reflected light microscope and a scanning electron microscope (SEM), equipped with energy-dispersive spectroscopy (EDS). The SEM-EDS back-scattered electron images (Figure 1) and analyses of chromite ores (Table 1) were carried out at the Faculty of Geology and Geoenvironment, National and Kapodistrian University of Athens (NKUA), using a JEOL JSM 5600 (Tokyo, Japan), scanning electron microscope, equipped with ISIS 300 OXFORD (Oxford shire, UK), automated energy dispersive analysis system. Analytical conditions were 20 kV accelerating voltage, 0.5 nA beam current, <2 μm beam diameter and 50 s count times. The following X-ray lines were used: FeKα, NiKα, CoKα, CuKα, CrKα, AlKα,TiKα, CaKα, SiKα, MnKα and MgKα. Cr, Fe, Mn, Ni, Co, Ti and Si, MgO for Mg and Al_2O_3 for Al. Contents of Fe_2O_3 and FeO were calculated on the basis of the spinel stoichiometry.

Table 1. Electron scanning electron microscope (SEM)/energy-dispersive spectroscopy (EDS) analyses of chromite from chromitites of Greece, Bulgaria and Kempirsai (Urals).

	Vourinos					Pindos							Skyros		
wt%	Vour. 1	Vour. 2	Vour. 3	Vour. 4	Vour. 5	Pi.1.	Pi.2	Pi. 3	Pi. 4	Pi. 5	Pi.6	Pi.7	Sky. 1	Sky. 2	Sky. 3
TiO_2	0.3	0.2	0.2	0.1	0.2	0.2	0.3	0.2	0.2	0.1	0.2	0.1	0.2	0.2	0.2
Al_2O_3	11.4	9.4	9.8	23.9	11.3	26.2	27.2	32.9	34.5	16.5	15.5	8.2	20.11	24.5	22.1
Cr_2O_3	59.7	63.2	61.3	44.8	60.4	41.5	39.5	35.9	34.8	52.8	52.1	61.4	48.8	45.3	47.1
MgO	13.6	12.1	13.6	16.2	12.4	14.5	13.6	16.4	16.7	12.4	10.1	11.7	13.9	15.7	10.2
FeO	13.1	14.1	12.7	11.1	14.6	13.5	15.7	12.3	12.5	15.6	18.7	14.9	13.6	11.8	19.7
Fe_2O_3	1.7	0.1	2.1	4.2	.5	3.4	3.1	2.3	2.6	2.9	2.7	2.8	2.2	2.3	0.3
MnO	0.3	0.2	0.2	0.2	0.2	0.3	n.d.	0.1	n.d.	0.2	0.3	0.3	0.1	0.1	0.2
NiO	0.2	0.1	n.d.	n.d.	0.2	0.3	n.d.	0.2	n.d.	0.2	0.1	0.2	0.1	0.2	0.2
Total	100.3	99.5	99.9	100.5	99.8	99.9	99.4	100.1	101.2	100.7	99.7	99.6	99.01	100.1	100
Cr/(Cr+Al)	0.77	0.81	0.81	0.56	0.78	0.52	0.49	0.42	0.4	0.70	0.69	0.81	0.62	0.55	0.58
$Mg/(Mg+Fe^{2+})$	0.65	0.62	0.66	0.72	0.60	0.65	0.61	0.71	0.72	0.59	0.49	0.58	0.64	0.69	0.51
$Fe^{3+}/(Cr+Al+Fe^{3+})$	0.031	0.000	0.025	0.036	0.0055	0.038	0.036	0.025	0.027	0.0413	0.028	0.036	0.027	0.026	0.0033
Numbers of Cations on the Basis of 32 Oxygens															
Ti	0.019	0.039	0.040	0.018	0.039	0.036	0.054	0.035	0.014	0.019	0.039	0.020	0.037	0.036	0.038
Al	3.426	2.889	2.997	6.736	3.451	7.415	7.758	8.966	9.241	4.650	4.728	2.555	5.896	6.922	6.532
Cr	12.037	13.031	12.523	8.471	12.383	7.891	7.558	6.563	6.254	10.645	10.662	12.835	9.602	8.586	9.340
Mg	5.170	4.703	5.250	5.774	4.790	5.210	4.906	5.652	5.658	4.712	3.896	4.611	5.156	5.610	3.813
Fe^{2+}	2.744	3.271	2.744	2.203	3.164	2.706	3.149	2.326	2.377	3.222	4.056	3.300	2.840	2.367	4.142
Fe^{3+}	0.498	0.001	0.400	0.757	0.087	0.620	0.575	0.401	0.436	0.659	0.533	0.570	0.427	0.420	0.053
Mn	0.065	0.044	0.045	0.040	0.044	0.061	0.000	0.020	0.000	0.043	0.065	0.067	0.021	0.020	0.042
Ni	0.157	0.021	0.000	0.000	0.042	0.058	0.000	0.037	0.000	0.041	0.020	0.042	0.020	0.039	0.040

	Rhodope Massif										Urals Kempirsai			
	Skyros		Greece				Bulgaria							
									Goliamo		Northern Part		Southern Part	
wt%	Sky. 4	Sky. 5	Soufli 1	Soufli 2	Gomati	Broucevci	Jacovitsa	Pletena	Kamenyane 1	Kamenyane 2	Batamshinsk		Main Ore Field	
TiO_2	0.1	0.2	n.d.	0.2	0.2	0.4	0.2	0.5	0.4	0.3	0.3	0.3	0.1	0.2
Al_2O_3	11.2	12.6	15.8	19.6	30.8	27.9	5.5	10.4	4.5	1.2	24.4	23.3	9.3	9.8
Cr_2O_3	59.1	58.1	53.9	47.8	35.5	37.3	58.8	49.8	33.2	24.9	46.6	47.1	61.3	60.2
MgO	13.8	14.4	13.4	14.1	15.3	16.6	10.2	8.2	11.9	7.2	14.3	14.2	15.5	15.2
FeO	12.1	11.5	13.4	13.5	13.9	10.4	16.2	20.7	13.6	20.2	14.1	14.1	9.9	10.3
Fe_2O_3	2.4	2.2	3.1	5.0	5.1	6.0	7.2	9.5	35.1	45.6	0.3	0.8	4.5	4.3
MnO	0.2	0.2	0.2	0.2	n.d.	0.2	0.6	0.5	n.d.	0.2	n.d.	n.d.	n.d.	n.d.

Table 1. *Cont.*

	Rhodope Massif										Urals Kempirsai			
	Skyros		Greece			Bulgaria								
									Goliamo		Northern Part		Southern Part	
wt%	Sky. 4	Sky. 5	Soufli 1	Soufli 2	Gomati	Broucevci	Jacovitsa	Pletena	Kamenyane 1	Kamenyane 2	Batamshinsk		Main Ore Field	
NiO	0.2	0.2	n.d.	0.2	n.d.	0.2	0.2	0.2	0.3	0.4	0.2	0.2	0.1	0.3
Total	99.1	99.7	100.2	99.8	100.3	99.1	99.1	100	99.2	99.9	100.4	99.9	100.8	100.3
Cr/(Cr+Al)	0.75	0.79	0.70	0.63	0.44	0.47	0.88	0.74	0.83	0.94	0.56	0.58	0.82	0.80
$Mg/(Mg+Fe^{2+})$	0.68	0.69	0.62	0.65	0.66	0.72	0.52	0.37	0.52	0.38	0.64	0.65	0.74	0.73
$Fe^{3+}/(Cr+Al+Fe^{3+})$	0.03	0.026	0.036	0.059	0.057	0.068	0.093	0.123	0.455	0.62	0.0015	0.009	0.053	0.052
Numbers of cations on the basis of 32 oxygens														
Ti	0.019	0.038	0.019	0.036	0.035	0.035	0.041	0.102	0.082	0.062	0.054	0.054	0.019	0.038
Al	3.414	3.813	4.636	5.591	8.475	7.828	1.767	3.277	1.452	0.386	6.987	6.,675	2.787	2.947
Cr	12.066	11.695	10.745	9.396	6.553	7.021	12.670	10.582	7.168	5.648	8.880	9.070	12.323	12.143
Mg	5.321	5.512	5.036	5.231	5.324	5.890	4.143	3.291	4.858	3.088	5.137	5.151	5.874	5.780
Fe^{2+}	2.613	2.442	2.940	2.805	2.711	2.067	3.715	4.653	3.114	4.839	2.879	2.864	2.125	2.197
Fe^{3+}	0.480	0.414	0.580	0.940	0.901	1.079	1.481	1.936	7.215	9,844	0.024	0.146	0.852	0.833
Mn	0.044	0.044	0.043	0.000	0.000	0.040	0.138	0.115	0.000	0.046	0.000	0.000	0.000	0.000
Ni	0.041	0.041	0.000	0.000	0.000	0.038	0.044	0.044	0.114	0.088	0.039	0.038	0.020	0.062

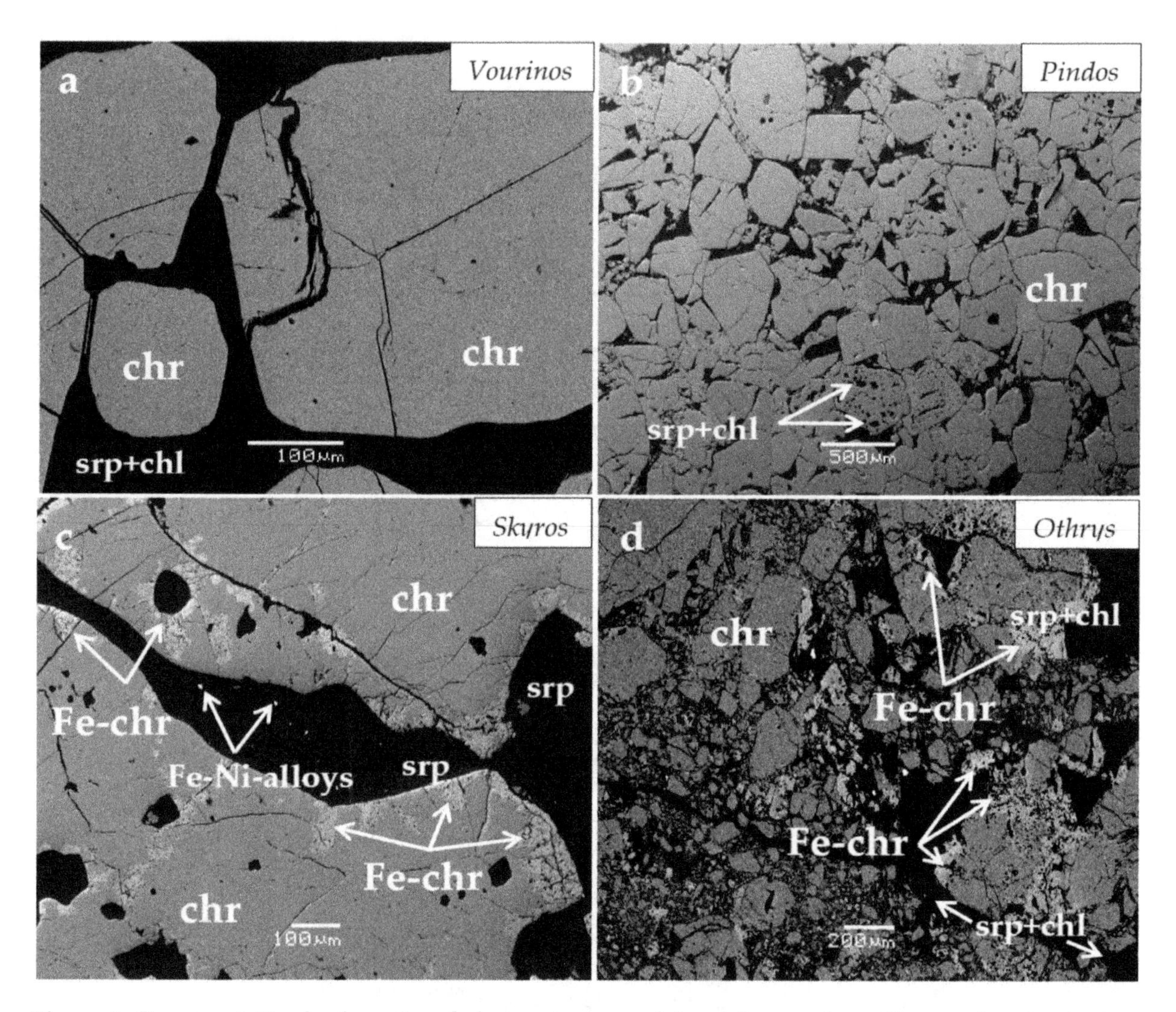

Figure 1. Representative back-scattered electron images of chromite ores from Greece, showing texture relationships between chromite and silicates, the presence of homogeneous chromite (**a**) and also abundant silicate inclusions (chlorite and serpentine) in the host chromite (**b**), porous texture and alteration to Fe-chromite (**c**,**d**). Abbreviations: chr = chromite; Fe-chr = iron-chromite; srp = serpentine; chl = chlorite.

2.2. *Whole Rock Analysis*

Major and trace elements in massive chromitite samples (more than 95 vol %) were determined by ICP-MS analysis, at the ACME Analytical Laboratories Ltd., Vancouver, BC, Canada (currently Bureau Veritas Commodities Canada Ltd.). The samples were dissolved using an acid mix (HNO_3–$HClO_4$–HF) digestion and then the residues were dissolved in concentrated HCl. The rare earth elements La, Ce, Pr, Nd, Sm, Eu, Gd, Tb, Dy, Ho, Er, Tm, Yb and Lu as well as Li, K, Ge, Sr, Y, Zr, Mo, Sb, Cs, W, Pb, Th, and U were lower than the detection limits of the analytical methods. The detection limits of the method for the presented elements are 1 ppm for Ga and V, 0.2 ppm for Co and Zn, 0.1 for Ni and 0.01 wt % for Fe. On the basis of the quality control report provided by the Analytical Labs, the results of analyses of the reference material in comparison to expected values, and the results from multistage analysis of certain samples, showed accuracy and precision in good agreement with accepted values for international standards. The analytical error, for Ga, for example, was <5%. Although the PGE data reported in Table 2 have been published previously (Table 2) a portion from the same samples was used for the presented trace element analyses.

Table 2. Trace element contents of high-Cr and high-Al chromitites.

Location	SEM/EDS		Trace Element (ppm)					wt%	ppb	
	Cr/(Cr+Al)	Mg/(Mg+Fe^{2+})	Ni	Co	V	Zn	Ga	Fe	ΣPGE *	Pd/Ir *
Vourinos 1	0.77	0.65	2000	240	500	260	14	8.1	140	0.06
Vourinos 2	0.81	0.62	1580	210	560	550	15	7.9	135	0.35
Vourinos 3	0.81	0.66	1900	200	400	300	13	8.72	92	0.55
Vourinos 4	0.56	0.72	1800	170	780	360	27	9.2	30	0.54
Vourinos 5	0.78	0.61	1800	180	620	280	14	7.76	109	0.63
Pindos 1	0.52	0.63	1300	140	600	400	23	7.84	51	1.0
Pindos 2	0.48	0.6	1500	260	710	410	32	9.77	143	1.0
Pindos 3	0.42	0.71	1500	260	710	520	34	7.8	117	6.33
Pindos 4	0.4	0.72	1630	240	580	460	32	7.3	6123	34.2
Pindos5	0.67	0.57	750	270	760	520	16	9.7	3875	12.2
Pindos 6	0.69	0.49	720	240	760	620	23	10.1	2098	7.2
Pindos 7	0.81	0.58	1450	290	560	490	17	10.2	181	0.22
Skyros 1	0.61	0.64	1300	250	1200	540	34	10.9	2300	0.08
Skyros 2	0.55	0.69	1600	200	1000	400	36	9.45	464	0.7
Skyros 3	0.58	0.51	1500	240	870	450	40	9.6	251	0.67
Skyros 4	0.75	0.69	1250	220	640	420	14	10.7	145	0.33
Skyros 5	0.79	0.69	1200	200	620	400	11	10.1	145	0.1
Othrys (n = 4)	0.54	0.69	1400	210	960	370	33	9.8	91	0.36
Rhodope Massif										
Greece										
Soufli1	0.70	0.62	1700	230	380	580	16	13.2	150	0.2
Soufli2	0.63	0.65	1150	220	460	280	12	10.6	82	1.0
Gomati	0.49	0.67	1030	130	730	280	24	9.6	104	0.25
Bulgaria										
Broucevci	0.47	0.72	1300	230	790	420	45	11	60	0.92
Jacovitsa	0.88	0.52	2250	310	240	760	9	13.9	197	0.46
Pletena	0.74	0.37	890	290	330	1030	13	17.9	563	0.07
Goliamo Kamenyane 1	0.83	0.52	1550	80	1000	450	12	11.6	87	0.14
Goliamo Kamenyane 2	0.94	0.38	2260	970	370	4030	6	64.3	40	1.93
Kemprsai (Urals)										
Northern	0.80	0.73	1600	210	160	160	14	9.1		
Batamshinsk	0.82	0.74	1600	230	200	190	16	9.5		
Southern	0.56	0.58	1500	240	680	480	48	12.6		
XL Let Kazakhstan	0.64	0.65	1700	230	730	340	49	10.1		

Symbol * = Data on PGE from literature [26–32].

3. A Brief Outline of Characteristics for the Studied Chromitites

All chromitite samples selected for the present study come from deposits and occurrences, which have been the subject of detailed geological, mineralogical and geochemical investigation [26–36] and references therein. The main ophiolite complexes of Greece (Vourinos, Othrys and Pindos) belong to the Upper Jurassic to Lower Cretaceous Tethyan ophiolite belt, and are characterized by heterogeneous deformation and rotation, during their original displacement and subsequent tectonic incorporation into continental margins [33]. The studied samples of chromitites are massive (Figure 1) and exhibit variations in the chromitite tonnage, the composition of chromite (Tables 1 and 2), the degree of transformation of ores and the associated ophiolites [27–37].

Chromite ores in the Vourinos complex occur in the mantle and cumulate sequences, with a tonnage estimated to approximately 10 Mt of high-Cr type, but at the central part of the complex there are high-Cr and high-Al ores in a spatial association, with low PGE contents [26]. The Othrys complex has a relatively high tonnage (approximately 3 Mt) of high-Al massive chromite ores and low PGE content [27].

The chromitite occurrences in the Pindos ophiolite complex are small (a few tens of m (x) a few tens of cm) and are hosted within completely serpentinized and weathered, intensively deformed dunite-harzburgite blocks, due to a strong plastic and brittle deformation that was superimposed on primary magmatic textures [17,18,34]. Chromitites throughout the Pindos complex are high-Cr and high-Al, often in a spatial association. The most salient feature of the Pindos chromitites is the enrichment in Pt and Pd at the area of Korydallos, at a level of 7 ppm PGE_{total} [28,29] and up to 29 ppm [35]. In the Achladones area on the Skyros island small massive chromitite bodies are of high-Al type and have elevated PGE contents, up to 3 ppm ΣPGE, although both high-Cr and high-Al types having low PGE content are found on the entire island [30].

Ophiolites associated with the Serbomacedonian massif (Gomati) and Rhodope massif including the ophiolites of Soufli (Greece), Dobromirci, Jacovitsa, Broucevci and Goliamo-Kamenjane (Bulgaria) host small (a few thousand tons) high-Cr and high-Al ores in a spatial are association, which occasionally contain elevated PGE concentrations. They are completely serpentinized, locally sheared and metamorphosed to antigorite-tremolite and/or talc schists. Detailed description of the characteristic mineralogy and texture of those chromitites have been published in previous studies [17,18,31,32,36].

The Kempirsai massif, covering an area of 2000 km^2, is divided by a shear zone into two parts: the southeastern part that is called Main Ore Field (MOF), hosting large high-Cr chromite deposits, and the northwestern area, the so-called Batamshinsk Ore Field (BOF), hosting much smaller high-Al chromite deposits [37]. An excellent description of the petrography and mineral chemistry, including mineral inclusions in the chromite of the giant chromite deposit of Kempirsai has been provided and discussed by Melcher et al [37]. These authors have interpreted their formation by a multistage process: High-Al chromitites may be derived from MORB-type tholeiitic melts, and high-Cr ones from boninitic magmas, during a second stage by interaction of hydrous high-Mg melts and fluids with depleted mantle in a supra-subduction zone setting.

4. Results

4.1. Compositional Variations in Chromite

The chromite samples from the central part of the Vourinos complex, the Pindos, Skyros island, Serbomacedonian, Rhodope and Kempirsai massifs show a wide variation in major elements from high-Cr, with the Cr/(Cr + Al) atomic ratio ranging from 0.81 to 0.69, to high-Al with the Cr/Cr + Al) ratio ranging from 0.63 to 0.4 (Table 1), falling in the range of metallurgical and refractorytype, respectively [21]. In addition, in the Bulgarian Rhodope massif (Jacovitsa, Pletena and Goliamo Kamenyane areas) altered chromite grains are dominant, having relatively high FeO and low Al_2O_3 and MgO contents (Table 1). As a consequence, the Cr/(Cr + Al) atomic ratios of those chromitites are significantly higher than those of high-Cr chromitites from the Vourinos complex (Table 1).

4.2. Distribution of Trace Elements in Chromitites

The geochemical data from whole rock analyses show a wide variation in major and trace element contents (Table 2). Gallium contents are lower in high-Cr chromitites (11 to 23 ppm) compared to high-Al ones (27–49 ppm), that seems to be independent on the degree of fractionation of parent magma, as exemplified by the Pd/Ir ratio [38]. The highest Co, Mn, Zn, Fe and lowest Ga were mainly recorded in strongly altered small chromite occurrences from the Rhodope massif in Bulgaria. They are in a good agreement with other chromitites [21–25] and are independent of the age of the associated ophiolites. Platinum-group elements (PGE) show total contents ranging from 30 to 6120 ppb and Pd/Ir ratios from 0.06 to 34, which are independent of the major element composition of chromitites (Table 2).

The results show a strong negative correlation (R ranges from −0.98 to −0.95) between the Cr/(Cr + Al) atomic ratio and Ga for the relatively large chromite deposits of Vourinos, Kempirsai massif (Urals) and the Skyros island. In addition, there is a less strong negative correlation for small chromite occurrences from the Pindos and Rhodope massifs ($R \geq -0.76$ and $R \geq -0.83$, respectively). Apart from Ga, the best correlation is found between Cr/(Cr + Al) and V for the Vourinos (R = −0.84), Skyros and Kempirsai ($R \geq -0.93$), whereas no significant relationship for chromitites from the Pindos and Rhodope massifs (Figure 2b) or between Cr/(Cr + Al) and other minor and trace elements is observed.

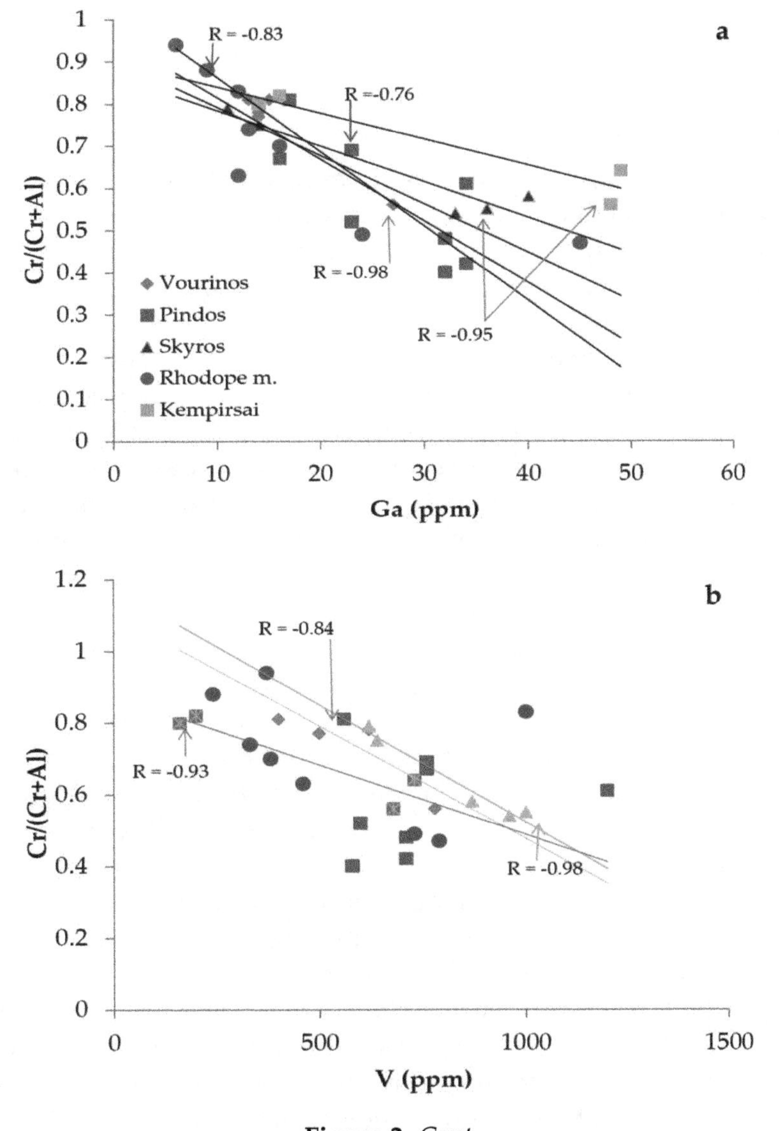

Figure 2. *Cont.*

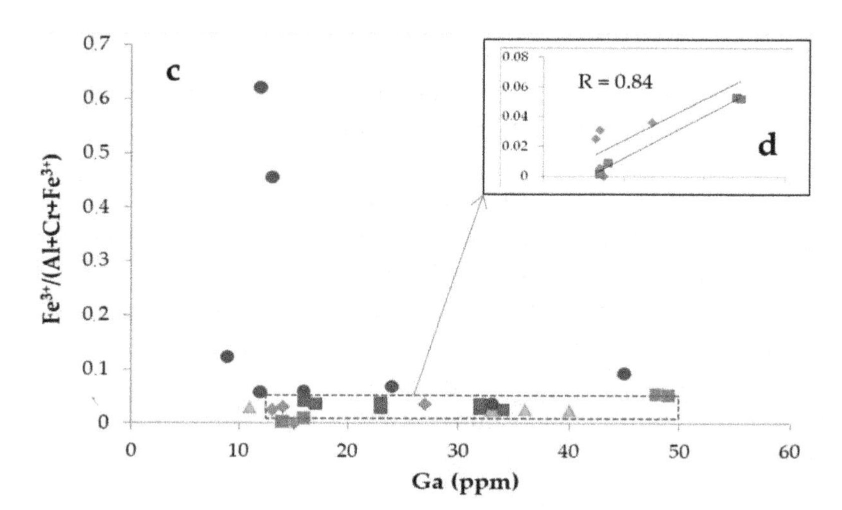

Figure 2. Plots of the Cr/(Cr + Al) atomic ratio versus Ga content (**a**); the Cr/(Cr + Al) ratio versus V content (**b**) and the $Fe^{3+}/(Al + Cr + Fe^{3+})$ ratio versus Ga content (**c**), including a detailed plot for the Vourinos and Kempirsai samples (**d**). Data from the Table 2.

5. Discussion

5.1. Factors Controlling the Spinel Chemistry

5.1.1. Magmatic Versus Post-Magmatic Processes

The wide variation of the Cr/(Cr + Al) atomic ratio for the chromitite samples from Greece, Bulgaria and Kempirsai massifs (Tables 1 and 2) fall in the range of metallurgical and refractory types. Differences in the trace element content (Table 2) may suggest trace element partitioning, depending on a number of factors, such as temperature, pressure, oxygen fugacity and the chemical composition of parent magmas [17–47].

As the partition coefficient of elements is defined as the ratio of the element content in a mineral and the melt [15], the Pd/Ir ratio can be used as an indicator of the degree of fractionation of parent magma for chromitites [38] and the presence of very low Pd/Ir values (low degree of fractionation) for both high-Cr and high-Al chromitites, suggest their origin from different magma sources. It has been argued that high-Cr chromitites, which have higher Sc, Mn, Co and Ni, and lower Ti, V, Zn and Ga contents may be derived from boninitic magmas, while high-Al ores may be derived from MORB-type tholeiitic magmas [17–24]. Experimental data at high temperature have shown that Ga is compatible in spinel with D values ranging between 0.9 and 11.2, and slightly lower D values in the most reducing experiments, while experimental data at temperatures >1300 °C and low oxygen fugacity have shown that there is no significant effect of temperature, composition and redox conditions [15]. However, the negative correlation between the Cr/(Cr + Al) atomic ratio and Ga content in natural chromitites points to the potential effect of the composition of the parent magma, while a positive trend between the $Fe^{3+}/(Al + Cr + Fe^{3+})$ atomic ratio and Ga content for large chromite deposits (Figure 2c,d) may suggest the effect of the redox conditions on the Ga distribution in chromitites. Specifically, high-Cr chromitites formed earlier from a primary magma (under relatively reducing conditions), compared to high-Al ones formed later from an evolved magma (and more oxidized conditions) magma [47].

In addition, differences in the negative correlations between the Cr/(Cr + Al) atomic ratio and Ga content and the slope of correlation lines for the different occurrences (Figure 2) may suggest that in addition to the composition of parent magmas, which is a major factor for large deposits (like Vourinos and Kempirsai massif, Urals) other factors such as temperature, pressure or redox conditions may be responsible for the observed deviation from linearity for small metamorphosed occurrences of chromitites, such as those from the Pindos and Rhodope massifs (Figure 2b). The lack of significant relationships between major and trace elements, in small chromitite occurrences from the Pindos and Rhodope massifs (Figure 2) may be related with post-magmatic processes. The elevated Mn, Co, Zn and Fe contents and depletion in Ga (Table 2) is consistent with the spinel chemistry in the Rhodope massif of Bulgaria, showing a trend of depletion in Ga, in the metamorphic Fe-chromite rims surrounding the cores of chromite grains, implying that most tetrahedral sites are still occupied by Fe^{2+} [17]. In addition, it has been suggested that the Mg cations can be replaced by Mn, Zn or Co, whereas Al and Fe^{3+} compete for the octahedral sites, hampering the entry of Ga [17,18,24].

Despite the recorded modification in trace elements by re-distribution during post-magmatic processes, as exemplified by bulk analysis (Tables 1 and 2) and spinel chemistry [17,18,24] limited only to relatively small chromitite occurrences, the well-established relationship ($R = -0.95$ to -0.98) between Cr/(Cr + Al) ratio and Ga (Table 2; Figure 1a) that is comparable to literature data for chromitites hosted in other ophiolite complexes [21–24,40–47] seems to be a salient feature.

5.1.2. Spinel Structure

The structure of spinel is a cubic close-packed array of 32 oxygen ions, with 64 tetrahedral vacancies and 32 octahedral vacancies in one unit cell each, containing 8 formula units, with the general formula: $A^{2+}B^{3+}{}_2O_4$, where A = Fe^{2+}, Mn^{2+}, Mg^{2+}, Co^{2+}, Zn^{2+}, Ni^{2+} and B = Fe^{3+}, Cr^{3+}, Al^{3+}, Ga^{3+}, V^{3+} [36]. Spinels are traditionally denoted as either "normal", where the A cation occupies T sites, the B cation occupies M sites, whereas in the "inverse" type cation B occupies the T site and the M site is occupied by both cations A and B [4,8–12]. The degree of inversion x characterizing the cation distribution can show values between $x = 0$ (normal spinel) and $x = 1$ (inverse spinel). The spinel structure is able to accommodate many cations (at least 36) by enlarging and decreasing its tetrahedral and octahedral bond distances, while the oxygen positional parameter (u) should be regarded as a measure of distortion of the spinel structure from cubic close packing or as the angular distortion of the octahedron [10,12]. The movements of cations between tetrahedral T and octahedral M sites, as a result of Mg^{2+} substitution, can be discussed based on structural parameters, such as bond lengths, cation-cation and cation-anion distances, bond angles and hopping lengths, which were calculated using experimental lattice constants and oxygen parameters [4,8–12].

5.1.3. Applications to Natural Spinels

Despite post-magmatic compositional changes in Fe-chromite within the chromitite ores, resulting in the remobilization of cations during metamorphism (700 °C to 450 °C) of chromitites [17,18,21,24,32], the structural incorporation of Ga into the chromite lattice is evidenced by the progressive and linear increase of Al or decreasing Cr/(Cr + Al) atomic ratio (Figure 2a). Experimental data on cation distribution versus temperature may provide valuable information related to the preferences of cations in the spinel lattice. $MgAl_2O_4$ is the most prominent example of a normal spinel because Mg^{2+} is much larger than Al^{3+} [2,6,48]. Gallium (Ga^{3+}) is smaller than Mg^{2+}, but significantly larger compared to Al^{3+}, leading to an ordering which is called mainly inverse, at least for the end-member composition $MgGa_2O_4$ [4].

The intra-crystalline exchange reaction in spinels has been modeled [49,50] and the order-disorder process has been described by the following exchange reaction: $^{T}Al + {}^{M}Mg = {}^{T}Mg + {}^{M}Al$ (where T = tetrahedral and M = octahedral site) in which the forward reaction implies an exchange of Mg with Al at the M site (ordering process), and backwards (disordering process). The cation distributions for both disordering and ordering experiments were obtained by measuring the oxygen positional parameter (u), the inversion parameter (x) (Al in T) site, using samples with varying composition. The Mueller kinetic model was satisfactory applied to the experimental data and allowed the calculation of the kinetic ordering constants K, linearly related to temperature by means of Arrhenius equations [48–51].

Martignago et al. [52] performed crystal structure refinements on three natural spinels, on low-Cr spinel containing small Fe^{3+} quantities and two other samples with high Cr (8.4 wt % Cr_2O_3) and low Cr (3.3 wt % Cr_2O_3) contents [52]. These experiments have shown that both parameters (u) and

(x) remained constant for the three different samples up to 600 °C, independently of Cr^{3+} contents. The distortion of spinels, started at higher temperature, near to 650 °C (Figure 3). The degree of distortion at the highest temperatures is inversely correlated with Cr^{3+} contents. Since Cr^{3+} has a tendency to be completely partitioned on the octahedral site, due to its electronic configuration and size [2,52–54], Al^{3+} cation is unable to substitute Cr^{3+}. The most salient feature derived from the above experimental data [52] is the abrupt increase of the inversion parameter (x) having the highest values for the sample with the lowest Cr content (L-Cr sample), and the lowest values for the sample with the highest Cr content (H-Cr) [52].

Such structural changes should cause modifications of the structural, physical and thermal properties of the spinels. Although the spinel structure is complicated and determination of several parameters may be required, the above results [52], which show that the Cr content in spinels affects the occupancy of Al in the tetrahedral site, may suggest that the co-existence of Ga^{3+} in high-Al chromitites (Table 2) [2,4–12] is related to the degree of disorder that is inversely correlated with Cr contents.

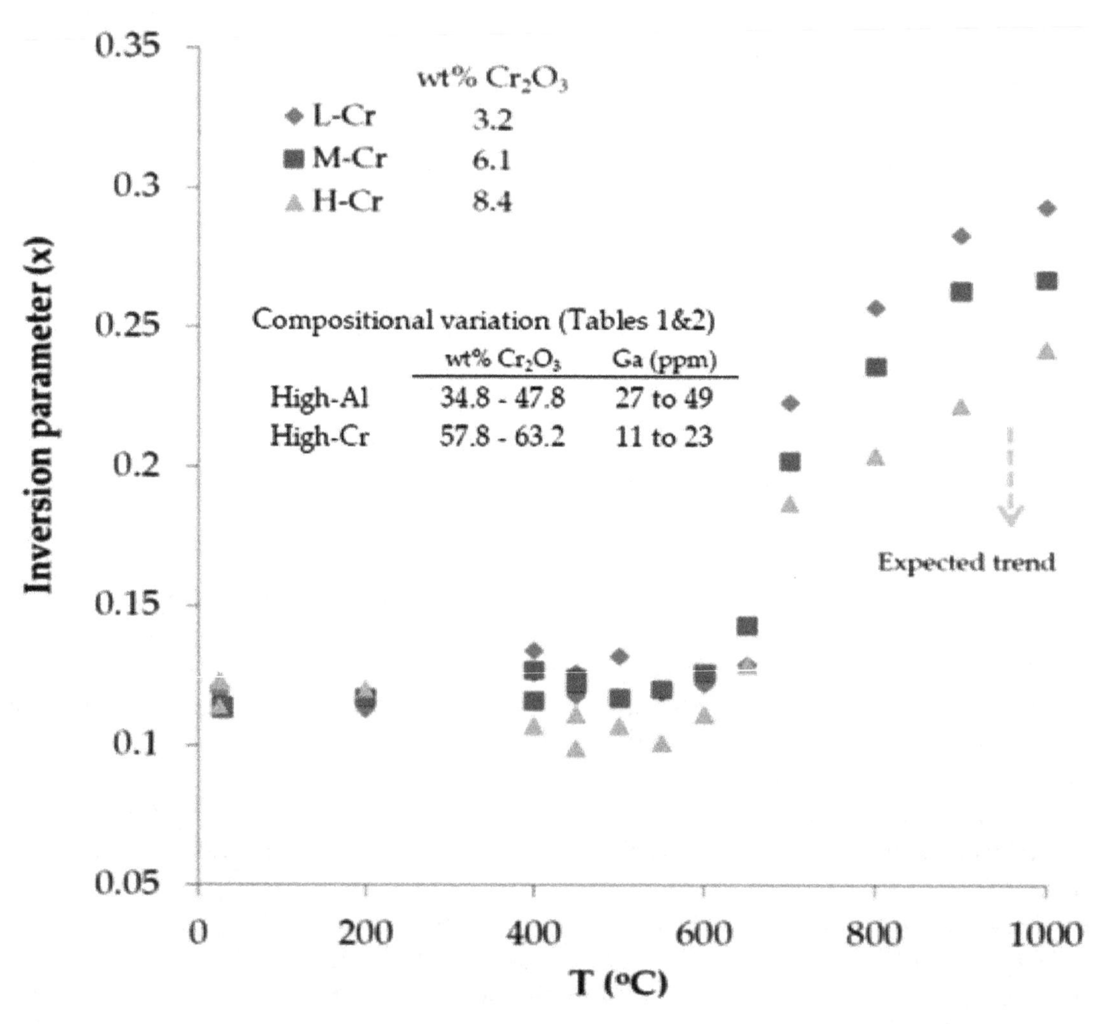

Figure 3. Plot of the Inversion parameter (x) *versus* temperature for spinels after [52]. A potential trend for chromitites with much higher Cr content (Table 1) is presented by the green arrow.

Therefore, potential controlling factors on the Ga preference in high-Al chromitites are (a) the composition of the parent magma (geotectonic setting), temperature and redox conditions, (b) the electronic configuration of Cr^{3+} resulting in occupation of M sites only, the ability of Al^{3+} to occupy T and M sites and the strong preference of Ga^{3+} to T sites, and (c) elevated values of the inversion parameter (x) that is inversely correlated with Cr^{3+} content and favors the co-existence of Ga^{3+} in high-Al chromitites.

6. Conclusions

The presented geochemical and mineral chemistry data on chromitites associated with ophiolite complexes, in conjunction with experimental literature data allowed us to draw the following conclusions:

(1) The lower Ga contents in high-Cr chromitites (11 to 23 ppm) compared to high-Al ones (27–49 ppm) suggest that the composition of the parent magma may be a major factor controlling the preference of Ga in high-Al chromitites.

(2) The positive trend between the $Fe^{3+}/(Al + Cr + Fe^{3+})$ atomic ratio and Ga content for large chromite deposits may suggest the effect of the redox conditions on the Ga distribution in chromitites.

(3) Plot of the Cr/(Cr + Al) atomic ratios versus Ga content exhibits differences in terms of the slope of correlation lines for the different occurrences, suggesting that, in addition to the composition of parent magmas, other factors such as temperature, pressure or redox conditions may affect the observed deviation from linearity for small metamorphosed chromitite bodies.

(4) The depletion of Ga and Al, and elevated Mn, Co, Zn and Fe contents in certain small chromitite occurrences, transformed during post-magmatic metamorphism, suggest potential change of the Ga content in Cr-spinel during sub-solidus reactions.

(5) Assuming that low-Cr spinel is characterized by the highest value of the inversion parameter (x) at higher than 650 °C, then the high Al content in spinels may be a driving force for the degree of inversion in the structure that facilitate the substitution of Al^{3+} for Ga^{3+} at magmatic conditions.

Author Contributions: I.-P.D.E. provided the SEM/EDS analyses. Both authors I.-P.D.E. and G.D.E. contributed to the elaboration and interpretation of the data, and carried out the final revision of the manuscript.

Acknowledgments: Many thanks are expressed to the National University of Athens for the donations the chromitite samples and the access to the analytical facilities (electron microprobe analysis). Many thanks are expressed to Heinz-Gunter Stosch, Karlsruhe Institute of Technology (KIT), Institute for Applied Geosciences, Germany, Maria Perraki, Technical University of Athens and the anonymous reviewers for the constructive criticism and suggestions on an earlier draft of the manuscript. The linguistic improvement of this work by H.G. Stosch is greatly appreciated.

References

1. Liu, Z.; Li, H. Metallurgical process for valuable elements recovery from red mud—A review. *Hydrometallurgy* **2015**, *155*, 29–43. [CrossRef]
2. Navrotsky, A.; Kleppa, O.J. The thermodynamics of cation distributions in simple spinels. *J. Inorg. Nucl. Chem.* **1967**, *29*, 2701–2714. [CrossRef]
3. Shannon, R.D. Revised effective ionic radii and systematic studies of interatomic distances in halides and chalcogenides. *Acta Crystallogr.* **1976**, *32*, 751–767. [CrossRef]
4. O'Neill, H.S.C.; Navrotsky, A. Simple spinels: Crystallographic parameters, cation radii, lattice energies, and cation distribution. *Am. Mineral.* **1983**, *68*, 181–194.
5. Sickafus, K.E.; Wills, J.M.; Grimes, N.W. Structure of spinel. *J. Am. Ceram. Soc.* **1999**, *82*, 3279–3292. [CrossRef]
6. Atkins, P.; Overton, T.; Rourke, J.; Weller, M.; Hagerman, M. *Inorganic Chemistry*, 5th ed.; W.H. Freeman and Company: New York, NY, USA, 2013.
7. Princivalle, F.; Della Giusta, A.; Carbonin, S. Comparative crystal chemistry of spinels from some suites of ultramafic rocks. *Mineral. Petrol.* **1989**, *40*, 117–126. [CrossRef]
8. Bosi, F.; Andreozzi, G.B.; Hålenius, U.; Skogby, H. Zn-O tetrahedral bond length variations in normal spinel oxides. *Am. Mineral.* **2011**, *96*, 594–598. [CrossRef]
9. Fregola, R.A.; Bosi, F.; Skogby, S.; Hålenius, U. Cation ordering over short-range and long-range scales in the $MgAl_2O_4$-$CuAl_2O_4$ series. *Am. Mineral.* **2012**, *97*, 1821–1827. [CrossRef]
10. Bosi, F. Chemical and structural variability in cubic spinel Oxides. *Acta Crystallogr.* **2019**, *B75*, 279–285. [CrossRef]

11. Lenaz, D.; Musco, M.E.; Petrelli, M.; Caldeira, R.; De Min, A.; Marzoli, A.; Mata, J.; Perugini, D.; Princivalle, F.; Boumehdi, M.A.; et al. Restitic or not? Insights from trace element content and crystal—Structure of spinels in African mantle xenoliths. *Lithos* **2017**, *278*, 464–476. [CrossRef]
12. Wei, C.; Feng, Z.; Scherer, G.G.; Barber, J.; Shao-Horn, Y.; Xu, Z.J. Cations in Octahedral Sites: A Descriptor for Oxygen Electrocatalysis on Transition-Metal Spinels. *Adv. Mater.* **2017**, *29*. [CrossRef] [PubMed]
13. Burton, J.D.; Culkin, F.; Riley, J.P. The abundances of gallium and germanium in terrestrial materials. *Geochim. Cosmochim. Acta* **1959**, *16*, 151–180. [CrossRef]
14. Paktunc, A.D.; Cabri, L.J. A proton- and electron-microprobe study of gallium, nickel and zinc distribution in chromian spinel. *Lithos* **1995**, *35*, 261–282. [CrossRef]
15. Wijbrans, C.H.; Klemme, S.; Berndt, J.; Vollmer, C. Experimental determination of trace element partition coefficients between spinel and silicate melt: The influence of chemical composition and oxygen fugacity. *Contrib. Mineral. Petrol.* **2015**, *169*, 1–33. [CrossRef]
16. Dare, S.A.S.; Pearce, J.A.; McDonald, I.; Styles, M.T. Tectonic discrimination of peridotites using fO2-Cr# and Ga–Ti–FeIII systematic in chrome-spinel. *Chem. Geol.* **2009**, *261*, 199–216.
17. Gervilla, F.; Padrón-Navarta, J.A.; Kerestedjian, T.; Sergeeva, I.; González-Jiménez, J.M.; Fanlo, I. Formation of ferrian chromite in podiform chromitites from the Golyamo Kamenyane serpentinte, Eastern Rhodopes, SE Bulgaria: A two-stage process. *Contrib. Mineral. Petrol.* **2012**, *164*, 643–657. [CrossRef]
18. Colás, V.; González-Jiménez, J.M.; Griffin, W.L.; Fanlo, I.; Gervilla, F.; O'Reilly, S.Y.; Pearson, N.J.; Kerestedjian, T.; Proenza, J.A. Fingerprints of metamorphism in chromite: New insights from minor and trace elements. *Chem. Geol.* **2014**, *389*, 137–152. [CrossRef]
19. Scowen, P.; Roeder, P.L.; Helz, R. Re-equilibration of chromite within Kilauea Iki lava lake, Hawaii. *Contrib. Mineral. Petrol.* **1991**, *107*, 8–20. [CrossRef]
20. Prasad, S.R.M.; Prasad, B.B.V.S.V.; Rajesh, B.; Rao, K.H.; Ramesh, K.V. Structural and dielectric studies of Mg^{2+} substituted Ni–Zn ferrite. *Mater. Sci. Pol.* **2015**, *33*, 806–815. [CrossRef]
21. Zhou, M.F.; Robinson, P.T.; Su, B.X.; Gao, J.F.; Li, J.W.; Yang, J.S.; Malpas, J. Compositions of chromite, associated minerals, and parental magmas of podiform chromite deposits: The role of slab contamination of asthenospheric melts in suprasubduction zone environments. *Gondwana Res.* **2014**, *26*, 262–283. [CrossRef]
22. Uysal, İ.; Tarkian, M.; Sadiklar, M.B.; Zaccarini, F.; Meisel, T.; Garuti, G.; Heidrich, S. Petrology of Al- and Cr-rich ophiolitic chromitites from the Muğla, SW Turkey: Implications from composition of chromite, solid inclusions of platinum-group mineral, silicate, and base-metal mineral, and Os-isotope geochemistry. *Contrib. Mineral. Petrol.* **2009**, *158*, 659–674. [CrossRef]
23. Proenza, J.; Gervilla, F.; Melgarejo, J.; Bodinier, J.L. Al- and Cr-rich chromitites from the Mayarí-Baracoa ophiolitic belt (eastern Cuba); consequence of interaction between volatile-rich melts and peridotites in suprasubduction mantle. *Econ. Geol.* **1999**, *94*, 547–566. [CrossRef]
24. González-Jiménez, J.M.; Locmelis, M.; Belousova, E.; Griffin, W.L.; Gervilla, F.; Kerestedjian, T.N.; Pearson, N.J.; Sergeeva, I. Genesis and tectonic implications of podiform chromitites in the metamorphosed Ultramafic Massif of Dobromirtsi (Bulgaria). *Gondwana Res.* **2015**, *27*, 555–574. [CrossRef]
25. Brough, C.P.; Prichard, H.M.; Neary, C.R.; Fisher, P.C.; McDonald, I. Geochemical variations within podiform chromitite deposits in the Shetland Ophiolite: Implications for petrogenesis and PGE concentration. *Econ. Geol.* **2015**, *110*, 187–208. [CrossRef]
26. Konstantopoulou, G.; Economou-Eliopoulos, M. Distribution of Platinum-group Elements and Gold in the Vourinos Chromitite Ores, Greece. *Econ. Geol.* **1991**, *86*, 1672–1682. [CrossRef]
27. Economou-Eliopoulos, M.; Parry, S.J.; Christidis, G. Platinum-group element (PGE) content of chromite ores from the Othrys ophiolite complex, Greece. In *Mineral Deposits: Research and Exploration. Where Do They Meet?* Papunen, H., Ed.; Balkema: Rotterdam, The Netherlands, 1997; pp. 414–441.
28. Economou-Eliopoulos, M.; Vacondios, I. Geochemistry of chromitites and host rocks from the Pindos ophiolite complex, northwestern Greece. *Chem. Geol.* **1995**, *122*, 99–108. [CrossRef]
29. Economou-Eliopoulos, M.; Sambanis, G.; Karkanas, P. Trace element distribution in chromitites from the Pindos ophiolite complex, Greece: Implications for the chromite exploration. In *Mineral Deposits*; Stanley, Ed.; Balkema: Rotterdam, The Netherlands, 1999; pp. 713–716.
30. Economou-Eliopoulos, M. On the origin of the PGE-enrichment in chromitites associated with ophiolite complexes: The case of Skyros island, Greece. In *9th SGA Meeting*; Andrew, C.J., Borg, G., Eds.; Digging Deeper: Dublin, Ireland, 2007; pp. 1611–1614.

31. Economou-Eliopoulos, M. Platinum-group element distribution in chromite ores from ophiolite complexes: Implications for their exploration. *Ore Geol. Rev.* **1996**, *11*, 363–381. [CrossRef]
32. Zhelyaskova-Panayiotova, M.; Economou-Eliopoulos, M. Platinum-group element (PGE) and gold concentrations in oxide and sulfide mineralizations from ultmmafic rocks of Bulgaria. *Ann. Univ. Sofia Geol. Congr.* **1994**, *86*, 196–218.
33. Rassios, A.; Dilek, Y. Rotational deformation in the Jurassic Meohellenic ophiolites, Greece, and its tectonic significance. *Lithos* **2009**, *108*, 192–206. [CrossRef]
34. Economou, M.; Dimou, E.; Economou, G.; Migiros, G.; Vacondios, I.; Grivas, E.; Rassios, A.; Dabitzias, S. *Chromite Deposits of Greece: Athens*; Theophrastus Publications: Athens, Greece, 1986; pp. 129–159.
35. Kapsiotis, A.; Grammatikopoulos, T.A.; Tsikouras, B.; Hatzipanagiotou, K. Platinum-group mineral characterization in concentrates from high-grade PGE Al-rich chromitites of Korydallos area in the Pindos ophiolite complex (NW Greece). *SGS Miner. Serv.* **2009**, *60*, 178–191. [CrossRef]
36. Bonev, N.; Moritz, R.; Borisova, M.; Filipov, P. Therma–Volvi–Gomati complex of the Serbo-Macedonian Massif, northern Greece: A Middle Triassic continental margin ophiolite of Neotethyan origin. *J. Geol. Soc.* **2018**. [CrossRef]
37. Melcher, F.; Grum, W.; Simon, G.; Thalhammer, T.V.; Stumpfl, E. Petrogenesis of the ophiolitic giant chromite deposit of Kempirsai, Kazakhstan: A study of solid and fluid inclusions in chromite. *J. Petrol.* **1997**, *10*, 1419–1458. [CrossRef]
38. Barnes, S.-L.; Naldrett, A.J.; Gorton, M.P. The origin of the fractionation of the platinum-group elements in terrestrial magmas. *Chem. Geol.* **1985**, *53*, 203–323. [CrossRef]
39. Economou-Eliopoulos, M. Apatite and Mn, Zn, Co-enriched chromite in Ni-laterites of northern Greece and their genetic significance. *J. Geochem. Exp.* **2003**, *80*, 41–54. [CrossRef]
40. Sack, R.O.; Ghiorso, M.S. Chromian spinels as petrogenetic indicators: Thermodynamic and petrological applications. *Am. Mineral.* **1991**, *76*, 827–847.
41. Pagé, P.; Barnes, S.-J. Using trace elements in chromites to constrain the origin of podiform chromitites in the Thetford Mines Ophiolite, Québec, Canada. *Econ. Geol.* **2009**, *104*, 997–1018. [CrossRef]
42. Uysal, I.; Sadiklar, M.; Tarkian, M.; Karsli, O.; Aydin, F. Mineralogy and composition of the chromitites and their platinum-group minerals from Ortaca (Muğla-SW Turkey): Evidence for ophiolitic chromitite genesis. *Mineral. Petrol.* **2005**, *83*, 219–242. [CrossRef]
43. Leblanc, M.; Violette, J.F. Distribution of aluminum-rich and chromium-rich chromite pods in ophiolite peridotites. *Econ. Geol.* **1983**, *78*, 293–301. [CrossRef]
44. Zhou, M.F.; Sun, M.; Keays, R.R.; Kerrich, R.W. Controls on platinum-group elemental distributions of podiform chromitites: A case study of high-Cr and high-Al chromitites from Chinese orogenic belts. *Geochim. Cosmochim. Acta* **1998**, *62*, 677–688. [CrossRef]
45. Zaccarini, F.; Garuti, G.; Proenza, J.A.; Campos, L.; Thalhammer, O.A.R.; Aiglsperger, T.; Lewis, J.F. Chromite and platinum group elements mineralization in the Santa Elena ultramafic nappe (Costa Rica): Geodynamic implications. *Geol. Acta* **2011**, *9*, 407–423.
46. Colás, V.; Padrón-Navarta, J.A.; González-Jiménez, J.M.; Griffin, W.L.; Fanlo, I.; O'reilly, S.Y.; Gervilla, F.; Proenza, J.A.; Pearson, N.J.; Escayola, M.P. Compositional effects on the solubility of minor and trace elements in oxide spinel minerals: Insights from crystal-crystal partition coefficients in chromite exsolution. *Am. Mineral.* **2016**, *101*, 1360–1372. [CrossRef]
47. Barnes, S.J.; Roeder, P.L. The range of spinel compositions in terrestrial mafic and ultramafic rocks. *J. Petrol.* **2001**, *42*, 2279–2302. [CrossRef]
48. Andreozzi, G.B.; Princivalle, F.; Skogby, H.; Della Giusta, A. Cation ordering and structural variations with temperature in MgAl2O4 spinel: An X-ray single-crystal study. *Am. Mineral.* **2000**, *85*, 1164–1171. [CrossRef]
49. Ma, Y.; Liu, X. Kinetics and Thermodynamics of Mg-Al Disorder in $MgAl_2O_4$-spinel: A Review. *Molecules* **2019**, *24*, 1704. [CrossRef] [PubMed]
50. Mueller, R.F. Kinetics and thermodynamics of intracrystalline distribution. *Mineral. Soc. Am. Spec. Pap.* **1969**, *2*, 83–93.
51. Mueller, R.F. Model for order-disorder kinetics in certain quasi-binary crystals of continuously variable composition. *J. Phys. Chem. Solids* **1967**, *28*, 2239–2243. [CrossRef]
52. Martignago, F.; Dal Negro, A.; Carbonin, S. How Cr3+ and Fe3+ affect Mg-Al order disorder transformation at high temperature in natural spinels. *Phys. Chem. Miner.* **2003**, *30*, 401–408. [CrossRef]

53. Princivalle, F.; Martignago, F.; Dal Negro, A. Kinetics of cation ordering in natural Mg (Al, Cr3+)$_2$O$_4$ spinels. *Am. Mineral.* **2006**, *91*, 313–318. [CrossRef]
54. Lavina, B.; Reznitskii, L.Z.; Bosi, F. Crystal chemistry of some Mg, Cr, V normal spinels from Sludyanka (Lake Baikal, Russia): The influence of V3+ on structural stability. *Phys. Chem. Miner.* **2003**, *30*, 599–605. [CrossRef]

8

Eliopoulosite, V_7S_8, A New Sulfide from the Podiform Chromitite of the Othrys Ophiolite, Greece

Luca Bindi [1,*], Federica Zaccarini [2], Paola Bonazzi [1], Tassos Grammatikopoulos [3], Basilios Tsikouras [4], Chris Stanley [5] and Giorgio Garuti [2]

1 Dipartimento di Scienze della Terra, Università degli Studi di Firenze, I-50121 Florence, Italy; paola.bonazzi@unifi.it
2 Department of Applied Geological Sciences and Geophysics, University of Leoben, A-8700 Leoben, Austria; federica.zaccarini@unileoben.ac.at (F.Z.); giorgio.garuti1945@gmail.com (G.G.)
3 SGS Canada Inc., 185 Concession Street, PO 4300, Lakefield, ON K0L 2H0, Canada; Tassos.Grammatikopoulos@sgs.com
4 Faculty of Science, Physical and Geological Sciences, Universiti Brunei Darussalam, BE 1410 Gadong, Brunei Darussalam; basilios.tsikouras@ubd.edu.bn
5 Department of Earth Sciences, Natural History Museum, London SW7 5BD, UK; c.stanley@nhm.ac.uk
* Correspondence: luca.bindi@unifi.it

Abstract: The new mineral species, eliopoulosite, V_7S_8, was discovered in the abandoned chromium mine of Agios Stefanos of the Othrys ophiolite, located in central Greece. The investigated samples consist of massive chromitite hosted in a strongly altered mantle tectonite, and are associated with nickelphosphide, awaruite, tsikourasite, and grammatikopoulosite. Eliopoulosite is brittle and has a metallic luster. In plane-reflected polarized light, it is grayish-brown and shows no internal reflections, bireflectance, and pleochroism. It is weakly anisotropic, with colors varying from light to dark greenish. Reflectance values of mineral in air (R_o, $R_{e'}$ in %) are: 34.8–35.7 at 470 nm, 38–39 at 546 nm, 40–41.3 at 589 nm, and 42.5–44.2 at 650 nm. Electron-microprobe analyses yielded a mean composition (wt.%) of: S 41.78, V 54.11, Ni 1.71, Fe 1.1, Co 0.67, and Mo 0.66, totali 100.03. On the basis of $\Sigma_{atoms} = 15$ apfu and taking into account the structural data, the empirical formula of eliopoulosite is $(V_{6.55}Ni_{0.19}Fe_{0.12}Co_{0.07}Mo_{0.04})_{\Sigma = 6.97}S_{8.03}$. The simplified formula is $(V, Ni, Fe)_7S_8$ and the ideal formula is V_7S_8, which corresponds to V 58.16%, S 41.84%, total 100 wt.%. The density, based on the empirical formula and unit-cell volume refined form single-crystal structure XRD data, is 4.545 g·cm^{-3}. The mineral is trigonal, space group $P3_221$, with a = 6.689(3) Å, c = 17.403(6) Å, V = 674.4(5) Å^3, Z = 3, and exhibits a twelve-fold superstructure ($2a \times 2a \times 3c$) of the NiAs-type subcell with V-atoms octahedrally coordinated by S atoms. The distribution of vacancies is discussed in relation to other pyrrhotite-like compounds. The mineral name is for Dr. Demetrios Eliopoulos (1947–2019), a geoscientist at the Institute of Geology and Mineral Exploration (IGME) of Greece and his widow, Prof. Maria Eliopoulos (nee Economou, 1947), University of Athens, Greece, for their contributions to the knowledge of ore deposits of Greece and to the mineralogical, petrographic, and geochemical studies of ophiolites, including the Othrys complex. The mineral and its name have been approved by the Commission of New Minerals, Nomenclature, and Classification of the International Mineralogical Association (No. 2019-96).

Keywords: eliopoulosite; sulfide; chromitite; Agios Stefanos mine; Othrys; ophiolite; Greece

---

1. Introduction

Only eight minerals containing V and S are in the list of valid species approved by the International Mineralogical Association (IMA). They include, in alphabetic order: colimaite, K_3VS_4 [1]; colusite,

$Cu_{13}VAs_3S_{16}$ [2]; germanocolusite, $Cu_{13}VGe_3S_{16}$ [3]; merelaniite, $Mo_4Pb_4VSbS_{15}$ [4]; nekrasovite, $Cu_{13}VSn_3S_{16}$ [5]; patronite, VS_4 [6]; stibiocolusite, $Cu_{13}V(Sb,Sn,As)_3S_{16}$ [7]; and sulvanite, Cu_3VS_4 [8]. All of them, with the exception of patronite, a mineral discovered in 1906 in the Minas Ragra of Peru [6], are sulfides or sulfosalts characterized by a complex composition including other metals besides vanadium.

During a recent investigation of the heavy mineral concentrates from a chromitite collected in the Othrys ophiolite (central Greece), three new minerals were discovered. Two of them are phosphides, namely tsikourasite, $Mo_3Ni_2P_{1+x}$ [9] and grammatikopoulosite, NiVP [10]. A chemical and structural study revealed the third mineral to be a new sulfide, trigonal in symmetry and having the ideal formula V_7S_8. The mineral and its name were approved by the Commission of New Minerals, Nomenclature and Classification of the International Mineralogical Association (No. 2019-096). The new mineral has been named after Dr. Demetrios Eliopoulos (1947–2019), a geoscientist at the Institute of Geology and Mineral Exploration (IGME) of Greece and his widow, Prof. Maria Eliopoulos (nee Economou, 1947), University of Athens, Greece, for their contributions to the knowledge of ore deposits of Greece and to the mineralogical, petrographic, and geochemical studies of ophiolites, including the Othrys complex. Holotype material is deposited in the Mineralogical Collection of the Museo di Storia Naturale, Università di Pisa, Via Roma 79, Calci (Pisa, Italy), under catalogue number 19911 (same type specimen of grammatikopoulosite).

2. Geological Background and Occurrence of Eliopoulosite

Eliopoulosite was discovered in a heavy-mineral concentrate obtained from massive chromitite collected in the mantle sequence of the Mesozoic Othrys ophiolite, located in central Greece (Figure 1A). The geology, petrography, and geodynamic setting of the Otrhys ophiolite have been discussed by several authors [11–24]. The studied sample was collected in the abandoned chromium mine of Agios Stefanos, located a few km southwest of the Domokos village (Figure 1B). In the studied area, a mantle tectonite composed of plagioclase lherzolite, harzburgite, and minor harzburgite-dunite occurs in contact with rocks of the crustal sequence (Figure 1C). The investigated chromitite was collected from the harzburgite-dunite that is cut across by several dykes of gabbro (Figure 1C).

The ophiolite of Othrys is a complete but dismembered suite (Mirna Group) and consists of three structural units: the uppermost succession with variably serpentinized peridotites, which is structurally bounded by an ophiolite mélange; the intermediate Kournovon dolerite, including cumulate gabbro and local rhyolite; and the lower Sipetorrema Pillow Lava unit also including basaltic flows, siltstones, and chert. The Mirna Group constitutes multiple inverted thrust sheets, which were eventually obducted onto the Pelagonian Zone, during the Late Jurassic-Early Cretaceous [11–14]. Three types of basalts with different geochemical signatures have been described: i) alkaline within-plate (WPB), occurring in the mélange, which is related to oceanic seamounts or to ocean-continent transition zones; ii) normal-type mid-ocean ridge (N-MORB); and iii) low-K tholeiitic (L-KT) rocks, which are erupted close to the rifted margin of an ocean-continent transition zone. Radiolarian data from interbedded cherts suggest Middle to Late Triassic ages [15]. The Othrys ophiolite is structurally divided into the west and east Othrys suites, which are thought to derive from different geotectonic environments. The former is related to an extension regime (back-arc basin or MORB [16–18]), while the latter formed in a supra-subduction zone (SSZ) setting [17–19]. This difference is reflected in the contrasting geochemical compositions of their ultramafic rocks, as well as in the diverse platinum group minerals (PGM) assemblages of their host chromitites [20,21], which suggest large mantle heterogeneities in this area. Conflicting views with regards to the evolution of the Othrys ophiolite suggest the involvement of Pindos [11,22,23] or Vardar (Axios) [15] oceanic domains.

The studied spinel-supergroup minerals that host eliopoulosite can been classified as magnesiochromite, although their composition is rather heterogeneous [9,10,25] with the amounts of Cr_2O_3 (44.96–51.64 wt.%), Al_2O_3 (14.18–20.78 wt.%), MgO (13.34–16.84 wt.%), and FeO (8.3–13.31 wt.%) varying significantly through the sampled rock. The calculated Fe_2O_3 is relatively high, and ranges from 6.72 to 9.26 wt.%. The amounts of MnO (0.33–0.60 wt.%), V_2O_3 (0.04–0.30 wt.%), ZnO (up to 0.07 wt.%), and NiO (0.03–0.24 wt.%) are low. The maximum TiO_2 content is 0.23 wt.%, which is typical for the podiform chromitites hosted in the mantle sequence of ophiolite complexes.

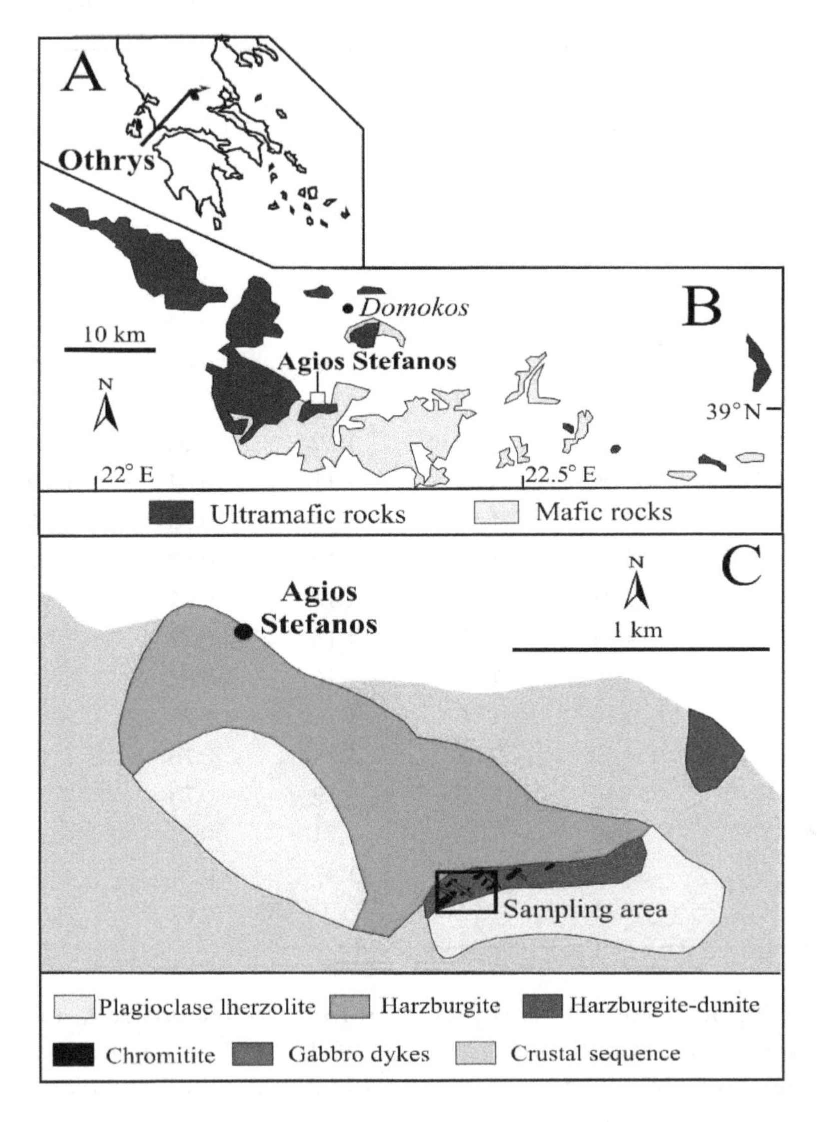

Figure 1. Location of the Othrys complex in Greece (**A**); general geological map of the Othrys ophiolite showing the location of the Agios Stefanos chromium mine (**B**) and (**C**) detailed geological setting of the Agios Stefanos area (modified after [10] and [5]).

3. Analytical Methods

The heavy minerals were concentrated at SGS Mineral Services, Canada, following the methodology described by several authors [9,10,21,25,26]. After the concentration process, the heavy minerals were embedded in epoxy blocks and then polished for mineralogical examination.

Quantitative chemical analyses and acquisition of back-scattered electron images of eliopoulosite were performed with a JEOL JXA-8200 electron-microprobe, installed in the E. F. Stumpfl laboratory, Leoben University, Austria, operating in wavelength dispersive spectrometry (WDS) mode. Major and minor elements were determined at 20 kV accelerating voltage and 10 nA beam current, with 20 s as the counting time for the peak and 10 s for the backgrounds. The beam diameter was about 1 μm in size. For the WDS analyses, the following lines and diffracting crystals were used: S = ($K\alpha$, PETJ), V, Ni, Fe, Co = ($K\alpha$, LIFH), and Mo = ($L\alpha$, PETJ). The following standards were selected: metallic vanadium for V, pyrite for S and Fe, millerite for Ni, molybdenite for Mo, and skutterudite for Co.

The ZAF correction method was applied. Automatic correction was performed for interference Mo-S. Representative analyses of eliopoulosite are listed in Table 1.

Table 1. Electron microprobe analyses (wt.%) of eliopoulosite.

Analysis	S	V	Ni	Fe	Co	Mo	Total
1	41.34	53.08	1.24	0.87	0.48	0.55	99.01
2	41.35	53.45	1.33	0.92	0.52	0.57	99.29
3	41.36	53.55	1.39	0.97	0.55	0.57	99.37
4	41.39	53.71	1.41	1.00	0.55	0.57	99.43
5	41.42	53.90	1.51	1.01	0.61	0.58	99.72
6	41.44	53.91	1.54	1.09	0.63	0.59	99.75
7	41.57	53.97	1.55	1.09	0.63	0.59	99.77
8	41.65	54.03	1.58	1.09	0.63	0.60	99.82
9	41.67	54.03	1.65	1.10	0.64	0.63	99.86
10	41.73	54.06	1.67	1.11	0.64	0.64	99.92
11	41.76	54.09	1.69	1.11	0.65	0.65	100.18
12	41.82	54.16	1.70	1.12	0.68	0.67	100.20
13	41.90	54.21	1.72	1.12	0.69	0.68	100.26
14	41.94	54.25	1.72	1.13	0.72	0.68	100.27
15	41.99	54.40	1.94	1.16	0.72	0.69	100.28
16	42.03	54.40	1.99	1.16	0.73	0.69	100.31
17	42.12	54.56	2.02	1.18	0.76	0.70	100.78
18	42.28	54.64	2.13	1.19	0.78	0.75	100.78
19	42.33	54.80	2.22	1.21	0.79	0.81	100.84
20	42.62	55.02	2.27	1.47	0.91	0.90	100.84
average	41.78	54.11	1.71	1.1	0.67	0.66	100.03

A small grain of eliopoulosite was hand-picked from the polished section under a reflected light microscope. The crystal (about 80 μm in size) was carefully and repeatedly washed in acetone and mounted on a 5 μm-diameter carbon fiber, which was, in turn, attached to a glass rod in preparation of the single-crystal X-ray diffraction measurements.

Single-crystal X-ray diffraction data were collected at the University of Florence (Italy) using a Bruker D8 Venture equipped with a Photon II CCD detector, with graphite-monochromatized Mo*K*α radiation (λ = 0.71073 Å). Intensity data were integrated and corrected for standard Lorentz-polarization factors with the software package *Apex*3 [27,28]. A total of 1289 unique reflections was collected up to 2θ = 62.24°.

The reflectance measurements on eliopoulosite were carried out using a WTiC standard and a J&M TIDAS diode array spectrophotometer at the Natural History Museum of London, UK.

4. Physical and Optical Properties

In the polished section, eliopoulosite occurs as tiny crystals (from 5 μm up to about 80 μm) and is anhedral to subhedral in habit. It consists of polyphase grains associated with other minerals, such as tsikourasite, nickelphosphide, awaruite, and grammatikopoulosite (Figure 2).

In plane-reflected polarized light, eliopoulosite is grayish-brown and shows no internal reflections, bireflectance, and pleochroism. It is weakly anisotropic, with colors varying from light to dark greenish. Reflectance values of the mineral in air (*R* in %) are listed in Table 2 and shown in Figure 3.

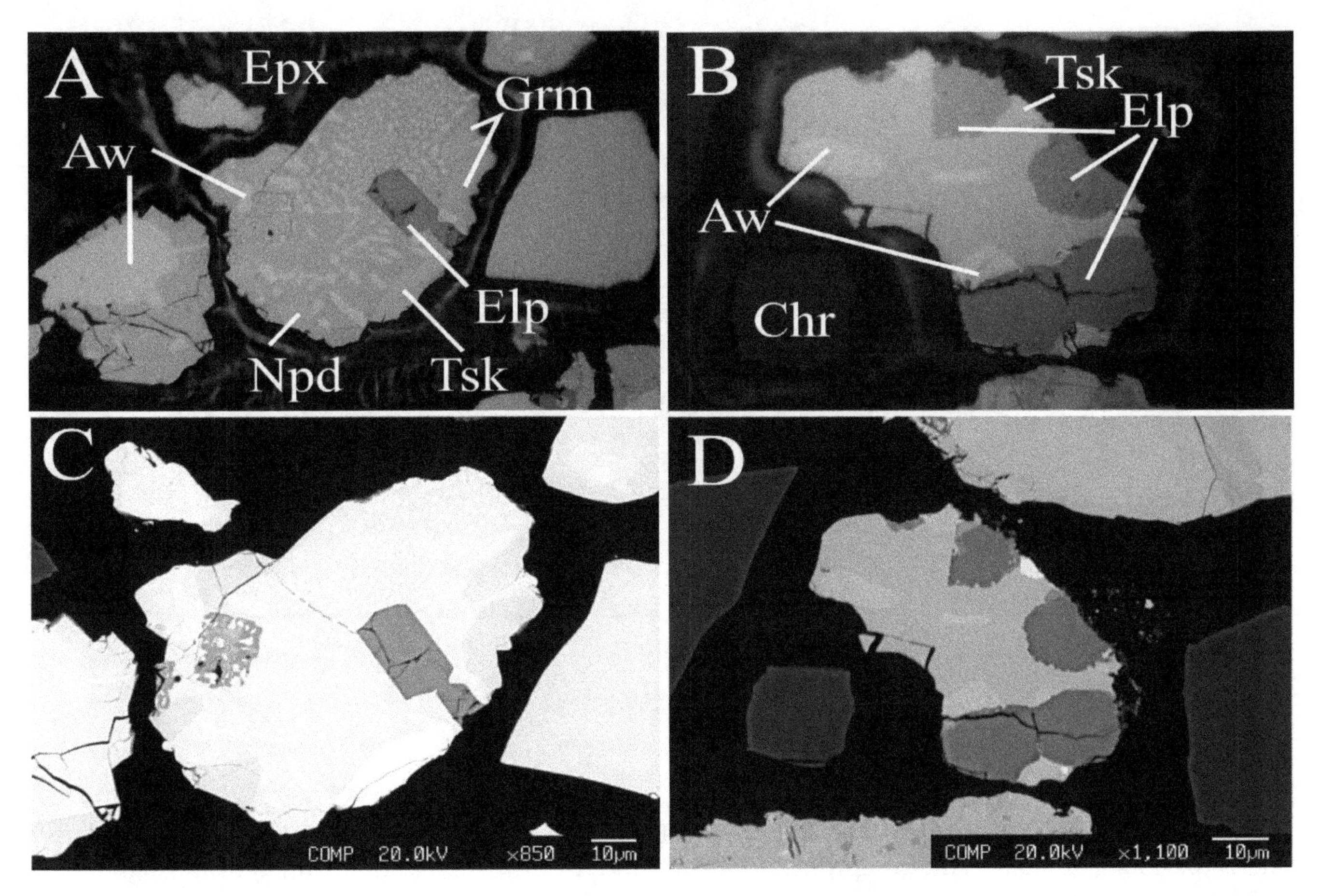

Figure 2. Digital image in reflected plane polarized light (**A**,**B**) and back-scattered electron image (**C**,**D**) showing eliopoulosite from the chromitite of Agios Stefanos. Abbreviations: Elp = eliopoulosite, Grm = grammatikopoulosite, Tsk = tsikourasite, Aw = awaruite, Npd = nickelphosphide, Chr = chromite, Epx = epoxy.

Table 2. Reflectance values of eliopoulosite, those required by the Commission on Ore Mineralogy (COM) are given in bold.

λ (nm)	R_o (%)	$R_{e'}$ (%)
400	32.0	33.1
420	32.7	33.7
440	33.5	34.5
460	34.3	35.3
470	**34.8**	**35.7**
480	35.2	36.1
500	36.1	37.0
520	36.9	37.8
540	37.7	38.7
546	**38.0**	**39.0**
560	38.6	39.7
580	39.5	40.8
589	**40.0**	**41.3**
600	40.5	41.9
620	41.2	42.9
640	42.1	43.8
650	**42.5**	**44.2**
660	42.8	44.6
680	43.5	45.5
700	44.3	46.2

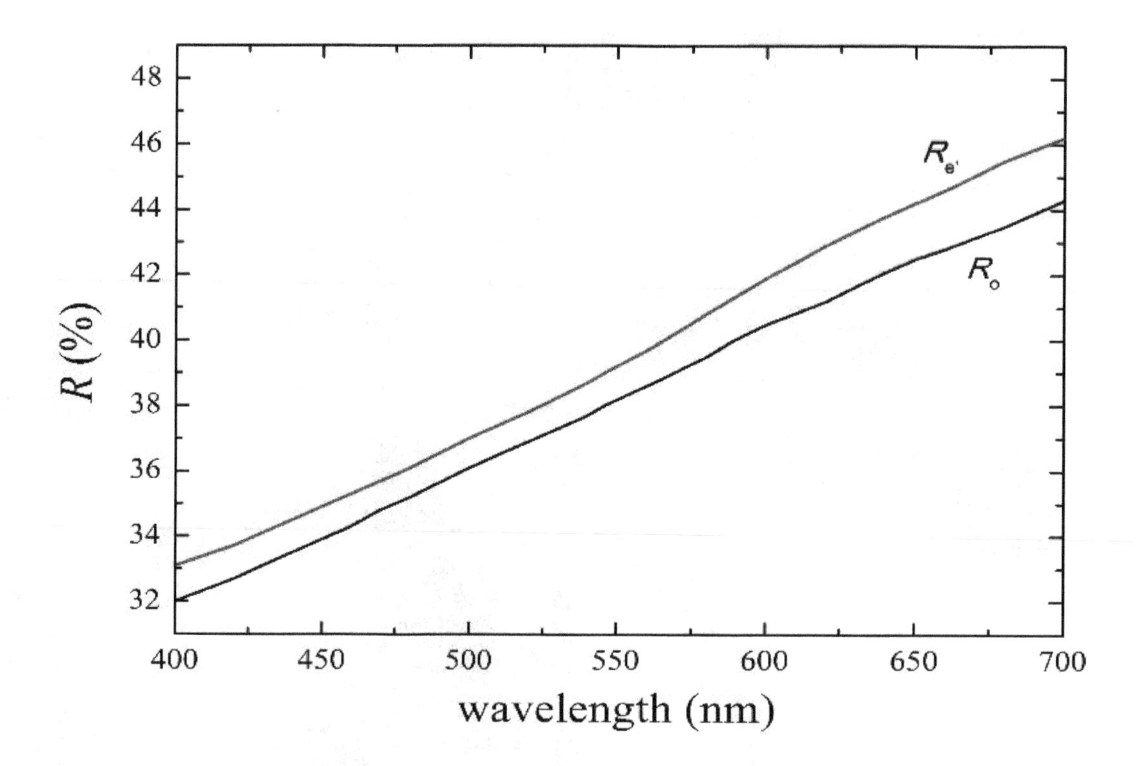

Figure 3. Reflectance data for eliopoulosite.

Due to the small amount of available material, density was not measured. The calculated density was = 4.545 g·cm^{-3}, based on the empirical formula and the unit-cell volume refined from single-crystal XRD data.

5. X-Ray Crystallography and Chemical Composition

The mineral is trigonal, with a = 6.689(3) Å, c = 17.403(6) Å, V = 674.4(5) Å^3, and Z = 3. The reflection conditions (*00l*: $l = 3n$), together with the observed R_{int} in the different Laue classes, point unequivocally to the choice of the space group $P3_221$. The structure solution was then carried out in this space group. The positions of most of the atoms (all the V positions and most of the S atoms) were determined by means of direct methods [29]. A least-squares refinement on F^2 using these heavy-atom positions and isotropic temperature factors produced an R factor of 0.123. Three-dimensional difference Fourier synthesis yielded the position of the remaining sulfur atoms. The program Shelxl-97 [30] was used for the refinement of the structure. The site occupancy factor (s.o.f.) at the cation sites was allowed to vary (V vs. structural vacancy) using scattering curves for neutral atoms taken from the International Tables for Crystallography [30]. Four V sites (i.e., V1, V2, V3, and V4) were found to be partially occupied by vanadium (75%), while V5 and V6 were found to be fully occupied by V and fixed accordingly. Sulfur atoms were found on four fully occupied general positions leading to an ideal formula V_7S_8. Given the almost identical partial occupancy of four V sites (i.e., 75%), we carefully checked either the possible presence of twinning or the acentricity of the model. The lack of the inversion center was confirmed using the Flack parameter in Shelxl (0.07(1)) and the trigonal model was double-checked with the Platon *addsymm* routine. At the last stage, with anisotropic atomic displacement parameters for all atoms and no constraints, the residual value settled at $R1$ = 0.0363 for 398 observed reflections ($Fo > 4\sigma(Fo)$) and 79 parameters and at $R1$ = 0.0537 for all 1289 independent reflections. Refined atomic coordinates and isotropic displacement parameters are given in Table 3, whereas selected bond distances are reported in Table 4. Crystallographic Information File (CIF) is deposited.

The calculated X-ray powder diffraction pattern (Table 5) was computed on the basis of the unit-cell data above and with the atom coordinates and occupancies reported in Table 3.

Table 3. Atoms, site occupancy, fractional atom coordinates (Å), and isotropic atomic displacement parameters (Å^2) for eliopoulosite.

Atom	Site Occupancy	x/a	y/b	z/c	U_{iso}
V1	$V_{0.749(6)}$	0.0001(6)	0	2/3	0.0088(6)
V2	$V_{0.762(6)}$	0.5005(6)	0	2/3	0.0095(6)
V3	$V_{0.755(6)}$	0.0006(7)	0	1/6	0.0113(6)
V4	$V_{0.749(7)}$	0.5010(7)	0	1/6	0.0187(8)
V5	$V_{1.00}$	0.4996(4)	0.4997(5)	0.16664(5)	0.0060(3)
V6	$V_{1.00}$	0.4997(5)	0.4996(5)	0.33327(5)	0.0120(4)
S1	$S_{1.00}$	0.1663(7)	0.3343(6)	0.08309(11)	0.0227(4)
S2	$S_{1.00}$	0.1659(7)	0.3328(7)	0.41639(11)	0.0226(4)
S3	$S_{1.00}$	0.1661(7)	0.3331(6)	0.74976(11)	0.0226(4)
S4	$S_{1.00}$	0.3333(6)	0.6673(7)	0.25009(11)	0.0228(4)

Table 4. Bond distances (Å) in the structure of eliopoulosite.

V1-S2	2.409(5) (×2)	V5-S1	2.413(4)
V1-S3	2.411(3) (×2)	V5-S4	2.416(3)
V1-S1	2.418(4) (×2)	V5-S1	2.417(4)
mean	2.413	V5-S2	2.418(4)
		V5-S4	2.419(3)
V2-S1	2.412(4) (×2)	V5-S3	2.419(4)
V2-S2	2.413(5) (×2)	mean	2.417
V2-S4	2.414(3) (×2)		
mean	2.413	V6-S1	2.407(4)
		V6-S4	2.411(4)
V3-S3	2.414(4) (×2)	V6-S3	2.413(4)
V3-S2	2.418(3) (×2)	V6-S3	2.413(4)
V3-S1	2.422(3) (×2)	V6-S2	2.415(4)
mean	2.418	V6-S4	2.417(4)
		mean	2.413
V4-S4	2.413(3) (×2)		
V4-S2	2.417(4) (×2)		
V4-S3	2.421(5) (×2)		
mean	2.417		
V1-V3	2.9006(10) (×2)		
V2-V5	2.9001(13) (×2)		
V4-V6	2.9018(14) (×2)		
V5-V6	2.8998(12) (×2)		

Table 5. Calculated X-ray powder diffraction data (d in Å) for eliopoulosite. The strongest observed reflections are given in bold.

hkl	d_{calc}	I_{calc}
200	**2.8964**	**29**
023	**2.5914**	**45**
026	**2.0495**	**100**
220	**1.6723**	**40**
029	1.6082	10
012	1.4503	8
400	1.4482	3
046	1.2957	20
$22\underline{12}$	1.0956	15
246	1.0242	12
600	0.9655	4

The chemical data (Table 1), yielding to a mean composition (wt.%) of S 41.78, V 54.11, Ni 1.71, Fe 1.1, Co 0.67, Mo 0.66, and total 100.03, were then normalized taking into account the structural results. The empirical formula of eliopoulosite on the basis of Σatoms = 15 apfu is $(V_{6.55}Ni_{0.19}Fe_{0.12}Co_{0.07}Mo_{0.04})_{\Sigma=6.97}S_{8.03}$. The simplified formula is $(V,Ni,Fe)_7S_8$ and the ideal formula is V_7S_8, which corresponds to V 58.16, S 41.84, and total 100 (wt.%).

The structure can be considered as a twelve-fold superstructure ($2a \times 2a \times 3c$) of the NiAs-type subcell with V atoms octahedrally coordinated by S atoms and sulfur located in a trigonal prism. There are six octahedral layers in the structure with rods of fully occupied V sites alternate to rods of partially occupied sites, so that every layer contains 3.5 vanadium atoms. In successive layers rods are directed along [110], [100], and [010] accordingly to the threefold screw axis. As shown in Figure 5, the distribution of fully (blue in color) and partially (pale blue) occupied sites determine the superstructure along the $\mathbf{a_1}$ and $\mathbf{a_2}$ axes. Differences in electron density between fully and partially occupied sites are indeed rather modest (22 vs. 16.5): Accordingly, *hkl* reflections with $h,k = 2n + 1$ are rather weak ($<I/\sigma(I)> = 27.1$ vs. 4890.3).

Since the octahedral vacancies' distribution within the layers at $z = 0$, $z = 1/6$, and $z = 1/3$ is perfectly replicated at layers at $z = 1/2$, $z = 2/3$, and $z = 5/6$ (Figure 5), no contribution to *hkl* reflections with $l = 2n + 1$ is given by metal atoms. However, their intensities ($<I/\sigma(I)> = 564.3$ vs. 3349.2) are even higher than those leading to doubling of **a**-axis, due to the different position of sulfur atoms related to the different octahedral orientation in the layer at $z = 0$ with respect to $z = 1/2$ (Figure 4).

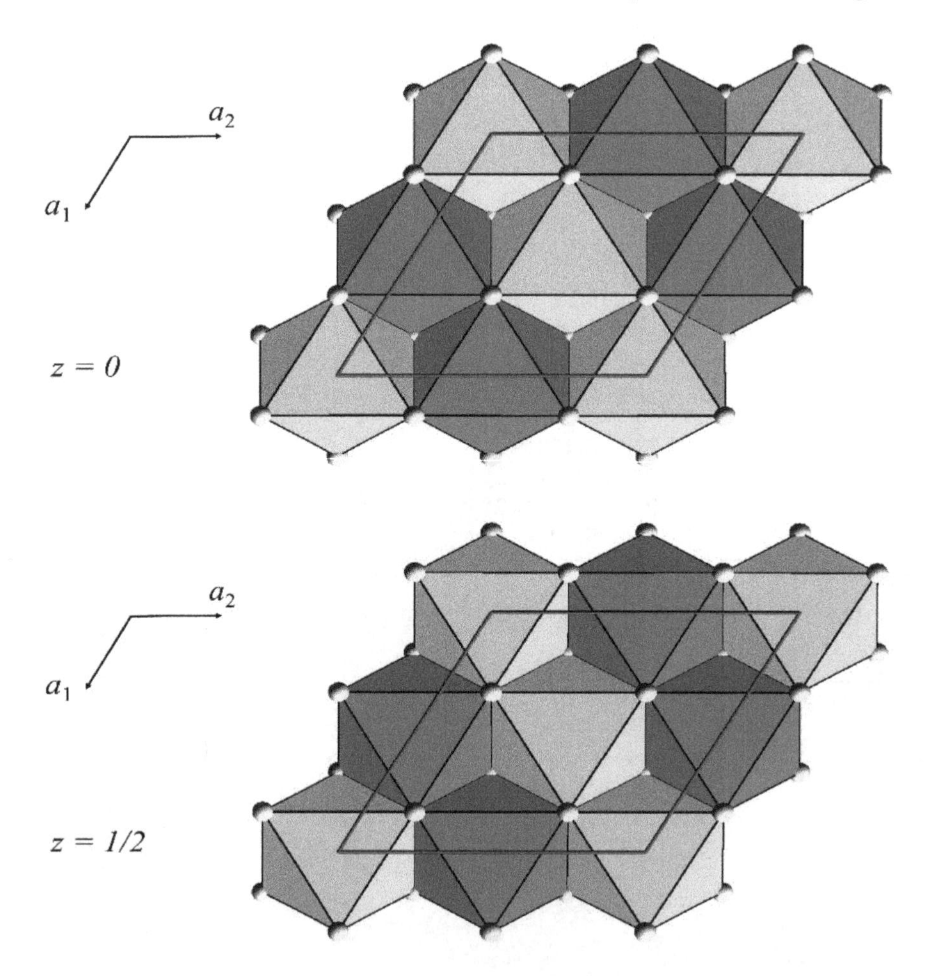

Figure 4. Octahedral layers in eliopoulosite projected down [001]. Layer at $z = 0$ (top of picture) repeats at $z = 1/3$ and $z = 2/3$; layer at $z = 1/2$ (bottom of picture) repeats at $z = 1/6$ and $z = 5/6$. Blue and pale blue colors represent full and partial occupancy. Sulfur atoms are depicted in yellow. The unit cell and the orientation of the structure are indicated.

The eliopoulosite structure shows strong analogies with the superstructures observed in the pyrrhotite-group of minerals. It is identical to the structure inferred for synthetic $VS_{1.125}$ [31]. The mean <V-S> bond distances in eliopoulosite (in the range 2.413–2.418 Å) are in excellent agreement with those found for synthetic vanadium sulfides. To achieve the charge balance, one should hypothesize the presence of both divalent and trivalent vanadium in eliopoulosite (i.e., $V^{2+}{}_5V^{3+}{}_2S_8$). The octahedral sites, however, do not show any significant difference symptomatic of a V^{2+}-V^{3+} ordering. Likewise, we did not find any evidence of ordering of the minor substituents (i.e., Ni and Fe) at a particular structural site. Furthermore, eliopoulosite shows, for the analogy of stoichiometry, unit cell and space group, close relationships with the metastable, trigonal form of 3C-Fe_7S_8 (a = 6.852(6), c = 17.046(2), $P3_121$) [32]. However, the distribution of vacancies is quite different (Figure 5b): As in other types of pyrrhotites [33], in 3C-Fe_7S_8 octahedral vacancies are completely ordered on one position alone so that the layers containing a void (i.e., three metal atoms) are alternating with fully occupied layers (i.e., four metal atoms).

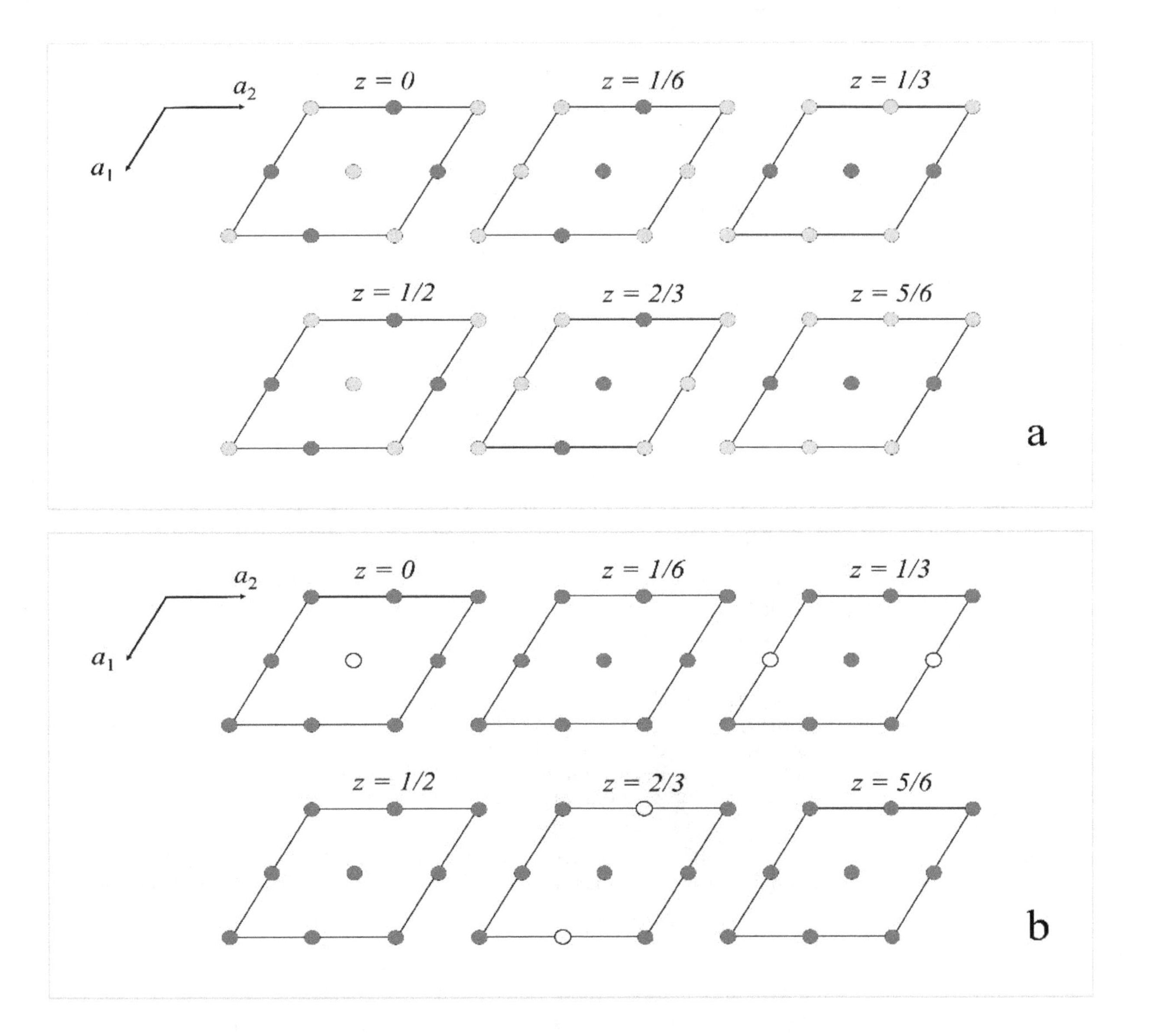

Figure 5. Metal distribution in eliopoulosite (**a**) and in synthetic 3C-Fe_7S_8 pyrrhotite (**b**). Blue and pale blue circles refer to metal positions with full and partial (75%) occupancy, respectively. White circles represent completely empty sites in synthetic 3C-Fe_7S_8 pyrrhotite [32]. The unit cell and the orientation of the structure are indicated.

Eliopoulosite does not correspond to any valid or invalid unnamed mineral [34]. According to literature data [1–8] eliopoulosite is the ninth mineral containing V and S that has been accepted by IMA. Noteworthy, eliopoulosite is the second V^{2+}-bearing mineral, the first being dellagiustaite, Al_2VO_4 [35]. The presence of divalent V requires extremely reducing conditions (well below the iron-wüstite buffer). Furthermore, based on its chemical composition, eliopoulosite is the second sulfide that contains only V as a major component discovered so far. Recently, Ivanova et al. [36] reported on the presence of a V,Cr,Fe-sulfide in the extremely reduced assemblage of a rare CB chondrite. Due to its small size (one micron) the grain was studied only by EBSD (Electron BackScattered Diffraction) and no crystallographic or optical data were provided. However, its ideal composition $(V,Cr,Fe)_4S_5$ seems different compared to that of eliopoulosite in terms of S content and V/Fe ratio and the presence of abundant Cr.

Interestingly, Selezneva et al. [37] recently showed that synthetic V_7S_8 shows differences in the magnetic behavior with respect to trigonal Fe_7S_8. Indeed, V_7S_8 is observed to exhibit a Pauli-paramagnetic behavior opposite to the classic ferrimagnetic ordering usually observed in pyrrhotite. Unfortunately, the small amount of the natural material precludes any possible measurement of both the magnetic susceptibility and magnetization. Such an experiment would have allowed us to verify if the amount of structural vacancies observed for eliopoulosite are enough to compensate the magnetic moments to avoid the ferromagnetic order.

6. Remarks on the Origin of Eliopoulosite

Eliopoulosite was found in the same sample in which tsikourasite and grammatikopoulosite were discovered, and the three minerals occur in the same mineralogical assemblage [9,10]. Therefore, we can argue that they formed under the same chemical-physical condition (i.e., in a strongly reducing environment). This assumption is fully supported by the chemical composition of eliopoulosite, that points to a low valence state for V. However, as already proposed for the origin of tsikourasite and grammatikopoulosite, it is still not possible to provide a conclusive model to explain exhaustively the genesis of eliopoulosite. The following hypothesis can be formulated, since all of them imply the presence of a reducing environment [9,10,38–43]: i) low-temperature alteration of chromitite during serpentinization process; ii) high-temperature reaction of the chromitites with reducing fluids, at mantle depth; iii) post-orogenic surface lightning strike; or iv) meteorite impact.

However, the probability of intercepting a fragment of a meteorite or a fulgurite in the Otrhys ophiolite during the sampling of the studied chromitite seems very unlikely, although a V-bearing sulfide has been reported in a meteorite [36].

Author Contributions: L.B., F.Z., and P.B. wrote the manuscript; F.Z. performed the chemical analyses; L.B. and P.B. performed the diffraction experiments; T.G. and B.T. provided the concentrate sample and information on the sample provenance and petrography of Othrys chromitite; G.G. discussed the chemical data and C.S. obtained the optical data. All the authors provided support in the data interpretation and revised the manuscript. All authors have read and agreed to the published version of the manuscript.

Acknowledgments: The authors acknowledge Ritsuro Miyawaki, Chairman of the CNMNC and its members for helpful comments on the submitted new mineral proposal. Many thanks are due to the editorial staff of Minerals.

References

1. Ostrooumov, M.; Taran, Y.; Arellano-Jimenez, M.; Ponse, A.; Reyes-Gasga, J. Colimaite, K_3VS_4—A new potassium-vanadium sulfide mineral from the Colima volcano, State of Colima (Mexico). *Rev. Mex. Cienc. Geol.* **2009**, *26*, 600–608.
2. Zachariasen, W.H. X-Ray examination of colusite, $(Cu,Fe,Mo,Sn)_4(S,As,Te)_{3-4}$. *Am. Mineral.* **1933**, *18*, 534–537.
3. Spiridonov, E.M.; Kachalovskaya, V.M.; Kovachev, V.V.; Krapiva, L.Y. Germanocolusite $Cu_{26}V_2(Ge,As)_6S_{32}$—a new mineral. *Vest. Moskov. Univers., Ser. 4, Geologiya* **1992**, *1992*, 50–54. (In Russian)

4. Jaszczak, J.A.; Rumsey, M.S.; Bindi, L.; Hackney, S.A.; Wise, M.A.; Stanley, C.J.; Spratt, J. Merelaniite, $Mo_4Pb_4VSbS_{15}$, a new molybdenum-essential member of the cylindrite group, from the Merelani Tanzanite Deposit, Lelatema Mountains, Manyara Region, Tanzania. *Minerals* **2016**, *6*, 115. [CrossRef]
5. Kovalenker, V.A.; Evstigneeva, T.L.; Malov, V.S.; Trubkin, N.V.; Gorshkov, A.I.; Geinke, V.R. Nekrasovite $Cu_{26}V_2Sn_6S_{32}$—A new mineral of the colusite group. *Mineral. Zh.* **1984**, *6*, 88–97.
6. Hillibrand, W.F. Vanadium sulphide, patronite, and its mineral associates from Minasragra, Peru. *Am. J. Sci.* **1907**, *24*, 141–151. [CrossRef]
7. Spiridonov, E.M.; Badalov, A.S.; Kovachev, V.V. Stibiocolusite $Cu_{26}V_2(Sb,Sn,As)_6S_{32}$: A new mineral. *Dokl. Akad. Nauk* **1992**, *324*, 411–414. (In Russian)
8. Trojer, F.J. Refinement of the structure of sulvanite. *Am. Mineral.* **1996**, *51*, 890–894.
9. Zaccarini, F.; Bindi, L.; Ifandi, E.; Grammatikopoulos, T.; Stanley, C.; Garuti, G.; Mauro, D. Tsikourasite, $Mo_3Ni_2P_{1+x}$ ($x < 0.25$), a new phosphide from the chromitite of the Othrys Ophiolite, Greece. *Minerals* **2019**, *9*, 248.
10. Bindi, L.; Zaccarini, F.; Ifandi, E.; Tsikouras, B.; Stanley, C.; Garuti, G.; Mauro, D. Grammatikopoulosite, NiVP, a new phosphide from the chromitite of the Othrys Ophiolite, Greece. *Minerals* **2020**, *10*, 131. [CrossRef]
11. Smith, A.G.; Rassios, A. The evolution of ideas for the origin and emplacement of the western Hellenic ophiolites. *Geol. Soc. Am. Spec. Pap.* **2003**, *373*, 337–350.
12. Hynes, A.J.; Nisbet, E.G.; Smith, G.A.; Welland, M.J.P.; Rex, D.C. Spreading and emplacement ages of some ophiolites in the Othris region (Eastern Central Greece). *Z. Deutsch Geol. Ges.* **1972**, *123*, 455–468.
13. Smith, A.G.; Hynes, A.J.; Menzies, M.; Nisbet, E.G.; Price, I.; Welland, M.J.; Ferrière, J. The stratigraphy of the Othris Mountains, Eastern Central Greece: A deformed Mesozoic continental margin sequence. *Eclogue Geol. Helv.* **1975**, *68*, 463–481.
14. Rassios, A.; Smith, A.G. Constraints on the formation and emplacement age of western Greek ophiolites (Vourinos, Pindos, and Othris) inferred from deformation structures in peridotites. In *Ophiolites and Oceanic Crust: New Insights from Field Studies and the Ocean Drilling Program*; Dilek, Y., Moores, E., Eds.; Geological Society of America Special Paper: Boulder, CO, USA, 2001; pp. 473–484.
15. Bortolotti, V.; Chiari, M.; Marcucci, M.; Photiades, A.; Principi, G.; Saccani, E. New geochemical and age data on the ophiolites from the Othrys area (Greece): Implication for the Triassic evolution of the Vardar ocean. *Ofioliti* **2008**, *33*, 135–151.
16. Barth, M.G.; Mason, P.R.D.; Davies, G.R.; Drury, M.R. The Othris Ophiolite, Greece: A snapshot of subduction initiation at a mid-ocean ridge. *Lithos* **2008**, *100*, 234–254. [CrossRef]
17. Barth, M.; Gluhak, T. Geochemistry and tectonic setting of mafic rocks from the Othris Ophiolite, Greece. *Contr. Mineral. Petrol.* **2009**, *157*, 23–40. [CrossRef]
18. Dijkstra, A.H.; Barth, M.G.; Drury, M.R.; Mason, P.R.D.; Vissers, R.L.M. Diffuse porous melt flow and melt-rock reaction in the mantle lithosphere at a slow-spreading ridge: A structural petrology and LA-ICP-MS study of the Othris Peridotite Massif (Greece). *Geochem. Geophys. Geosyst.* **2003**, *4*, 278. [CrossRef]
19. Magganas, A.; Koutsovitis, P. Composition, melting and evolution of the upper mantle beneath the Jurassic Pindos ocean inferred by ophiolitic ultramafic rocks in East Othris, Greece. *Int. J. Earth Sci.* **2015**, *104*, 1185–1207. [CrossRef]
20. Garuti, G.; Zaccarini, F.; Economou-Eliopoulos, M. Paragenesis and composition of laurite from chromitites of Othrys (Greece): Implications for Os-Ru fractionation in ophiolite upper mantle of the Balkan Peninsula. *Mineral. Dep.* **1999**, *34*, 312–319. [CrossRef]
21. Tsikouras, B.; Ifandi, E.; Karipi, S.; Grammatikopoulos, T.A.; Hatzipanagiotou, K. Investigation of platinum-group minerals (PGM) from Othrys chromitites (Greece) using superpanning concentrates. *Minerals* **2016**, *6*, 94. [CrossRef]
22. Robertson, A.H.F. Overview of the genesis and emplacement of Mesozoic ophiolites in the Eastern Mediterranean Tethyan region. *Lithos* **2002**, *65*, 1–67. [CrossRef]
23. Robertson, A.H.F.; Clift, P.D.; Degnan, P.; Jones, G. Palaeogeographic and palaeotectonic evolution of the Eastern Mediterranean Neotethys. *Palaeogeogr. Palaeoclim. Palaeoecol.* **1991**, *87*, 289–343. [CrossRef]
24. Economou, M.; Dimou, E.; Economou, G.; Migiros, G.; Vacondios, I.; Grivas, E.; Rassios, A.; Dabitzias, S. Chromite deposits of Greece. In *Chromites, UNESCO's IGCP197 Project Metallogeny of Ophiolites*; Petrascheck, W., Karamata, S., Eds.; Theophrastus Publ. S.A.: Athens, Greece, 1986; pp. 129–159.

25. Ifandi, E.; Zaccarini, F.; Tsikouras, B.; Grammatikopoulos, T.; Garuti, G.; Karipi, S.; Hatzipanagiotou, K. First occurrences of Ni-V-Co phosphides in chromitite of Agios Stefanos mine, Othrys ophiolite, Greece. *Ofioliti* **2018**, *43*, 131–145.
26. Zaccarini, F.; Ifandi, E.; Tsikouras, B.; Grammatikopoulos, T.; Garuti, G.; Mauro, D.; Bindi, L.; Stanley, C. Occurrences of new phosphides and sulfide of Ni, Co, V, and Mo from chromitite of the Othrys ophiolite complex (Central Greece). *Per. Mineral.* **2019**, 88. [CrossRef]
27. Bruker. *APEX3*; Bruker AXS Inc.: Madison, WI, USA, 2016. Available online: https://www.bruker.com/products/x-ray-diffraction-and-elemental-analysis/single-crystal-x-ray-diffraction/sc-xrd-software/apex3.html (accessed on 6 March 2020).
28. Bruker. *SAINT and SADABS*; Bruker AXS Inc.: Madison, WI, USA, 2016. Available online: https://www.bruker.com/products/x-ray-diffraction-and-elemental-analysis/single-crystal-x-ray-diffraction/sc-xrd-software/apex3.html (accessed on 6 March 2020).
29. Sheldrick, G.M. A short history of SHELX. *Acta Crystallogr.* **2008**, *A64*, 112–122. [CrossRef]
30. Wilson, A.J.C. *International Tables for Crystallography: Mathematical, Physical, and Chemical Tables*; International Union of Crystallography: Chester, UK, 1992; Volume 3.
31. Grønvold, F.; Haraldsen, H.; Pedersen, B.; Tufte, T. X-ray and magnetic study of vanadium sulfides in the range V_5S_4 to V_5S_8. *Rev. Chim. Minéral.* **1969**, *6*, 215.
32. Nakano, A.; Tokonami, M.; Morimoto, N. Refinement of 3C pyrrhotite, Fe_7S_8. *Acta Crystallogr.* **1979**, *B35*, 722–724. [CrossRef]
33. Morimoto, N. Crystal structure of a monoclinic pyrrhotite. *Rec. Progr. Nat. Sci. Japan* **1978**, *3*, 183–206.
34. Smith, D.G.W.; Nickel, E.H. A system for codification for unnamed minerals: report of the Subcommittee for Unnamed Minerals of the IMA Commission on New Minerals, Nomenclature and Classification. *Can. Mineral.* **2007**, *45*, 983–1055. [CrossRef]
35. Cámara, F.; Bindi, L.; Pagano, A.; Pagano, R.; Gain, S.E.M.; Griffin, W.L. Dellagiustaite: A novel natural spinel containing V^{2+}. *Minerals* **2019**, *9*, 4. [CrossRef]
36. Ivanova, M.A.; Ma, C.; Lorenz, C.A.; Franchi, I.A.; Kononkova, N.N. A new unusual bencubbinite (cba), Sierra Gorda 013 with unique V-rich sulfides. *Met. Plan. Sci.* **2019**, *54*, 6149.
37. Selezneva, N.V.; Ibrahim, P.N.G.; Toporova, N.M.; Sherokalova, E.M.; Baranov, N.V. Crystal structure and magnetic properties of pyrrhotite-type compounds $Fe_{7-y}V_yS_8$. *Acta Phys. Polon.* **2018**, *A133*, 450–452. [CrossRef]
38. Malvoisin, B.; Chopin, C.; Brunet, F.; Matthieu, E.; Galvez, M.E. Low-temperature wollastonite formed by carbonate reduction: a marker of serpentinite redox conditions. *J. Petrol.* **2012**, *53*, 159–176. [CrossRef]
39. Etiope, G.; Tsikouras, B.; Kordella, S.; Ifandi, E.; Christodoulou, D.; Papatheodorou, G. Methane flux and origin in the Othrys ophiolite hyperalkaline springs, Greece. *Chem. Geol.* **2013**, *347*, 161–174. [CrossRef]
40. Etiope, G.; Ifandi, E.; Nazzari, M.; Procesi, M.; Tsikouras, B.; Ventura, G.; Steele, A.; Tardini, R.; Szatmari, P. Widespread abiotic methane in chromitites. *Sci. Rep.* **2018**, *8*, 8728. [CrossRef]
41. Xiong, Q.; Griffin, W.L.; Huang, J.X.; Gain, S.E.M.; Toledo, V.; Pearson, N.J.; O'Reilly, S.Y. Super-reduced mineral assemblages in "ophiolitic" chromitites and peridotites: The view from Mount Carmel. *Eur. J. Mineral.* **2017**, *29*, 557–570. [CrossRef]
42. Pasek, M.A.; Hammeijer, J.P.; Buick, R.; Gull, M.; Atlas, Z. Evidence for reactive reduced phosphorus species in the early Archean ocean. *Proc. Nat. Acad. Sci. U.S.A.* **2013**, *110*, 100089–100094. [CrossRef]
43. Ballhaus, C.; Wirth, R.; Fonseca, R.O.C.; Blanchard, H.; Pröll, W.; Bragagni, A.; Nagel, T.; Schreiber, A.; Dittrich, S.; Thome, V.; et al. Ultra-high pressure and ultra-reduced minerals in ophiolites may form by lightning strikes. *Geochem. Perspec. Lett.* **2017**, *5*, 42–46. [CrossRef]

9

Hydrometallurgical Recovery and Process Optimization of Rare Earth Fluorides from Recycled Magnets

Prince Sarfo *, Thomas Frasz, Avimanyu Das and Courtney Young *

Metallurgical and Materials Engineering, Montana Tech, 1300 West Park Street, Butte, MT 59701, USA; tfrasz@mtech.edu (T.F.); adas@mtech.edu (A.D.)

* Correspondence: psarfo@mtech.edu (P.S.); cyoung@mtech.edu (C.Y.)

Abstract: Magnets containing substantial quantities of rare earth elements are currently one of the most sought-after commodities because of their strategic importance. Recycling these rare earth magnets after their life span has been identified to be a unique approach for mitigating environmental issues that originate from mining and also for sustaining natural resources. The approach is hydrometallurgical, with leaching and precipitation followed by separation and recovery of neodymium (Nd), praseodymium (Pr) and dysprosium (Dy) in the form of rare earth fluorides (REF) as the final product. The methodology is specifically comprised of sulfuric acid (H_2SO_4) leaching and ammonium hydroxide (NH_4OH) precipitation followed by reacting the filtrate with ammonium bifluoride ($NH_4F{\cdot}HF$) to yield the REF. Additional filtering also produces ammonium sulfate ($(NH_4)_2SO_4$) as a byproduct fertilizer. Quantitative and qualitative evaluations by means of XRD, ICP and TGA-DSC to determine decomposition of ammonium jarosite, which is an impurity in the recovery process were performed. Additionally, conditional and response variables were used in a surface-response model to optimize REF production from end-of-life magnets. A REF recovery of 56.2% with a REF purity of 62.4% was found to be optimal.

Keywords: rare earth elements; magnets; recycling; recovery; fluorides; modelling

---

1. Introduction

Disquiet around the sustainability of rare-earth elements (REE) provisions has stimulated determination not only to recycle but also to improve the proficiency of the materials they are used to make [1]. With respect to the amount of REEs produced, neodymium (Nd) usage in the production of neodymium-Iron-Boron (NdFeB) magnets from mine output is about 13%. Out of that, 34% of the magnets produced are used in the manufacturing of actuator hard disk drives [1–3]. With their life span centered on their application, rare earth materials in hard disk drives are also applied in parts such as printed circuit boards (PCB), spindles, and so on [4].

In the production of these magnets, specific additions of elements are also used to adjust their properties [5,6]. For example, cobalt (Co) is used to substitute REE and iron (Fe) materials (up to over 5%) to increase the Curie temperature [7,8]. Dysprosium (Dy) addition increases the temperature characteristics such that the compound has a better stability against demagnetization. It also decreases the residual induction of the magnet, which leads to lower magnetic field properties [6,9]. To lower the production costs, Pr is now used as a substitute of Nd (up to 20–25%) in the production of magnets. Additionally, the magnet material is coated with a protective layer, such as copper (Cu), aluminum (Al), or nickel (Ni), as well as polymeric material to hold the magnet on the steel plate [4]. Although this shows that recycling of end-of-life magnets can help reduce the criticality of these REE in the near future, commercial recycling of REE is low, at less than 1%. This is mainly due to inefficient collection, technological difficulties, and high cost of processing [10–15].

Separation stage(s) have always been an important step in recycling. For rare earth magnets, since the element of interest is found with two or more others, they may require different extraction technology. Hydrometallurgical approaches have been shown to be a very efficient way to separate the REEs in which chemical separation by leaching is performed [16,17]. For REE leaching, lixiviants are directly added with or without heat treatment to dissolve the solid materials. Once the materials are in solution, various processes such as precipitation, solvent extraction, and ion exchange can be used to economically produce individual REE in the required form.

As a part of the impurities encountered in REE recovery, some of the leached Fe is also recovered as ammonium jarosite in the final REF product. To remove this impurity, the final product form is subjected to high temperature (250–500 °C) to decompose the ammonium jarosite [18–21]

For solvent extraction and ion exchange, different cationic, anionic and solvating extractants such as di (2-ethyl-hexyl) phosphoric acid (D2EHPA), dialkyl phosphonic acid (Cyanex 272), 2-ethyl-hexyl phosphonic acid mono-2-ethyl-hexyl ester (PC 88A), neodecanoic acid (Versatic 10), tributyl phosphate (TBP), and tricaprylylmethylammonium chloride (Aliquat 336) have been reported for the separation of REEs from solution with D2EHPA being more commonly used with nitrate, sulfate, chloride and perchlorate solutions, PC 88A with chloride solutions, and TBP with nitrate solutions [22–35]. Interestingly, many of the same chemical types used as solvent extractants are also used in solid form as ion-exchange resins from the same type of leaching solutions [23,24,28,32]. For both solvent extraction and ion exchange, the REE-loaded material must then be selectively stripped. The resulting solutions are then predominantly processed to precipitate the individual REEs, often as REOs, but not always [36–39].

This paper presents the recycling Nd magnet scrap using a novel hydrometallurgical process involving H_2SO_4 leach, NH_4OH precipitation, and $NH_4F{\cdot}HF$ reaction. The latter step transforms the precipitate into rare earth fluorides (REF) which should be appropriate feedstock for subsequent pyrometallurgical processing into metal in molten fluoride electrolysis [39]. The application of hydrofluoric acid (HF) was completely avoided and the process was optimized through statistical analysis and modelling.

2. Experimental

2.1. Materials

Samples of scrap magnets were obtained from end-of-life computer hard drives from the IT Department of Montana Tech. Ammonium hydroxide (NH_4OH) dissolved in water at 28–30% concentration was obtained from VWR International LLC. Sulfuric acid (98% H_2SO_4) from Pharmco Products Inc. and ammonium bifluoride (98.8% $NH_4F{\cdot}HF$) manufactured by J.T. Baker were the other reagents used in the work.

2.2. Sample Preparation

2.2.1. Demagnetization

Nd magnets were obtained by disassembling the actuators in various hard drives. After loading in a ThermoScientific Lindberg/Blue M box furnace, the furnace was programmed to heat up to 500 °C at 5 °C per minute under ambient air. This was done in order to cause demagnetization and weaken the adhesive used to hold the magnet on the steel plate. The temperature was held at 500 °C for 60 min. Afterwards, the demagnetized magnets were air-cooled and sorted from the steel plates.

2.2.2. Comminution

Comminution was done to liberate the NdFeB part of the demagnetized sample so that ground mass could be easily leached with the lixiviant in view of the increased surface area of the material. Using a disc pulverizer from Bico Inc. (Burbank, CA, USA), the samples were initially comminuted with a set of 3.1 mm. That product was further comminuted with a set of 0.3 mm. A sieve analysis was performed to determine the size distribution of the comminuted material and is discussed later.

2.3. *Hydrometallurgical Processing*

REEs were leached from the comminuted sample in a 2M H_2SO_4 acid solution with sample to solution ratio of 1g:10mL [4]. This process was done under a fume hood for 2 h with the acid solution being added in small amounts because of the aggressiveness of the reaction as shown in Equation (1).

$$Nd_2Fe_{14}B + 45H^+ + 3H_2O \rightarrow 2Nd^{3+} + 14Fe^{3+} + BO_3{}^{3-} + 25.5H_2 \quad (1)$$

Filtration was performed after leaching to separate the REE-acidic pregnant solution from the residue. NH_4OH was then added to the filtrate in a ratio of 1 mL NH_4OH to 20 mL rare earth rich pregnant solution to adjust the pH to 1.2, as shown in Equation (2).

$$NH_4OH + Nd^{3+} + 2SO_4{}^{2-} + H^+ + 2H_2O \rightarrow (NH_4)Nd(SO_4)_2(H_2O)_3 \quad (2)$$

Upon addition, the solution was stirred at 90 rpm to completely dissolve back into solution anything that formed when the NH_4OH was added. After that, the solution was allowed to sit for 12 h so the REE-rich precipitate could fully form and settle. Finally, filtration was performed so that the REE-rich residue could be collected and allowed to air dry. The dry REE-rich precipitate is then added into a mixture of $NH_4F{\cdot}HF$ and deionized water in a ratio of 1g:1.5g:10g and stirred for 45 min to enhance the formation of REF. The residue obtained after filtration was air dried and analyzed for REF content. The process flow sheet is shown in Figure 1. Various stages are also identified in the figure to facilitate further discussion later in this paper.

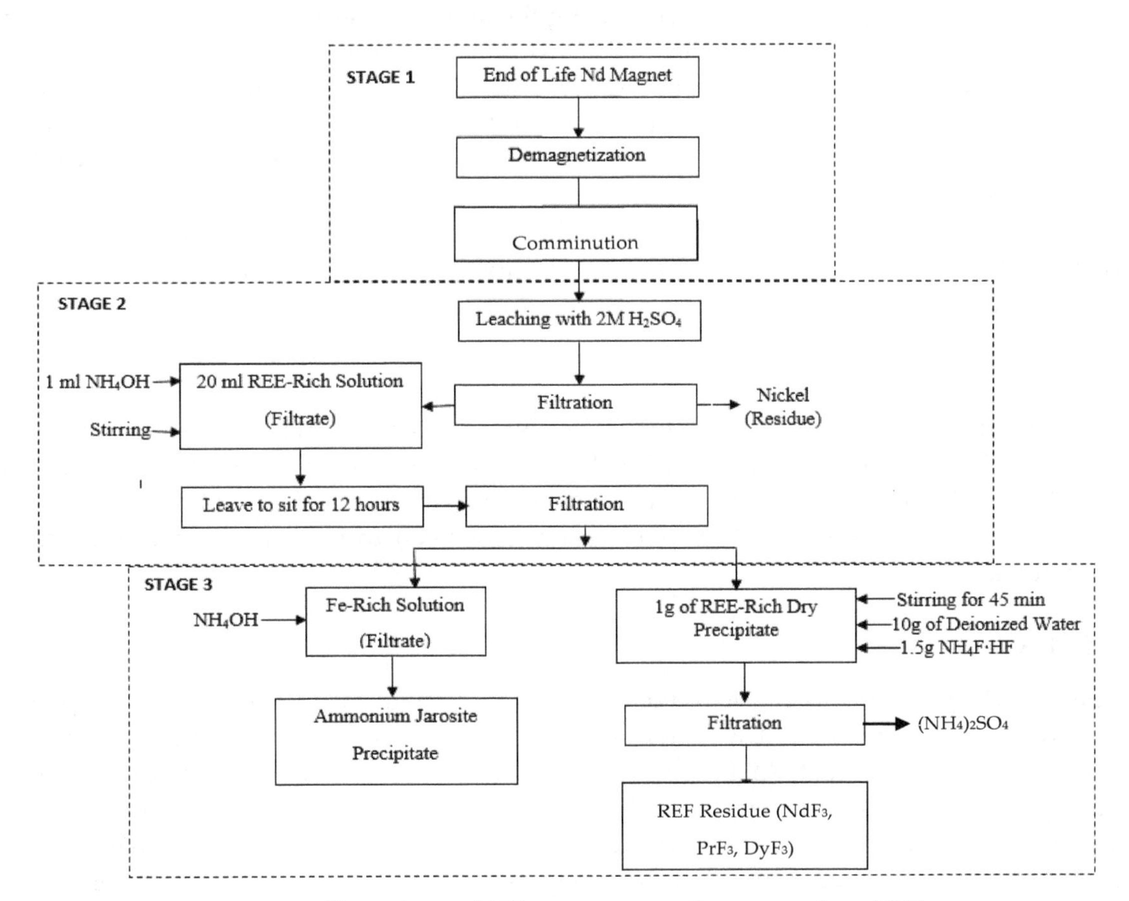

Figure 1. Flow sheet of Nd magnet recycling to produce REF.

Recovery of Fe in the form of ammonium jarosite, as shown in Equation (3), can also be done by the addition of more NH_4OH into the Fe-rich solution (as shown in Stage 3). With Equation (3) representing the reaction, the pH of the solution increased until a pH of 2 where maximum ammonium jarosite precipitated [17].

$$NH_4OH + 3Fe^{3+} + 2SO_4^{2-} + 5H_2O \rightarrow (NH_4)Fe_3(OH)_6(SO_4)_2 + 5H^+ \quad (3)$$

2.4. Modelling of REF Recovery

Stage 1 and Stage 2 of the flow sheet in Figure 1 were identified as less critical compared to Stage 3 of the process. Hence, it was decided to optimize Stage 3 in the present work. To optimize the REF recovery, Response Surface Methodology (RSM) was pragmatic in the analysis of the experiments. RSM is a mathematical and statistical technique that employs empirical models to fit the experimental data with reference to the Design of Experiments (DOE). With the process responses not following a linear model, the Box-Behnken design was employed for designing the experimental matrix to delineate the response surfaces generated by the condition variables [39–44]. This design selects points in the experimental domain for a three-level factorial arrangement in such a way that permits proficient approximation of the first and second order coefficients for the mathematical model [42]. The user identifies a high level and a low level of each condition variable and the mid-point is automatically identified for the point selection. Several experiments (usually 3 or 5) are conducted at the mid-point of all variables to estimate the inherent variability associated with the experimental technique.

In this work, the objectives were to maximize the amount of REF recovered from the precipitate along with their purity. This RSM was used at the point in the experiment where the REE-rich precipitate is added to $NH_4F{\cdot}HF$ to produce the REF in Stage 3. The experiments were performed as per the RSM design of experiments developed using the statistical software Design Expert 9 procured from Stat-Ease Inc., Minneapolis, MN, USA [43]. During the scoping tests, several condition variables were identified that affected the amount of REF recovered from the precipitate as well as its purity. To limit the number of experiments, the most important three condition variables were chosen, namely, the amount of deionized water (mL), amount of $NH_4F{\cdot}HF$ (g) and degree of stirring (min). Of course, the identified responses were the amount of REF recovered from the precipitate and the purity of the REF. The selected points in the experimental domain and the responses obtained are discussed later, along with the response surfaces, model equations, optimization and the interaction of the condition variables.

2.5. Material Characterization

After demagnification, the feed materials as well as intermediate and final products from this study were characterized using Scanning Electron Microscopy with Energy Dispersive X-ray (SEM-EDX), X-ray Diffractometry (XRD), Inductively Coupled Plasma–Optical Emission Spectrometry (ICP-OES), and a thermal analyzer with thermogravimetric (TG) and differential scanning calorimetry (DSC) capabilities.

2.5.1. SEM/EDX

The SEM-EDX analyzer was employed to determine the chemical compositions of all phases in the demagnetized material. The SEM-EDX system uses a TESCAN TIMA with a tungsten filament and an EDAX Z2 analyzer (TESCAN ORSAY HOLDING, a.s., Kohoutovice, Czech Republic). Cross-sectioned sides of a representative sample were hand-separated and cold-mounted in epoxy using molds approximately 25 mm in diameter and 10 mm in thickness. Resulting mounts were ground and polished to a smooth finish and then conductively coated with carbon to obtain SEM images by backscattered electron (BSE) detection. EDX analyses helped determine the chemical compositions of all products.

2.5.2. XRD

X-ray diffraction was carried out with a Rigaku Ultima IV X-ray Diffractometer (XRD) (RIGAKU AMERICAS CORPORATION, Woodland, TX, USA) using Cu-Kα radiation at 40 kV and 40 mA. This was used for quantitative analysis and also to determine the various phases of the precipitates and the products obtain after $NH_4{\cdot}HF$ addition.

2.5.3. ICP-OES

A ratio of 3:1 *v/v* HCl to HNO_3 was used to digest a representative sample of the final product obtained at the end of each experiment and resulting solutions were analyzed by ICP-OES to determine the elemental content of the sample (Thermo Electron Duo View iCAP 6500, Waltham, MA, USA).

2.5.4. TGA-DSC

Thermal decomposition of the ammonium jarosite, obtained in the final stage, was carried out in a TA Instruments SDT Q650 simultaneous thermal analyzer with thermogravimetric (TG) and differential scanning calorimetry (DSC) capabilities (TA Instruments, New Castle, DE, USA). Analyses were performed using 10 mg sample with a heating rate of 10 °C/min to 400 °C under argon atmosphere. The furnace temperature was regulated precisely to provide a uniform rate of decomposition.

3. Results and Discussions

3.1. Granulometry

Regarding Stage 1 of Figure 1 and using sieves ranging in aperture size from 0.149 to 4.699 mm, the size distribution of the comminuted sample is shown in Figure 2. It can be seen from this figure that nearly 50% of the material passed through the 0.6 mm sieve size. The 80% passing size (d80) of the comminuted mass is about 1.7 mm.

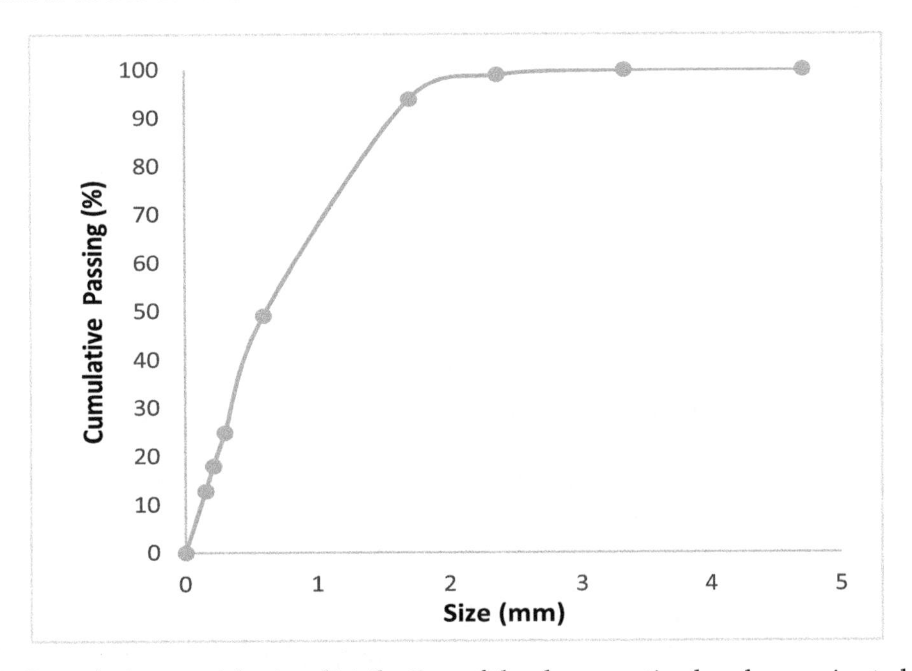

Figure 2. Cumulative particle size distribution of the demagnetized and comminuted sample.

3.2. SEM-EDX Analysis

Image from the Backscatter Electron (BSE) detector on the SEM-EDX demonstrate some dissimilar phases in the demagnetized and comminuted sample (see Stage 1 of Figure 1). "Three Dots" analysis was performed using the EDX. Of the three indicated points shown in Figure 3, each has a distinctive chemistry: point (a) appears to be Ni, which is the outer layer; point (b) is a combination of Pr, Nd and

Fe, where Pr is used to substitute Nd (up to 20–25%) to lower production costs; and point (c) shows a mixture of Nd and Fe. It is noted that boron (B) was not detected due to its low intensity, resulting from it being a light element (i.e., element #5 on the periodic table), as well as from the inherent disability of the BSE detector that was used [45].

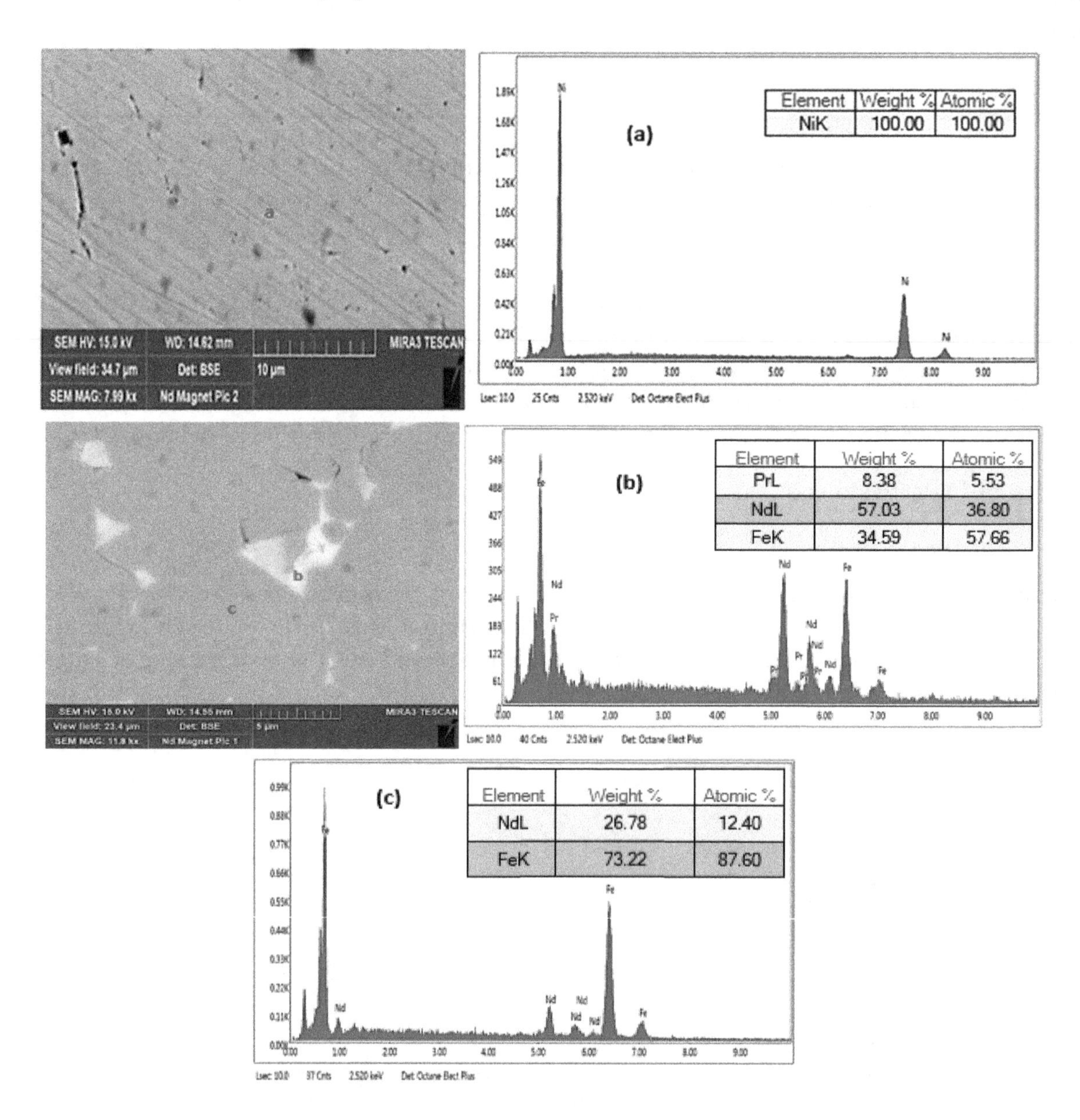

Figure 3. SEM image and EDX composition of the demagnetized and comminuted sample (see text for details). (**a**) Ni; (**b**) Pr, Nd, Fe; (**c**) Nd, Fe.

3.3. XRD Results

Figures 4 and 5 show the phase compositions of the precipitates (Stage 2 feed to Stage 3 rich, dry precipitate, as per Figure 1) and the final product (Stage 3 REF residue, as per Figure 1), respectively. From Figure 4, it can be said that most of the Fe remains in solution and that the dry precipitate is made up of mainly $(NH_4)Nd(SO_4)_2(H_2O)_3$. This compound is an ammonium neodymium sulfate double salt and is noted to be similar to ammonium jarosite [17] with Nd substituting for Fe. Figure 5 also shows the phases of the product made after adding and stirring the dry precipitate in $NH_4F{\cdot}HF$ for 45 min, which is then heated under argon atmosphere to 400 °C and allowed to cool. Equation (4) depicts a possible mechanism for REF formation with the REE being Nd assuming ammonium (NH_4^+) is in excess:

$$2(NH_4)Nd(SO_4)_2(H_2O)_3 + 3NH_4F{\cdot}HF + 3NH_4^+ \rightarrow 2NdF_3 + 4(NH_4)_2SO_4 + 6H_2O + 3H^+ \quad (4)$$

The product is clearly a REF containing REEs of Nd, Dy and Pr. It can therefore be concluded that REF can be produced by this process and the resulting fluorides can be used later as feedstock for pyrometallurgical metal production [39].

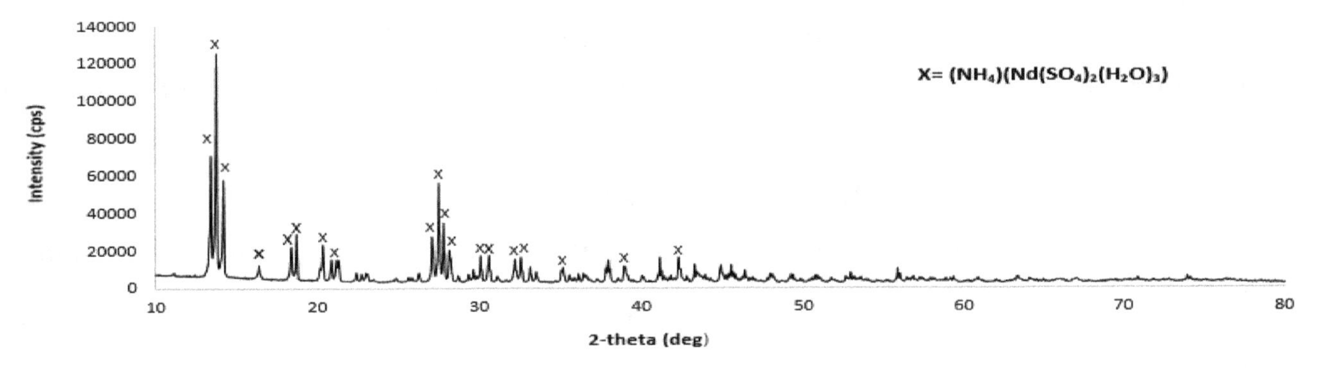

Figure 4. Crystal phases in the REE rich precipitate after adding NH_4OH.

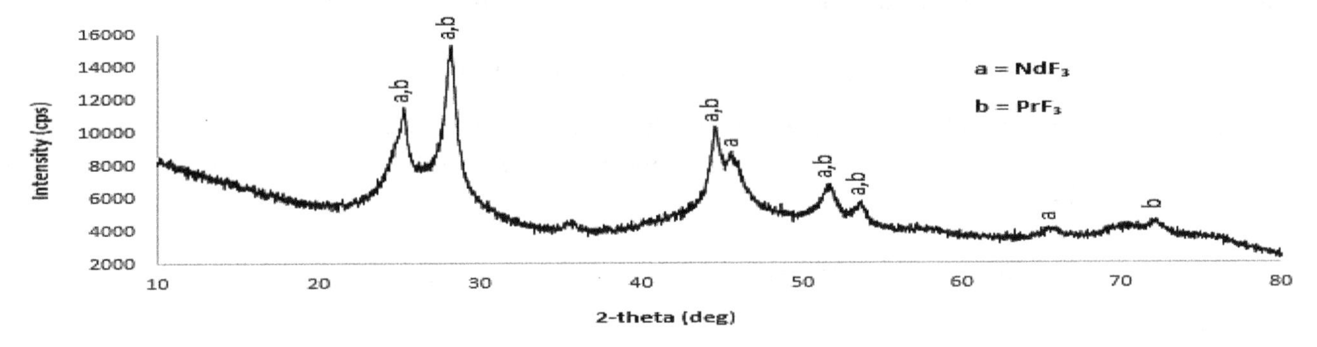

Figure 5. Crystal phases in the REF filtrate after 45 min in $NH_4F{\cdot}HF$ and heated to 400 °C.

3.4. ICP Results

Table 1 shows the chemical composition of REEs in the final REF product indicating Dy and Fe are present, apparently in amounts below the XRD detection limit. Together, the total REE amount sums to 63.25%. Based on stoichiometry and assuming the REEs exist as REF (i.e., NdF_3, PrF_3 and DyF_3), the F-content would be 24.94%. This leaves a balance of 10.59%, which is expected to be ammonium sulfate [$(NH_4)_2SO_4$] and ammonium jarosite [$(NH_4)Fe_3(SO_4)_2(OH)_6$], as per Equation (4). Furthermore, based on the stoichiometry of the jarosite, the Fe content would equate to 3.44% of these other components, suggesting there is 7.15% ammonium sulfate. Both ammonium sulfate and ammonium jarosite can be monohydrated [46] which will thermally decompose at low temperature. In this regard, a TGA-DSC study was undertaken on the final REF residue from Stage 3. Because the product was heated to 400 °C under argon, a subsequent TGA-DSC scan did not exceed 400 °C. It is important to note that this REF residue, as discussed above, is not pure, and therefore contains some ammonium sulfate and ammonium jarosite.

Table 1. The chemical composition of the final product.

Sample Composition	Dy	Fe	Nd	Pr	Others
Weight (%)	2.89	1.22	52.95	7.41	35.53

3.5. TGA-DSC Results

The TGA-DSC graphs of the REF residue are shown in Figure 6. The plot shows that mass loss occurs in five steps which agrees with the literature [46–49]. In this regard, the first two steps equate simply to crystalline water loss of the ammonium sulfate (Equation (5)) and ammonium jarosite (Equation (7)), respectively:

25–50 °C

$$(NH_4)_2SO_4{\cdot}H_2O \rightarrow (NH_4)_2SO_4 + H_2O \quad (5)$$

75–125 °C

$$(NH_4)Fe_3(SO_4)_2(OH)_6{\cdot}H_2O \rightarrow (NH_4)Fe_3(SO_4)_2(OH)_6 + H_2O \quad (6)$$

The third step corresponds to the decomposition of dehydrated ammonium sulfate to ammonium bisulfate:

140–210 °C

$$(NH_4)_2SO_4 \rightarrow NH_4HSO_4 + NH_3 \quad (7)$$

The fourth step is likely caused by complete thermal decomposition of the ammonium bisulfate:

220–250 °C

$$NH_4HSO_4 \rightarrow 1/3NH_3 + 1/3N_2 + SO_2 + 2H_2O \quad (8)$$

Finally, the fifth step appears to be the dehydroxylation of the dehydrated ammonium jarosite, thereby accounting for the continued loss in weight as the temperature increased to 380 °C:

>260 °C

$$(NH_4)Fe_3(SO_4)_2(OH)_6 \rightarrow (NH_4)(FeO)_3(SO_4)_2 + 3H_2O \quad (9)$$

If the temperature had been increased to 600 °C, three additional steps involving the sequential reactions of the dehydroxylated ammonium jarosite would be observed such that all of the Fe ultimately becomes hematite [46–49]:

>385 °C

$$2(NH_4)(FeO)_3(SO_4)_2 \rightarrow 2NH_3 + H_2O + 2Fe_3O_{2.5}(SO_4)_2 \quad (10)$$

>510 °C

$$2Fe_3O_{2.5}(SO_4)_2 \rightarrow 2Fe_2O_3 + Fe_2(SO_4)_3 + SO_3 \quad (11)$$

>540 °C

$$Fe_2(SO_4)_3 \rightarrow Fe_2O_3 + 3SO_3 \quad (12)$$

where $Fe_3O_{2.5}(SO_4)_2$ is essentially equivalent to a solid solution of 2/3 $Fe_2(SO_4)_3$ and 5/6 Fe_2O_3.

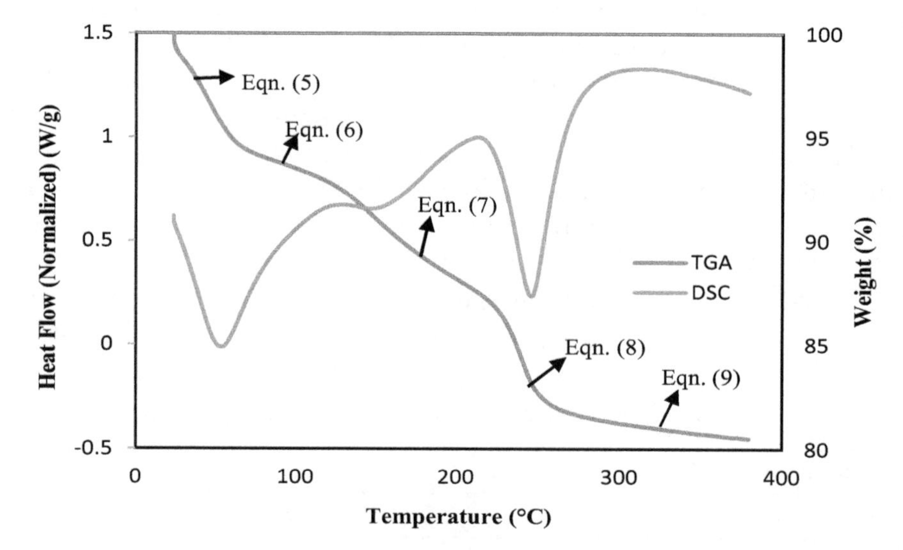

Figure 6. TGA-DSC curve for REF residue from Stage 3.

4. Model Development and Process Optimization Using RSM

Table 2 shows the Design of Experiments (DOE) along with the condition variables and responses for Stage 3 processing. In all, 17 experiments were performed, each using 2 g of REE-rich dry precipitate from Stage 2. The following ranges of the condition variables were employed: deionized water volume was 10–30 mL; 2–4 g of $NH_4F{\cdot}HF$ was added; the degree of stirring was 15–45 min. Responses were REE recovery and REF purity where recovery refers to the amount of REE from Stage 2 being converted to REF residue and purity refers to the quality of the REF residue based on ICP analysis.

Table 2. DOE conditions and their responses.

Standard	Run	Volume of Deionized Water (mL)	Amount of $NH_4{\cdot}HF$ (g)	Degree of Stirring (mins)	REE Recovery (%)	Purity of REF Residue (%)
2	1	30	2	30	46.76	60.59
8	2	30	3	45	55.58	63.26
10	3	20	4	15	53.15	54.01
7	4	10	3	45	63.03	46.12
4	5	30	4	30	49.63	58.38
17	6	20	3	30	47.16	56.93
3	7	10	4	30	64.09	49.66
1	8	10	2	30	50.44	45.52
12	9	20	4	45	55.41	51.84
16	10	20	3	30	49.67	53.08
5	11	10	3	15	57.48	51.19
11	12	20	2	45	54.65	48.59
15	13	20	3	30	50.09	54.94
14	14	20	3	30	52.74	51.39
6	15	30	3	15	52.13	50.85
9	16	20	2	15	53.56	50.65
13	17	20	3	30	51.38	57.42

Furthermore, five experiments were completed with all condition variables being at their midpoints. The results were analyzed to develop a statistically significant model for the responses of the REE recovery and REF purity as shown in Equations (13) and (14):

$$\textit{REE Recovery (\%)} = 50.14 - 3.87A + 0.087B + 0.84C - 2.70AB - 0.53AC + 2.68A^2 + 4.15C^2 + 4.04A^2B + 1.41A^2C \quad (13)$$

$$\textit{REF Purity (\%)} = 54.75 + 4.20A + 1.65B - 1.06C - 1.59AB + 4.37AC + 0.18\,A^2 - 1.40B^2 - 2.08C^2 - 1.17A^2B + 2.89A^2C \quad (14)$$

where A denotes the volume of deionized water used, B denotes the amount of $NH_4F{\cdot}HF$ added, and C denotes the degree of stirring. In both cases, a cubic model represented the data best and the R^2 values were 0.94 and 0.93 for recovery and purity, respectively.

Using Equations (13) and (14), 3-D surface plots in Figures 7 and 8 were generated to illustrate the effects of the condition variables on the responses. From Figure 7a,b, it can be said that $NH_4F{\cdot}HF$ addition appears to have more prominent effect on the process efficiency. The impact of deionized water addition is more prominent at higher stirring rates and higher $NH_4F{\cdot}HF$ additions. With respect to stirring, it can be observed from Figure 7a that the amount of REF recovery increases with stirring time for both 10 and 30 mL of water used. However, recovery is higher for lower volumes of water than for higher ones, and this same pattern is observed with respect to $NH_4F{\cdot}HF$ addition as illustrated in Figure 7b when using Equation (13).

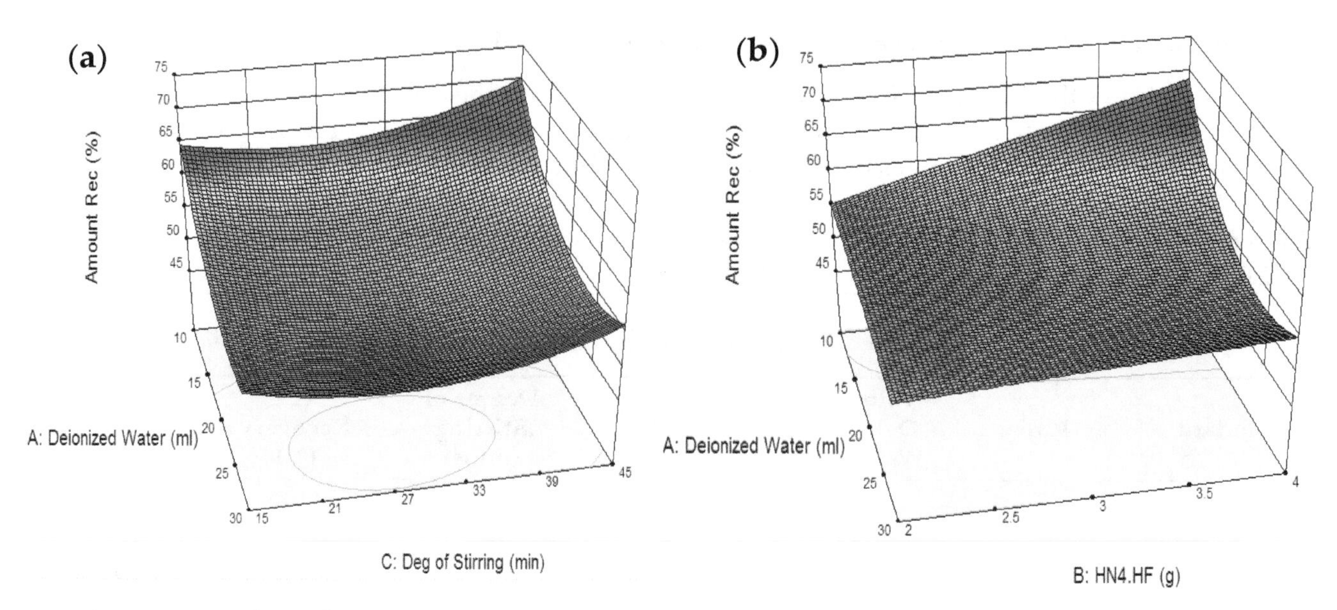

Figure 7. 3D plots of REF recovery with respect to process variables at (**a**) 4 g of $NH_4F{\cdot}HF$ and (**b**) 43 min for degree of stirring.

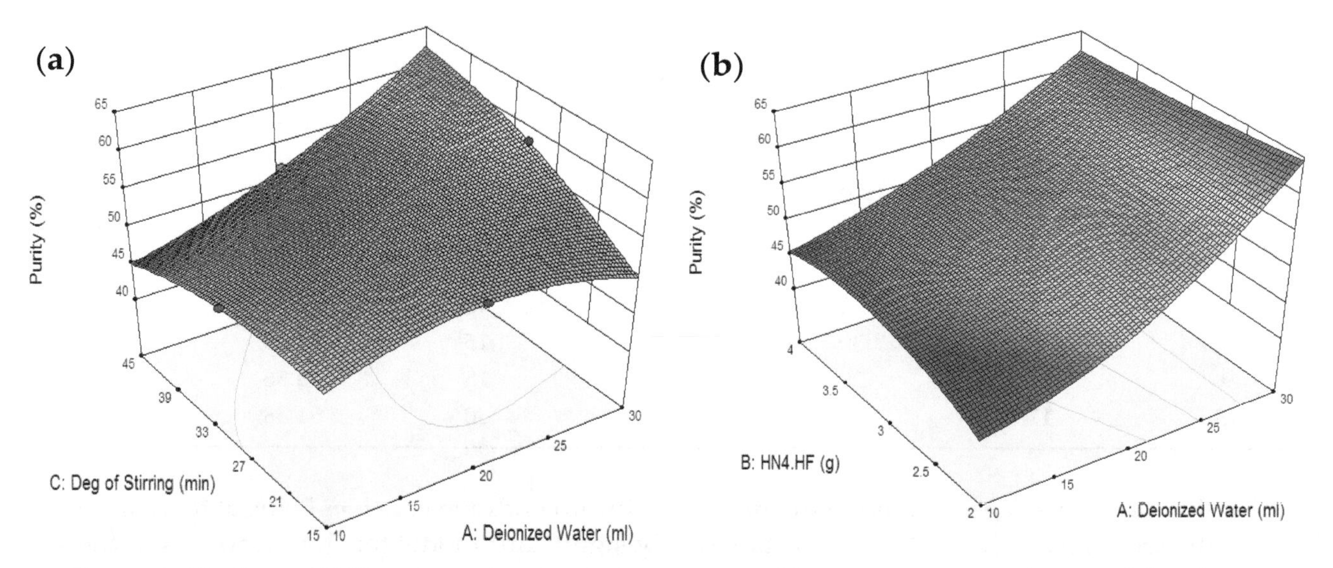

Figure 8. 3D plots of REF purity with respect to process variables at (**a**) 4 g of $NH_4F{\cdot}HF$ and (**b**) 44 min for degree of stirring.

Figure 8a,b were generated from Equation (14) and illustrates the response surfaces for REF purity in relations to the process variables. From these plots, it is observed that the amount of deionized water again did influence purity of REF recovery to a significant extent at all $NH_4F{\cdot}HF$ additions and also at higher stirring degrees. The degree of stirring, as observed in Figure 8a, seems to affect the REF insignificantly at low water additions. However, REF purity increased sharply with stirring time at a high volume of water addition. Figure 8b shows that $NH_4F{\cdot}HF$ addition has less impact on the purity of REF. Sharp increase in the purity of the REF is seen with increase in the volume of water.

It may also be noted that, in Equations (13) and (14), the process variables are all in terms of the coded values and differ in the range +1 t o−1, with the mid-point having a value of zero. Thus, for B, the maximum amount of $NH_4F{\cdot}HF$ used (4 g) corresponds to +1 and the lowest (2 g) corresponds to –1, whereas the midpoint of 3 g corresponds to 0. To use these statistically significant equations, the actual variable value needs to be converted to the coded form lying between +1 and −1. Using the above model equations optimization was carried out to find the conditions that maximized both REF recovery and REF purity. The identified conditions and their maximum corresponding responses are as follows:

Amount of Deionized Water Used: 30 mL
$NH_4F{\cdot}HF$ *Added*: 4 g
Degree of Stirring: 45 min

REF Recovery: 56.23%
REF Purity: 62.42%

5. Conclusions

From this study, it is established REF can be produced from end-of-life Nd magnet from hard disk drive by hydrometallurgical techniques using $NH_4F{\cdot}HF$. The final product (REF) can also be used as part of the salt bath for molten salt electrolysis using fluoride salt [39]. This process is preferred over those reported in the literature because the use of hazardous HF was avoided. Employing statistical design of experiments (DOE), process parameters such as volume of deionized water, amount of $NH_4F{\cdot}HF$ and degree of stirring were defined in detail, and a conducive process regime for the recovery of the REFs was identified. It was established that at higher volumes of water, $NH_4F{\cdot}HF$ addition, and degree of stirring, optimum REF recovery with reasonably high purity can be achieved.

Author Contributions: Conceptualization, P.S., C.Y., and A.D.; methodology, P.S., T.F., and A.D.; software, P.S.; validation, P.S., A.D., and C.Y.; formal analysis, P.S., and T.F.; investigation, T.F.; resources, P.S.; data curation, P.S., and T.F.; writing—original draft preparation, P.S. and T.F.; writing—review and editing, P.S., A.D., and C.Y.; visualization, P.S., and T.F.; supervision, C.Y., and A.D..; project administration, C.Y.; funding acquisition, C.Y. All authors have read and agreed to the published version of the manuscript.

Acknowledgments: The views and conclusions contained in this document are those of the authors and should not be interpreted as representing the official policies, either expressed or implied, of the Army Research Laboratory or the U.S. Government. The U.S. Government is authorized to reproduce and distribute reprints for Government purposes notwithstanding any copyright notation herein. The authors acknowledge Grant Wallace and Jan Chorney for their support.

References

1. Yang, Y.; Walton, A.; Sheridan, R.; Güth, K.; Gauß, R.; Gutfleisch, O.; Buchert, M.; Binnemans, K. REE recovery from end-of-life NdFeB permanent magnet scrap: A critical review. *J. Sustain. Metall.* **2017**, *3*, 122–149. [CrossRef]
2. Walters, A.; Lusty, P. *Rare Earth Elements*; British Geological Survey: Nottingham, UK, 2010.
3. Ueberschaar, M.; Rotter, V.S. Enabling the recycling of rare earth elements through product design and trend analyses of hard disk drives. *J. Mater. Cycles Waste Manag.* **2015**, *17*, 266–281. [CrossRef]
4. Önal, M.A.R.; Borra, C.R.; Guo, M.; Blanpain, B.; Van, G.T. Hydrometallurgical recycling of NdFeB magnets: Complete leaching, iron removal and electrolysis. *J. Rare Earths* **2017**, *35*, 574–584. [CrossRef]
5. Rademaker, J.H.; Kleijn, R.; Yang, Y. Recycling as a strategy against rare earth element criticality: A systemic evaluation of the potential yield of NdFeB magnet recycling. *Environ. Sci. Technol.* **2013**, *47*, 10129–10136. [CrossRef]
6. Zepf, V. *Rare Earth Elements: A New Approach to the Nexus of Supply, Demand and Use: Exemplified along the Use of Neodymium in Permanent Magnets*; Springer: Heidelberg, Germany, 2013.
7. Goldman, A. *Handbook of Modern Ferromagnetic Materials*; Kluwer Academic Publishers: Dordrecht, The Netherlands, 1999; p. 649.
8. Rotter, V.S.; Chancerel, P.; Ueberschaar, M. Recycling-oriented product characterization for electric and electronic equipment as a tool to enable recycling of critical metals. In *Rewas*; Springer: Berlin, Germany, 2013.
9. Nguyen, R.T.; Diaz, L.A.; Imholte, D.D.; Lister, T.E. Economic assessment for recycling critical metals from hard disk drives using a comprehensive recovery process. *JOM* **2017**, *69*, 1546–1552. [CrossRef]
10. Meyer, L.; Bras, B. Rare earth metal recycling, sustainable systems and technology. In Proceedings of the IEEE International Symposium, Dana Point, CA, USA, 16–18 May 2011.

11. Tanaka, M.; Oki, T.; Koyama, K.; Narita, H.; Oishi, T. Recycling of rare earths from scrap. In *Handbook on the Physics and Chemistry of Rare Earths*; Bunzli, J.C.G., Pecharsky, V.K., Eds.; Elsevier: Amsterdam, The Netherlands, 2013; Volume 43, pp. 159–211.
12. Anderson, C.D.; Anderson, C.G.; Taylor, P.R. A survey of recycled rare earths metallurgical processing. In Proceedings of the 51st Annual Conference of Metallurgists of CIM, Niagara Falls, ON, Canada, 30 Septembere–3 October 2012; pp. 411–422.
13. Graedel, T.E.; Allwood, J.; Birat, J.P.; Reck, B.K.; Sibley, S.F.; Sonnemann, G.; Buchert, M.; Hagelüken, C. *Recycling Rates of Metals—A Status Report*; United Nations Environment Programme: Nairobi, Kenya, 2011.
14. Reck, B.K.; Graedel, T.E. Challenges in metal recycling. *Science* **2012**, *337*, 690–695. [CrossRef]
15. Eggert, R.; Wadia, C.; Anderson, C.; Bauer, D.; Fields, F.; Meinert, L.; Taylor, P. Rare earths: Market disruption, innovation, and global supply chains. *Annu. Rev. Environ. Resour.* **2016**, *41*, 199–222. [CrossRef]
16. Goonan, T.G. *Rare Earth Elements—End Use and Recyclability*; US Geological Survey: Reston, VA, USA, 2011.
17. Lyman, J.W.; Palmer, G.R. *Recycling of Neodymium Iron Boron Magnet Scrap*; University of Michigan Library: Ann Arbor, MI, USA, 1993.
18. Weber, R.J.; Reisman, D.J. *Rare Earth Elements: A Review of Production, Processing, Recycling, and Associated Environmental Issues*; US EPA Region: Washington, DC, USA, 2012.
19. Majzlan, J.; Stevens, R.; Boerio-Goates, J.; Woodfield, B.F.; Navrotsky, A.; Burns, P.C.; Crawford, M.K.; Amos, T.G. Thermodynamic properties, low-temperature heat-capacity anomalies, and single-crystal X-ray refinement of hydronium jarosite,$(H_3O)Fe_3(SO_4)_2(OH)_6$. *Phys. Chem. Miner.* **2004**, *31*, 518–531. [CrossRef]
20. Alonso, M.; López-Delgado, A.; López, F.A. A kinetic study of the thermal decomposition of ammoniojarosite. *J. Mater. Sci.* **1998**, *33*, 5821–5825. [CrossRef]
21. Dutrizac, J.E. Converting jarosite residues into compact hematite products. *JOM* **1990**, *42*, 36–39. [CrossRef]
22. Thakur, N.V.; Jayawant, D.V.; Iyer, N.S.; Koppiker, K.S. Separation of neodymium from lighter rare earths using alkyl phosphonic acid, PC 88A. *Hydrometallurgy* **1993**, *34*, 99–108. [CrossRef]
23. Preston, J.S. *The Recovery of Rare Earth Oxides from A Phosphoric Acid Byproduct. Part 4. The Preparation of Magnet-Grade Neodymium Oxide from The Light Rare Earth Fraction*; Mintek: Randburg, South Africa, 1996.
24. Lu, D.; Horng, J.S.; Hoh, Y.C. The separation of neodymium by quaternary amine from didymium nitrate solution. *J. Less Common Met.* **1989**, *149*, 219–224. [CrossRef]
25. Morais, C.A.; Ciminelli, V.S.T. Process development for the recovery of high-grade lanthanum by solvent extraction. *Hydrometallurgy* **2004**, *73*, 237–244. [CrossRef]
26. Radhika, S.; Nagaphani, K.B.; Lakshmi, K.M.; Ramachandra, R.B. Solvent extraction and separation of rare earths from phosphoric acid solutions with TOPS 99. *Hydrometallurgy* **2001**, *110*, 50–55. [CrossRef]
27. Panda, N.; Devi, N.; Mishra, S. Solvent extraction of neodymium (III) from acidic nitrate medium using Cyanex 921 in kerosene. *J. Rare Earths* **2012**, *30*, 794–797. [CrossRef]
28. El-Hefny, N.E. Kinetics and mechanism of extraction and stripping of neodymium using a Lewis cell. *Chem. Eng. Process. Process. Intensif.* **2007**, *46*, 623–629. [CrossRef]
29. Lee, M.-S.; Lee, J.-Y.; Kim, J.-S.; Lee, G.-S. Solvent extraction of neodymium ions from hydrochloric acid solution using PC88A and saponified PC88A. *Sep. Purif. Technol.* **2006**, *46*, 72. [CrossRef]
30. Banda, R.; Jeon, H.; Lee, M. Solvent extraction separation of Pr and Nd from chloride solution containing La using Cyanex 272 and its mixture with other extractants. *Sep. Purif. Technol.* **2012**, *98*, 481–487. [CrossRef]
31. Guo, C.; Zhi, Y. Effective solvent extraction of La, Ce and Pr from hydrochloric acid with a novel extractant N, N-dihexyloxamic acid. *J. Chem. Technol. Biotechnol.* **2016**, *92*, 1596–1600. [CrossRef]
32. Chen, J.; Guo, L.; Deng, Y.; Li, D. Application of [A336][P507]/[P204] on High Selective Extraction and Separation of Rare Earths (III) from Mechanism to Techniques. In Proceedings of the 52nd Conference of Metallurgists (COM), Montreal, QC, Canada, 27–31 October 2013; pp. 367–374.
33. Rout, A.; Kotlarska, J.; Dehaen, W.; Binnemans, K. Liquid–liquid extraction of neodymium(iii) by dialkylphosphate ionic liquids from acidic medium: The importance of the ionic liquid cation. Electronic supplementary information (ESI) available: NMR spectra of the ionic liquids. *Phys. Chem. Chem. Phys.* **2013**, *15*, 16533–16541. [CrossRef]
34. Esmaeil, J.; Malek, S. The production of rare earth elements group via tributyl phosphate extraction and precipitation stripping using oxalic acid. *Arab. J. Chem.* **2016**, *9*, S1532–S1539.
35. Belova, V.V.; Voshkin, A.A.; Kholkin, A.I.; Payrtman, A.K. Solvent extraction of some lanthanides from chloride and nitrate solutions by binary extractants. *Hydrometallurgy* **2009**, *97*, 198–203. [CrossRef]

36. London, I.M.; Goode, J.R.; Moldoveanu, G.; Rayat, M.S. Rare Earth Elements Symposium. In Proceedings of the 52nd Conference of Metallurgists (COM), Montreal, QC, Canada, 27–31 October 2013.
37. Yan, C.; Liao, C.; Jia, J.; Wang, M.; Li, B. Comparison of economical and technical indices on rare earth separation processes of bastnasite by solvent extraction. *J. Rare Earths* **1999**, *17*, 58–63.
38. Krishnamurthy, N.; Gupta, C.K. *Extractive Metallurgy of Rare Earths*; CRC Press: Boca Raton, FL, USA, 2015; p. 869.
39. Sarfo, P. Recovery of Rare Eearth Elements by Advanced Processing Technologies. Ph.D. Thesis, Montana Tech, Butte, MT, USA, October 2019.
40. Sarfo, P.; Wyss, G.; Ma, G.; Das, A.; Young, C. Carbothermal reduction of copper smelter slag for recycling into pig iron and glass. *Miner. Eng.* **2017**, *107*, 8–19. [CrossRef]
41. Sarfo, P.; Das, A.; Wyss, G.; Young, C. Recovery of metal values from copper slag and reuse of residual secondary slag. *Waste Manag.* **2017**, *70*, 272–281. [CrossRef]
42. Box, G.E.P.; Behnken, D.W. Some new three level designs for the study of quantitative variables. *Technometrics* **1960**, *2*, 455–475. [CrossRef]
43. Bezerra, M.A.; Santelli, R.E.; Oliveira, E.P.; Villar, L.S.; Escaleira, L.A. Response surface methodology (RSM) as a tool for optimization in analytical chemistry. *Talanta* **2008**, *76*, 965–977. [CrossRef]
44. Stat-Ease, Inc. Stat-Ease® DOE Software. 2003. Available online: https://www.hearne.software/Resources/Design-Ease/softoverview.aspx (accessed on 10 October 2018).
45. Ingemarsson, L.; Halvarsson, M. *Sem/edx Analysis of Boron*; High Temperature Corrosion Centre-Chalmer University of Technology: Gothenburg, Sweden, 2011.
46. Das, G.K.; Anand, S.; Acharya, S.; Das, R.P. Preparation and decomposition of ammoniojarosite at elevated temperatures in H_2O–$(NH_4)_2SO_4$–H_2SO_4 media. *Hydrometallurgy* **1995**, *38*, 263–276. [CrossRef]
47. Frost, R.L.; Wills, R.-A.; Kloprogge, J.T.; Martens, W. Thermal decomposition of ammonium jarosite $(NH_4)Fe_3(SO_4)_2(OH)_6$. *J. Therm. Anal. Calorim.* **2006**, *84*, 489–496. [CrossRef]
48. Kunda, W.; Veltman, H. Decomposition of jarosite. *Metall. Trans. B* **1979**, *10*, 439–446. [CrossRef]
49. Spratt, H.; Rintoul, L.; Avdeev, M.; Martens, W. The thermal decomposition of hydronium jarosite and ammoniojarosite. *J. Therm. Anal. Calorim.* **2014**, *15*, 101–109. [CrossRef]

10

Podiform Chromitites and PGE Mineralization in the Ulan-Sar′dag Ophiolite East Sayan, Russia

Olga N. Kiseleva [1,*], Evgeniya V. Airiyants [1], Dmitriy K. Belyanin [1,2] and Sergey M. Zhmodik [1,2]

[1] Sobolev Institute of Geology and Mineralogy, Siberian Branch Russian Academy of Science, pr. Academika Koptyuga 3, Novosibirsk 630090, Russia; jenny@igm.nsc.ru (E.V.A.); bel@igm.nsc.ru (D.K.B.); zhmodik@igm.nsc.ru (S.M.Z.)

[2] Faculty of Geology and Geophysics, Novosibirsk State University, Novosibirsk 630090, Russia

* Correspondence: kiseleva_on@igm.nsc.ru

Abstract: In this paper, we present the first detailed study on the chromitites and platinum-group element mineralization (PGM) of the Ulan-Sar′dag ophiolite (USO), located in the Central Asian Fold Belt (East Sayan). Three groups of chrome spinels, differing in their chemical features and physical–chemical parameters, under equilibrium conditions of the mantle mineral association, have been distinguished. The temperature and log oxygen fugacity values are, for the chrome spinels I, from 820 to 920 °C and from (−0.7) to (−1.5); for chrome spinels II, 891 to 1003 °C and (−1.1) to (−4.4); and for chrome spinels III, 738 to 846 °C and (−1.1) to (−4.4), respectively. Chrome spinels I were formed through the interaction of peridotites with mid-ocean ridge basalt (MORB)-type melts, and chrome spinels II were formed through the interaction of peridotites with boninite melts. Chrome spinels III were probably formed through the interaction of andesitic melts with rocks of an overlying mantle wedge. Chromitites demonstrate the fractionated form of the distribution of the platinum-group elements (PGE), which indicates a high degree of partial melting at 20–24% of the mantle source. Two assemblages of PGM have been distinguished: The primary PGE assemblage of Os-Ir-Ru alloys-I, $(Os,Ru)S_2$, and IrAsS, and the secondary PGM assemblage of Os-Ir-Ru alloys-II, Os^0, Ru^0, RuS_2, OsS_2, IrAsS, RhNiAs with Ni, Fe, and Cu sulfides. The formation of the secondary phases of PGE occurred upon exposure to a reduced fluid, with a temperature range of 300–700 °C, log sulfur fugacity of (−20), and pressure of 0.5 kbar. We have proposed a scheme for the sequence of the formation and transformation of the PGMs at various stages of the evolution of the Ulan-Sar′dag ophiolite.

Keywords: ophiolite; chromitites; PGE mineralogy; geodynamic setting

---

1. Introduction

Podiform chromitites are associated with restite peridotites in ophiolite complexes [1–5]. The concentration, platinum-group element distribution, and mineral form and composition of chrome spinels in podiform chromitites are sensitive indicators of mantle processes: the partial melting degree, initial melt sources, and melt saturation with volatile components (S, H_2O, etc.) [6–14]. Thus, the geochemical features of chromitites, their distribution and the mineral composition of PGE can be used to evaluate the physical–chemical parameters, mineral equilibria parameters, initial melt composition and geodynamic setting of ophiolite formations [14–22]. The use of an olivine–chrome spinel and olivine geothermometer, as well as experimental data on the temperature equilibria of the primary PGMs, provides important information on the formation temperatures of the mantle olivine spinel PGM assemblage and deformation processes [23–25]. The chemical composition of chrome spinels and PGMs, the microstructural features of this assemblage, and accessory sulfide mineralization give important information for the reconstruction of magmatic, hydrothermal, and metamorphic processes. In this paper, we describe the chemistry of chrome spinels, the geochemical features of chromitites and

PGMs from chromitites, the physical–chemical parameters of mineral equilibria, and the alteration of PGE–chromite mineralization in the mantle peridotites of the Ulan-Sar'dag ophiolite.

2. Materials and Methods

Twenty-one samples of chromitites from the western part of Ulan-Sar'dag were studied. The chemical composition of chrome spinels and olivines and the Al_2O_3 impurities in olivine were determined by wavelength-dispersive analysis using a Camebax-micro electron microprobe from the Sobolev Institute of Geology and Mineralogy, Russian Academy of Science, Novosibirsk, Russia (Analytical Center for multi-elemental and isotope research SB RAS). A JEOL JXA-8100 analysis methodology specific to olivines to obtain their trace elements has been elaborated: The accelerating voltage is 20 kV, the probe current is 400 nA, and the counting time per line and background measurement are both 10 s. The number of measurements per analysis is 25 [26]. For an error of 10 relative %, the quantitation limit of Al_2O_3 is within 45–100 ppm. The analytical conditions for olivines and chrome spinels are as follows: The accelerating voltage is 20 kV, the probe current is 30 nA, the beam size is 2 μm, and the signal accumulation time is 10 s. The standards used were natural and synthetic silicates and oxides. The detection limit for oxides was 0.01–0.03 wt. %.

The content of PGE in chromitites and dunites was determined in the rocks using the ICP-MS microprobe and kinetic methods, along with pre-concentration (nickel matte), at the TsNIGRI analytical laboratory (Moscow, Russia). The ICP-MS microprobe method with pre-concentration (nickel matte) involves melting the probe (50 g) at 1100 °C, during which two phases are formed: A sulfide (nickel matte) and an oxide (skimming) phase. These phases are segregated according to the concentration of all noble elements in nickel matte. The collector is an alloy of sulfides and a metal alloy, in which the bulk of the noble metals is collected [27]. The received matte is cleaned through skimming and crushed for 10–15 s. A portion of the matte is dissolved in hydrochloric acid, while the Ni and Cu of the matte are almost completely dissolved. Platinum metals in the nonsolute residue are dissolved in a mixture of hydrochloric and nitric acids. The mass concentration of Pt, Pd, Rh, Ru, and Ir in the solution is determined by ICP-MS. Osmium is separated from other noble metals by distillation. The method is based on the ability of osmium to form volatile tetraoxides under oxidizing conditions [28]. The determination of the osmium concentration was carried out using a kinetic method. The method is based on the catalytic effect of osmium in indicator oxidation-reduction reactions. The reaction rate depends on the concentration of the catalyst [28–30]. The detection limits were 2 ppb for Os, Ir, Ru, and Rh; and 5 ppb for Pt and Pd.

The studied platinum-group minerals were polished plates (21 pieces) and heavy fractions, extracted from chromite ores. The selected PGM grains were mounted in epoxy blocks and polished with a diamond paste for further analysis. Microtextural observations of PGM were performed by means of reflected-light microscopy. The chemical composition of PGMs was determined using a MIRA 3 LMU scanning electron microscope, with an attached INCA Energy 450 XMax 80 microanalysis energy dispersive system, at the Sobolev Institute of Geology and Mineralogy, Russian Academy of Science, Novosibirsk, Russia (Analytical Center, IGM SB RAS). Pure metals were used as the standards to determine the chemical composition of the PGE, Ni, and Cu; arsenopyrite was used for As; and pyrite was used for Fe. The minimum detection limits of the elements (wt. %) were found to be 0.1–0.2 for S, Fe, Co, Ni, and Cu; 0.2–0.4 for As, Ru, Rh, Pd, Sb, and Te; and 0.4–0.7 for Os, Ir, and Pt.

3. Geological Setting

The ophiolite complexes are widely distributed in the south-eastern part of Eastern Sayan (Siberia, Russia) (Figure 1).

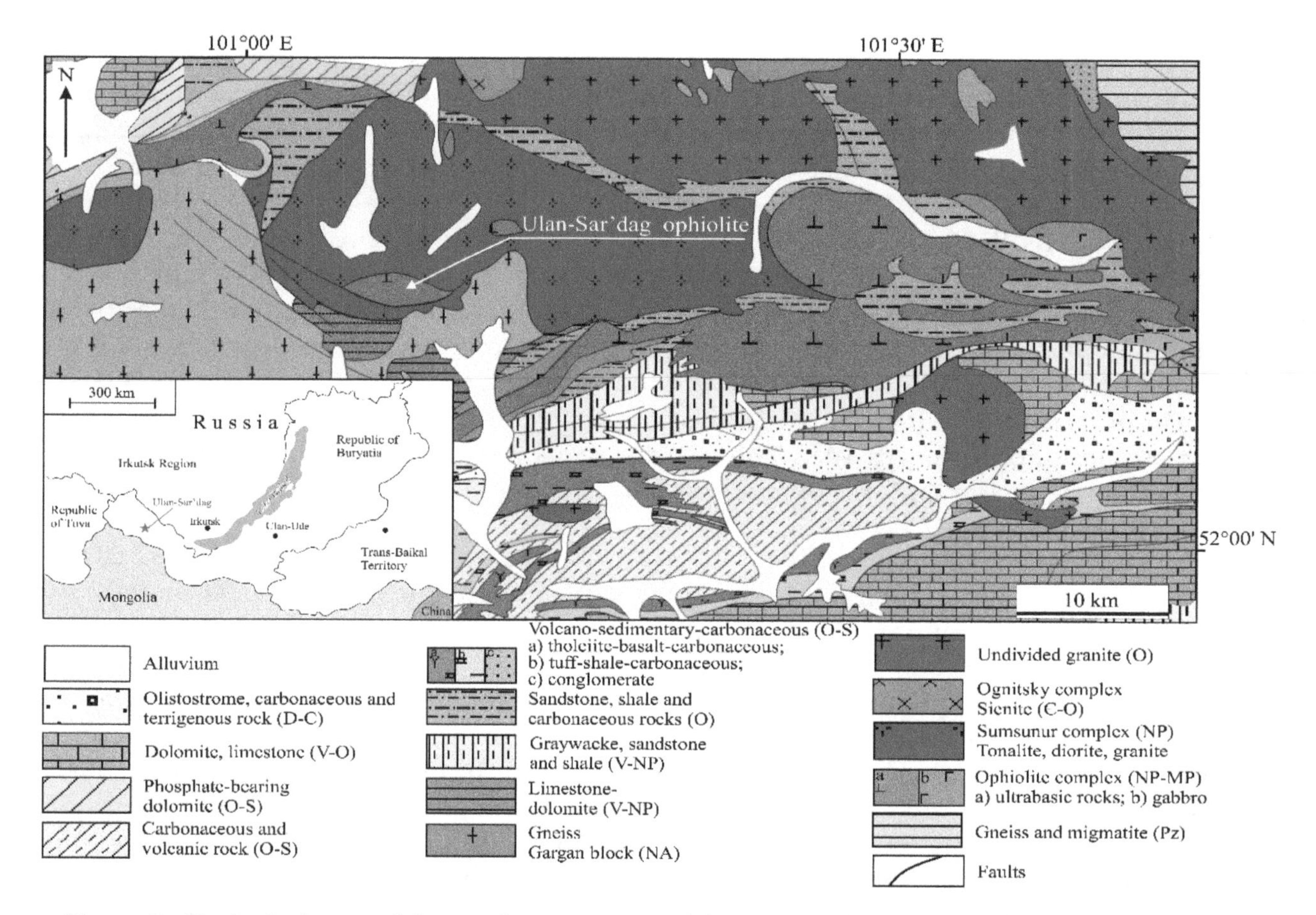

Figure 1. Geological map of the southeastern part of the Eastern Sayan region [31]. NB—northern branch, SB—southern branch.

The ophiolites are localized in the form of extended branches in the south (Il'chir) and north (Holbin-Hairhan). The features of these ophiolites are as follows: rock associations, geochemical and mineralogical characteristics, a geological structure resembling individual "massifs", geodynamic conditions and a formation time, which are actively studied for a considerable time. Dobretsov and Zhmodik et al. obtained data on the heterogeneity of ophiolites of the south-eastern part of Eastern Sayan (SEPES) [32,33]. They indicate that the ophiolites of the southern branch were formed in a mid-ocean ridges setting, and ophiolites of the northern branch were formed in an island arc setting [32–38]. The Ulan-Sar'dag ophiolite (USO) occupies a special structural position in the Eastern Sayan and Central Asian Fold Belt (Dunzhugur island arc). The USO is a tectonic plate, which is located between ophiolites of the southern and northern branches of the Dunzhugur island arc, near the contact zone of the Gargan "block" gneisses, with granites of the Sumsunur complex. The USO is underlaid by volcanogenic and sedimentary rocks of the Ilchir suite and limestones of the Irkut suite. Ophiolites include mantle restites (dunites, harzburgites), podiform chromitites, cumulates (metawehrlite, metapyroxenite, metagabbro), a basic dike and a volcanic complex (Ilchir suite) (Figures 2 and 3). USO has a lenticular body that is elongated in the east–west direction and is 1.5×5 km^2. It is composed of dunites and harzburgites. Harzburgites predominate in the central part, dunites and serpentinites prevail in the margin, and the latter lie in contact with the volcanogenic-sedimentary sequence of the Ilchir suite. The rocks of the cumulative series are metamorphosed under the conditions of epidote-amphibolite and amphibolite facies. The volcanogenic-sedimentary sequence is composed of volcanic, volcanogenic-sedimentary, and metasedimentary rocks (black schists and marbles). Volcanogenic-sedimentary rocks are sulfurized, and sulfide mineralization is confined to the schistosity zones. High-Ti basalts (MORB tholeiites), boninites, and island-arc andesites are distinguished in the volcanic complex. All rocks in the ophiolite nappe bottom are intensively deformed and mark the thrust zone. They include numerous crushing zones, signs of shear displacement and gliding planes. In the contact zone of serpentinites and rocks of the Ilchir suite, the talcum powder zones and zones of actinolite-tremolite composition are widely manifested.

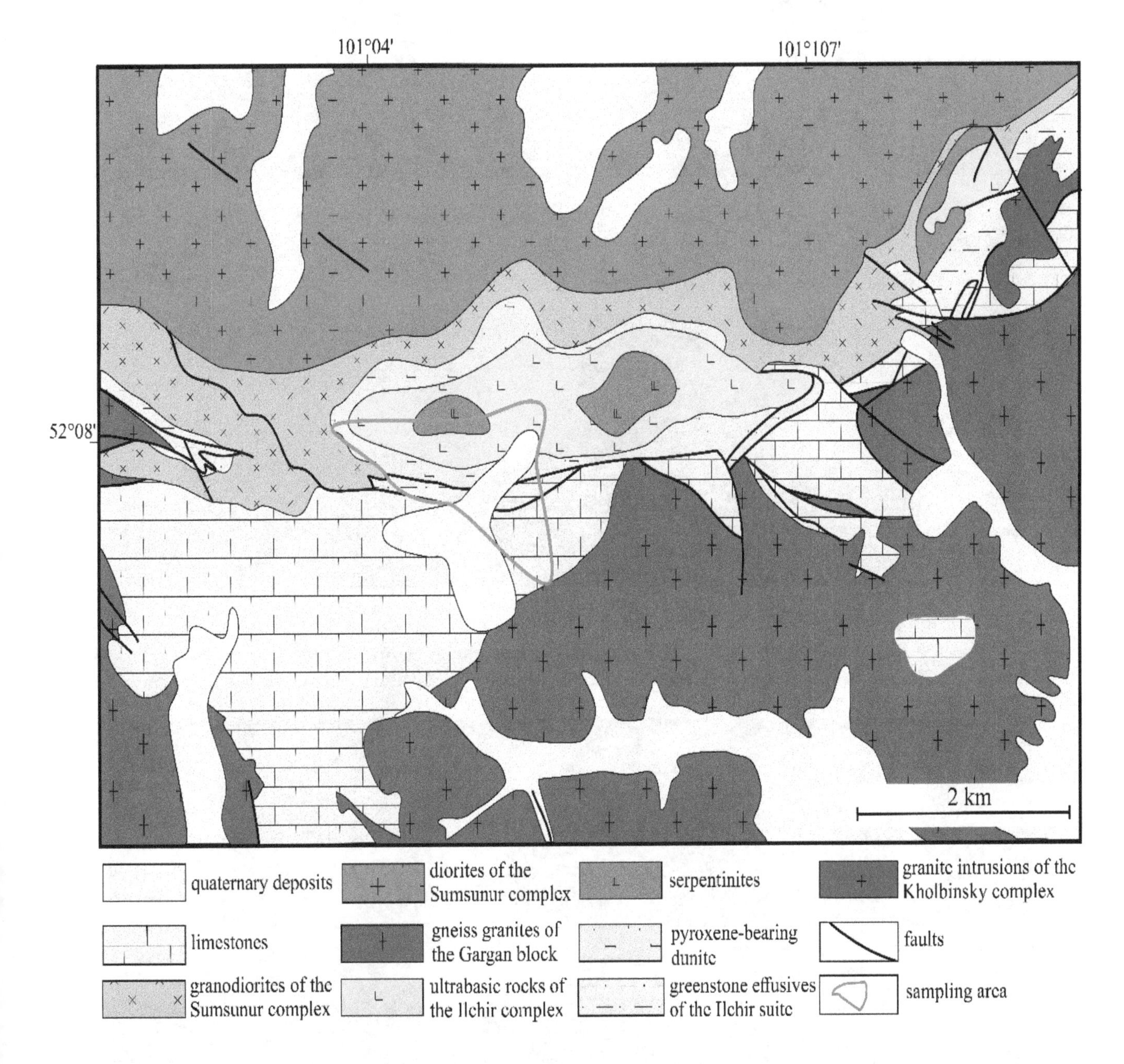

Figure 2. Geological scheme of the Ulan-Sar'dag ophiolite [39] with author additions.

The types of schlieren, lenticular, and vein-like chromite bodies are localized in dunites and serpentinites (Figure 4). The length of schlieren and massive chromitite bodies ranges from a few centimeters to several meters, and their width varies from 5 cm to 0.5 m. The predominant structural-textural type is a massive chromitite. In some schlieren chromitites, the tectonic flow structures (Figure 4e) and the "snowball" type (Figure 4f) are observed.

Figure 3. Photographs of the field relationship between mantle peridotites and volcanic-sedimentary rocks (Ilchir suite), limestone thickness (Irkut suit), Gargan gneisses, and Susunur tonalities: (**a**) view of the south side of the Ulan-Sar'dag ophiolite, (**b**) view of the north side of the Ulan-Sar'dag ophiolite.

Figure 4. Structural features of chromitite pods: (**a**) chromitite seams in dunites; (**b**) schlieren; (**c**) massive pods; (**d**) transformation of the schlieren type into the massive type due to deformation processes; (**e**) folded schlieren-type chromitites; (**f**) structure of the "snowball".

4. Results

4.1. Podiform Chromitites

Massive chromitites are the predominant petrographic variety of rocks in the mantle peridotites of the USO; disseminated, schlieren, and rhythmically-banded ores are less common. Chrome spinels make up 80–95 vol. % of massive chromitites. The intergranular space is filled with olivine and secondary silicates: serpentine, chlorite, and, rarely, talc. The structure of massive chromitites varies from fine-grained to coarse-grained. The crystal bodies vary from subidiomorphic grains (0.1–0.5 cm) to allotrimorphic aggregates of grains of up to 1.5 cm in size. Chrome spinels have a cataclastic texture, and they are partially fragmented. The grains are dissected by cracks filled with secondary silicates. In the central part, chrome spinels are unchanged, and in cracks and on the grain rims, they are often replaced by chrome magnetite. Some grains contain olivine inclusions. A wide range of minerals show accessory mineralization in chromite ores: PGMs, sulfides, sulfarsenides, and sulfosalts of base metals (Figure 5).

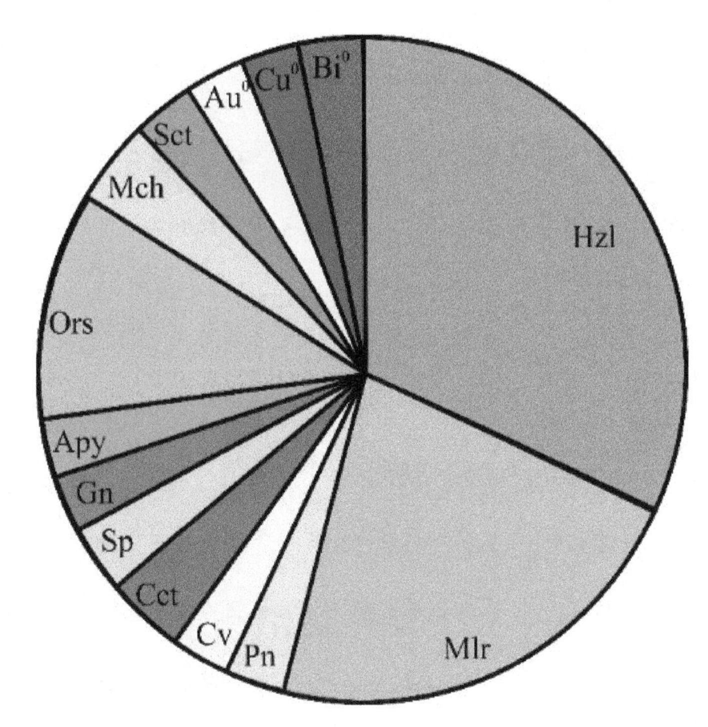

Figure 5. Accessory mineral association in chromitites. Hzl—heazlewoodite Ni_3S_2; Mlr—millerite NiS; Pn—pentlandite $(Fe,Ni)_9S_8$; Cv—covellite CuS; Cct—chalcocite Cu_2S; Orc—orcelite Ni_5As_2; Mh—maucherite Ni11As_8; Sct—scutterudite $Co_4(As_4)_3$; Apy—arsenopyrite FeAsS; Sp—sphalerit ZnS; Gn—galena PbS; Au^0, Cu^0, Bi^0—native gold, copper and bismuth.

4.1.1. Chrome Spinels

According to the chemical composition, ore chrome spinels are divided into three groups and are represented by chrome picotite, alumo-chromite, chromite, chrome-magnetite and magnetite (Table 1, Figure 6). In one sample, chrome spinels of all three groups may be present. Chrome spinels of groups I, II, and III have the following compositions (wt. %): Al_2O_3 = (17–43), (7–16), and (9–18); Cr_2O_3 = (26–54), (46–66), and (47–57); MgO = (10–20), (9–13), and (8–10); and FeO = (6–14), (1–18), and (18–20), respectively. In chromitites of the USO, chrome picotites are found, which is typical only of this massif, in contrast to chrome spinels from other ophiolite complexes of the southeastern part of East Sayan [38]. Chrome spinels have a homogeneous composition. Altered chrome spinels demonstrate increased contents of FeO, MnO, NiO, ZnO, and Fe_2O_3 and decreased contents of Cr_2O_3 and Al_2O_3. Chrome spinels are often replaced by chrome magnetite and magnetite along the cracks.

Table 1. Representative chemical composition of Cr spinels and the calculated parental melt composition (wt. %) of Cr spinels. Thermometric and oxybarometric data for Cr spinels.

No.	1	2	3	4	5	6	7	8	9	10	11	12	13	14	15	16	17	18	19	20	21	22
	I Group					II Group													III Group			
N Sample	37	41	74	77	101	130	132	120	10	48	52	47	53	5	112	88	9	60	71	11	22	18
TiO_2	0.1	0.1	0.1	0.1	0.1	0.10	0.1	0.1	0.1	0.05	0.07	0.1	0.1	0.1	0.2	0.2	0.1	0.05	0.04	0.07	0.05	0.07
Al_2O_3	40.3	41.0	24.6	24.1	19.2	12.19	7.97	15.3	14.6	13.52	16.53	14.5	14.8	13.9	12.5	12.4	14.6	14.6	13.0	14.0	16.5	19.6
Cr_2O_3	29.7	28.5	44.2	43.7	49.2	59.57	63.94	56.4	54.2	56.51	53.70	55.6	55.6	56.5	56.4	57.0	57.3	54.0	55.4	49.0	48.0	48.3
MnO	0.5	0.5	0.2	0.3	0.3			0.3	0.3	0.32	0.56	0.3	0.4	0.3	0.5	0.4	0.4	0.3	0.4	0.4	0.7	0.3
FeO	7.9	7.9	14.7	14.8	14.4	15.4	19.1	13.3	14.8	15.8	15.3	15.4	16.4	15.6	16.1	16.3	17.7	19.0	18.3	18.6	20.9	19.8
Fe_2O_3	2.0	2.3	2.1	2.6	3.3	2.4	0.7	1.2	2.1	1.7	2.1	1.6	2.3	1.7	2.7	2.7	1.1	1.8	2.3	5.6	5.8	2.5
MgO	19.7	19.8	13.6	13.3	12.9	12.6	10.8	13.7	12.4	11.7	12.4	12.0	11.7	11.9	11.4	11.3	10.9	9.6	9.8	9.6	8.3	9.6
V_2O_5	0.8	0.8	0.1	0.1	0.1			0.1	0.1	0.13		0.1		0.1	0.1	0.1	0.1	0.1	0.07	0.11	0.12	0.09
NiO	0.3	0.3	0.2	0.1	0.0			0.0	0.1	0.02		0.0		0.0	0.0	0.0	0.1	0.0	0.03	0.06	0.07	0.02
ZnO	0.0	0.0	0.0	0.2	0.7			0.0	0.0	0.14		0.1		0.2	0.3	0.2	0.0	0.3	0.19	0.10	0.46	0.46
Total	101.3	101.1	99.8	99.2	100.2	102.3	234.6	100.4	98.6	99.8	100.7	99.7	101.2	100.2	100.1	100.5	102.2	99.8	99.5	97.6	100.9	100.7
Al′	56	57	35	34	27	23	26	20	20	19	23	20	20	19	20	20	20	20	18	20	23	28
Cr′	41	40	62	62	69	74	84	80	80	79	74	80	80	80	80	80	80	80	78	71	68	69
Fe′	2.7	3	3	3.7	4	2.9	0.9	1.7	3.0	2.3	2.9	2.2	3.2	2.3	3.7	3.7	1.6	2.6	3	8	8	4
Mg′	72	71	48	47	47	59	50	49	54	57	55	56	58	57	59	59	62	66	35	34	29	33
f′	28	29	52	53	53	41	50	49	54	57	55	56	58	57	59	59	62	66	65	66	71	67
TiO_2melt	0.1	0.1	0.1	0.1	0.1	0.18	0.11	0.1	0.2	0.10	0.13	0.1	0.1	0.1	0.4	0.3	0.2	0.10	0.09	0.14	0.10	0.14
Al_2O_3 melt	18.3	18.5	15.0	14.9	13.5	11.1	9.3	12.3	12.2	11.7	12.7	12.1	12.1	11.8	11.3	11.3	12.0	12.1	11.5	12.0	12.7	13.6
(Fe/Mg)m	0.3	0.3	0.5	0.6	0.5	0.5	0.7	0.4	0.5	0.6	0.5	0.6	0.6	0.6	0.6	0.6	0.7	0.9	0.78	0.80	1.07	0.96
T°C (Ol-Sp)	919.1	893.9	823.4	820.7	885.7	938.62	991.6	1003.9	948.9	931.8	891.6	931.5	921.2	933.2	942.4	913.6	861.2	812.6	846.3	821	742.7	738.1
fO_2	−0.988	−0.765	−1.498	−1.246	−1.1	−2.48	−1.3	−2.1	−1.3	−1.8	−2.5	−1.9	−1.6	−1.9	−1.1	−2.8	−4.4	−3.3	−3.01	−1.49	−1.73	−2.78
T°C (Al in Ol)												893–1332				1073–1225						

Notes: Al′ = Al/(Al + Cr + Fe^{3+}); Cr′ = Cr/(Cr + Al + Fe^{3+}); Fe′ = Fe^{3+}/(Cr + Al + Fe^{3+}); Mg′ = Mg/(Mg + Fe^{2+}); f′ = Fe^{2+}/(Mg + Fe^{2+}); Sample No. 1–5—Cr-spinels I; 6–18—Cr-spinels II; 19–22—Cr-spinels III; T°C (Ol-Sp)—olivine-spinel geothermometer; $f(O_2)$—oxygen fugacity; T°C (Al in Ol)—olivine geothermometer.

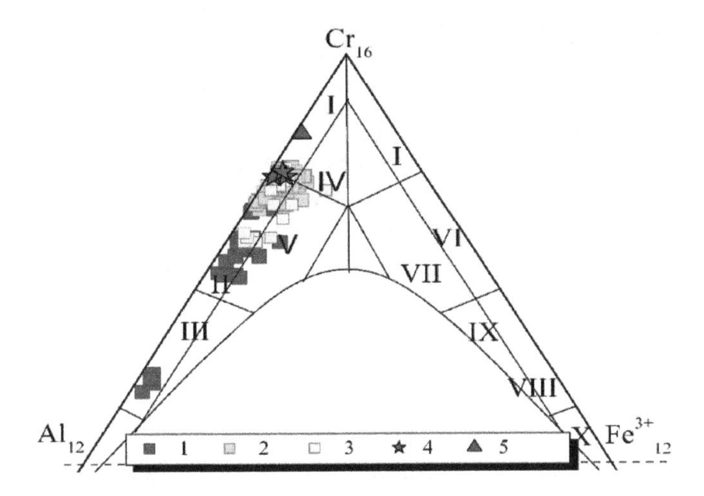

Figure 6. Classification diagram of chrome spinels from the chromitites of Ulan-Sar'dag, based on the structural formula of spinels. Composition fields: I—chromite; II—alumo-chromite; III—chrome picotite; IV—subferrichromite; V—subferrialumochromite; VI—ferrichromite; VII—subferrialumoferrichromite; VIII—chrome magnetite; IX—subalumochrome magnetite; X—magnetite [40]. Notes: 1—chrome spinels-I; 2—chrome spinels-II; 3—chrome spinels-III; 4—chrome spinels-II, with inclusions $(Os,Ru)S_2$; 5—chrome spinels-II, with olivine inclusion.

4.1.2. Olivine

Olivine in chromitites occupies the intergranular space of chrome spinels; it is often replaced by serpentine or chlorite. The data, which correspond to chrysotile and forsterite, are shown in Table 2. The MgO content is 49–54 wt. %, and the NiO variations are insignificant and amount to 0.37–0.42 wt. %, which corresponds to olivines from depleted peridotites. The fraction of the Fo component is 92–97.

Table 2. Chemical composition of the olivine from chromitites (wt. %).

N-Sample	SiO_2	MgO	FeO	NiO	MnO	CaO	Al_2O_3	Cr_2O_3	Total	Mg#
298-1	41.28	52.59	4.98	0.41	0.11	0.001		0.002	99.38	95
298-2	41.83	52.25	4.50	0.42	0.09	0.001		0.002	99.08	95
298-3	41.26	52.20	5.68	0.42	0.12	0.002		0.004	99.68	94
298-4	41.31	52.10	5.71	0.42	0.13	0.001	0.0118	0.002	99.68	94
298-5	41.41	52.68	5.34	0.44	0.11	0.002	0.0052	0.002	99.98	95
3-6	41.08	51.19	6.95	0.37	0.11	0.001	0.0191	0.002	99.73	93
3-7	41.10	51.04	6.98	0.37	0.11	0.003	0	0.001	99.60	93
3-8	41.15	51.03	6.85	0.38	0.12	0.004	0.0119	0	99.54	93
3-9	40.83	50.85	7.13	0.38	0.13	0.015	0.0104	0.002	99.34	93
3-10	41.13	51.17	6.92	0.38	0.10	0.003	0.0015	0	99.69	93
3-11	41.22	51.02	6.89	0.37	0.11	0.004	0.0117	0.001	99.62	93
3-12	41.09	51.16	6.90	0.38	0.13	0.006	0.0005	0.001	99.66	93
3-13	41.07	51.19	6.89	0.37	0.13	0.005	0.0006		99.65	93
3-14	41.10	51.08	6.93	0.37	0.10	0.006	0.0009		99.58	93
3-15	41.24	51.26	6.79	0.38	0.13	0.009	0.003	0.002	99.80	93
305-16	41.33	53.61	3.28					0.61	98.83	97
305-17	41.38	53.86	3.45					0.69	99.38	97
2-18	41.05	51.71	6.47						99.23	93
973-19	40.25	50.24	8.14					0.01	98.64	92
976-20	41.43	51.26	6.90					0.02	99.61	93
977-21	40.09	50.39	6.85			0.02		0.00	97.35	93
939-22	40.94	49.59	7.28			0.01		0.03	97.85	92
939-23	40.74	49.06	9.61			0.04		0.36	99.81	90
970-24	40.85	49.93	7.87			0.00		0.02	98.67	92
980-25	42.39	53.59	5.50			0.01		0.02	101.51	95

Notes: (Samples 16–18)—inclusions in chromite.

4.2. Geochemistry of Platinum-Group Elements

The content of PGE in mantle peridotites and podiform chromitites was determined. The content of PGE in dunites and serpentinites is 94–180 ppb; in chromitites, it ranges from 242 to 992 ppb (Table 3). The PGE content increases with an increase in the volume percentage of chromite in the rock. In addition, depending on the proportion of the chromite and silicate components in the rock, the IPGE/PPGE ratio changes (Figure 7a,b) (IPGE: Os, Ir, and Ru; PPGE: Rh, Pt, and Pd). For example, in massive ores (85–95 vol. % of chromite), IPGE > PPGE, and IPGE/PPGE = 0.86–2.15. In schlieren lenticular chromitites and chromitites with deformational textures, the PPGE proportion increases and amounts to IPGE/PPGE = 0.03–0.67. In chromitites I, the IPGE/PPGE is higher than in chromitites II (Table 3).

Table 3. PGE abundance for the dunites and chromitites of the Ulan-Sar'dag ophiolite (ppb).

No.	1	2	3	4	5	6	7	8	9	10	11
N Sample	6	305	6 mas	7 mas	294 mas	307 mas	6 mas	3 mas	17 mas	20 mas	2 shl
Os	6	8	45	51	58	117	49	37	81	46	7
Ir	21	6	26	58	43	82	57	20	35	20	6
Ru	7	14	68	53	116	221	121	59	44	46	20
Rh	7	3	12	10	8	19	24	16	9	11	21
Pt	35	15	39	49	31	54	49	64	46	41	35
Pd	104	48	177	182	104	122	478	97	87	78	903
Total	180	94	367	403	360	615	778	293	302	242	992
Pt/Ir	1.67	2.5	1.5	0.84	0.72	0.66	0.86	3.20	1.31	2.05	5.83
∑IPGE	34	28	139	162	217	420	227	116	160	112	33
∑PPGE	146	66	228	241	143	195	551	177	142	130	959
IPGE/PPGE	0.23	0.42	0.61	0.67	1.52	2.15	0.41	0.66	1.13	0.86	0.03
∑PGE	180	94	367	403	360	615	778	293	302	242	992
Cr# (Crt)			68	67	65	57	70	69	84	83	75
Al#(Crt)			30	30	42	34	26	27	14	15	22

Notes: 1,2—dunites; 3–11—chromitites. mas—massive pods, shl—shlieren pods; 3–6—group I chromitites; 7–10—group II chromitites; 11—group III chromitite.

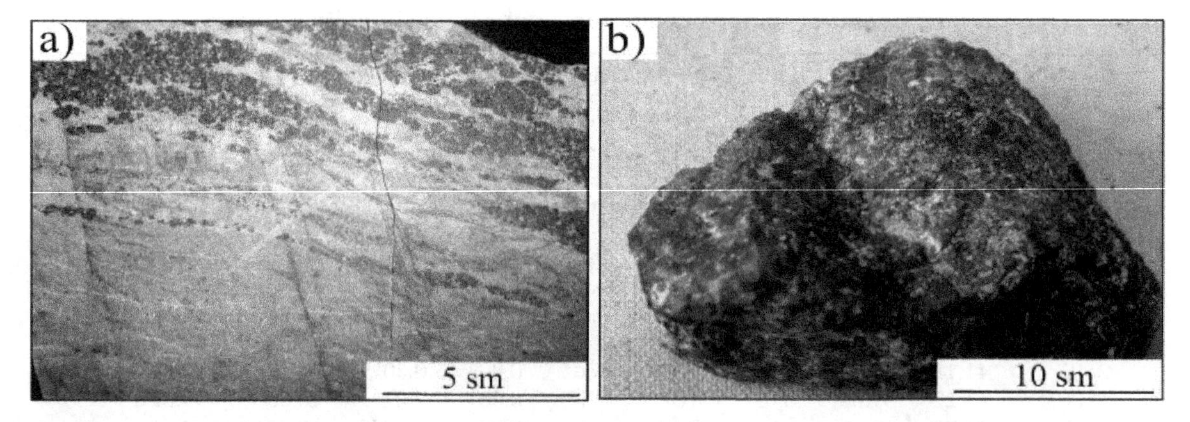

Figure 7. Photographs of chromitites with different values of IPGE/PPGE: (**a**) schlieren densely disseminates in serpentinizated dunite (∑PGE = 903ppb, and IPGE/PPGE = 0.03); (**b**) massive (∑PGE = 615 ppb, and IPGE/PPGE = 2.15).

4.3. Mineralogy of Platinum-Group Elements

The first data on PGE mineralization (PGM) in chromitites of USO have been obtained. The chemical composition of the minerals is presented in Table 4, and classification diagrams are presented in Figure 8a,b. Primary and secondary PGMs have been distinguished. The most common platinum-group minerals in the chromitites of the USO are PGE sulfides: laurite-erlichmanite $(Ru,Os)S_2$, with different Ru/(Ru + Os) ratios. Other PGE phases are represented by high-temperature Os-Ir-Ru alloys-I, phases of variable-composition Os-Ir-Ru alloys-II, native Os, Ru, sulfarsenides of (Os, Ir, Ru), and zaccarinite RhNiAs. The primary and secondary PGMs will be described separately.

Table 4. Representative composition of platinum-group elements in the chromitite from the Ulan-Sar′dag ophiolitic massif.

No. an	Os	Ir	Ru	Rh	Fe	Ni	S	As	Sb	O	Total	Os	Ir	Ru	Rh	Fe	Ni	S	As	Sb	O	Ru/(Ru + Os)
	wt. %											apfu										
1	75.01	21.95	3.38		0.72						101.06	0.39	0.11	0.03		0.01						
2	74.84	20.6	5.16		0.33			0			100.93	0.39	0.11	0.05		0.01						
3	79.75	20.74	2.35		0.35						103.19	0.41	0.1	0.02		0.01						
4	23.87	5.23	34.23		0.59		33.36				97.28	0.24	0.05	0.65		0.02		2				0.59
5	23.21	5.18	39.55	0.6			34.33				102.87	0.23	0.05	0.73	0.01	0.00		2				0.63
6	33.29	5.5	27.26		0.58		31.27				97.9	0.36	0.06	0.55		0.02		2				0.45
7	22.92	4.3	39.85				35.23				102.3	0.22	0.05	0.73				2				0.63
8	20.65	6.86	39.48				33,15				100.14	0.21	0.07	0.75				2				0.66
9	8.94	5.76	50.87				37.82				103.39	0.08	0.05	0.85				2				0.85
10	33.62	2.93	34.3				32	0.66			103.51	0.35	0.03	0,67				1.98	0.02			0.51
11	49.97	2.18	20.8				30.12				103.07	0.56	0.02	0.44				2.00				0.29
12		3.5	58.59				38.19	1.67			101.95		0.03	0.95				1.96	0.04			1
13			62.81				39.43				102,24			1.01				2				1
14	1.91	4.45	57.45		0,32	0.49	36.31	0.92			101,85	0.02	0.04	0,99		0,01	0.01	1.98	0.02			0.97
15	92.73	0	8.57			0.47	1.2	0.19			103.16	0.79		0.14		0,00	0.01	0.06				
16	37.98	31.17	29.55		0.88	1.61					101.19	0.29	0.23	0.42		0.02	0.04					
17	28.79	27.29	37.06	4.4	0.71		1.15				99.4	0.21	0.2	0.51	0.06	0.02						
18	33.33	11.19	53.61							2.12	100.25	0.19	0.06	0.6							0.14	
19	34.25	9.51	54.3				1.57			3.19	102.82	0.17	0.05	0.54				0.05			0.19	
20	88.01	11.34			1.09	2.01		0.2		0.66	103.31	0.8	0.1			0.03	0.06					
21	59.46	16.78	21.62		0.58	3.06					101.5	0.46	0.13	0.32		0.01	0.07					
22	2.52	2.11	93.14			2.55					100.32	0.01	0.01	0.93			0.04					
23	3.01	3.67	52.84			0.99	36.78	3.3			100.59	0.03	0.03	0.88			0.03	1.93	0.07			0.95
24			60.54		0.32	0.46	38.33				99.65			1		0.01	0.01	2				1
25	47.44	4.97	22.55		0.37		27.28				102.61	0.59	0.06	0.52		0.02		2				0.32
26			60.93				38.64				99.57			1				2				1
27	9.25	3.47	54.02				34.92				101.66	0.09	0.03	0.98				2				0.85
28	60.28	6.17	6.88		0.37	1.04	23.79				98.53	0.85	0.09	0.18		0.02	0.05	2				0.1
29			59.89		0.46		37.67				98.02			1.01		0.01		2				1
30		58.69	2.1	2.31			12.51	23.25			98.86		0.29	0.02	0.02			0.37	0.30			
31	1.18	59.23	1.44	0.35		0.81	11.38	24.82			99.21	0.03	0.41	0.1				0.63	0.37			
32	1.8	56.58	5.84	0.6			12.85	19.07	1		97.74	0.01	0.29	0.06	0.01			0.39	0.25	0.01		
33		48.23	11.47	3.91			16.81	20.93			101.35		0.21	0.09	0.03			0.44	0.23			
34		61.04	2.53	0.87			11.81	23.78	1.16				0.30	0.02	0.01			0.35	0.30	0.01		
35	4.21	3.15	10	37.77		22.35		25.66			103.14	0.02	0.01	0.08	0.3		0,31		0.28			
36	21.34	10.58	26.92	0.57		28.95	6.5	14.23			109.09	0.17	0.08	0.40	0.01		0.74	0.31	0.29			

Notes: Primary high temperature PGM: Os-Ir-Ru alloys-I—(1–3); laurite I—(4–7); altered laurite II—(8–14); secondary PGM: Os-Ir-Ru alloys-II—(15–22); laurite III—(23–29); irarsite—(30–34); RhNiAs—35; (Ru,Ni,Os,Ir,Rh)AsS—36.

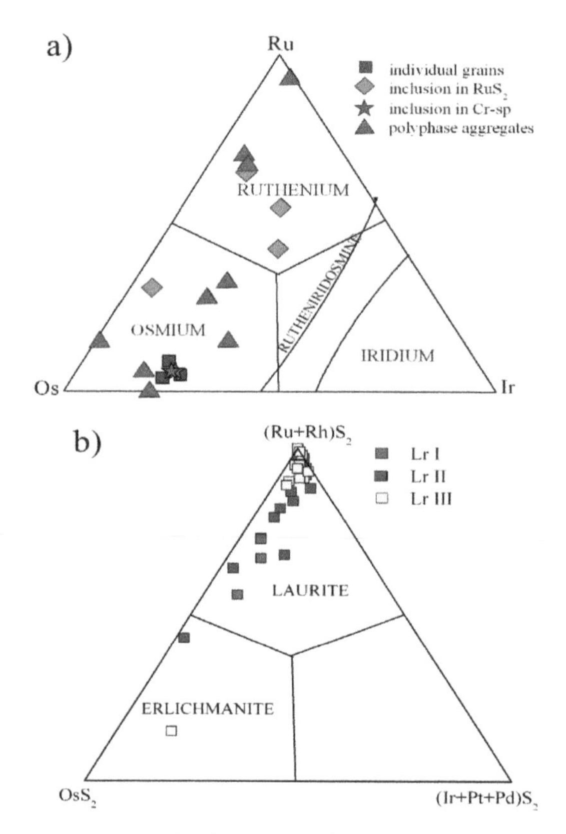

Figure 8. Diagram of the compositions of PGE minerals: (**a**) Os-Ir-Ru alloys; (**b**) Laurite-erlichmanite: Lr I—primary, inclusion in chrome spinels or isolated grains, laurite II—primary altered sulfides, and laurite III—secondary sulfides. The fields shown in the diagrams are drawn from [41].

4.3.1. Primary Platinum-Group Minerals

High-temperature Os-Ir-Ru alloys-I are in the form of idiomorphic inclusions in chrome spinels, and xenomorphic grains are intergrown with laurite. Single grains contain inclusions of $(Ru,Os)S_2$ of less than 10 μm in size. The alloys are enriched with Os, and its content varies from 71 to 79 wt. %; the content of Ir ranges from 20 to 28 wt. %; and the content of Ru is very low and ranges from 2 to 5 wt. %. There are some impurities of Fe and Ni. In the classification diagram, the compositions of the primary alloys correspond to osmium with low contents of Ir and Ru (Figure 8a). Dissolution microstructures are observed in some grains (Figure 9).

Laurite-Erlichmanite $(Ru,Os)S_2$

Sulfides can be divided into two groups, according to their composition and microstructural features (Table 4). The first group, laurite I, contains laurite-erlichmanite with a homogeneous composition.

Laurite-erlichmanite occurs as an inclusion in chrome spinels (10–15 μm), and in some cases, it can be intergrown with amphibole (magnesio-hastingsite hornblende) in chrome spinel or be found as individual grains (Figure 9c,d). The ratio, Ru′ = Ru/(Ru + Os), is 0.61–0.78. The content of Os is 20–33 wt. %, and Ru is 27–40 wt. %. In the classification diagram, they are in the laurite field (Figure 8b). The second group, laurite II, contains laurites of a heterogeneous composition, containing micro inclusions of (Os-Ir-Ru) II alloys and laurites, which are replaced by PGE sulfarsenides (Figure 9e–g). Laurite II (Lr II) is characterized by wide variations in the contents of Os of 1.2–49.9 wt. %, Ru of 20.8-62.2 wt. %, and the Ru/(Ru + Os) ratio of 0.44–0.99. In the classification diagram, they are in the fields of erlichmanite and laurite. In the chemical composition of laurite, a sulfur deficiency is often registered (Table 4). Insignificant amounts of irarsite IrAsS are found in the chromitites of the USO, where it replaces laurite. Irarsite forms corrosive, looped structures to replace laurite (Figure 9g,h).

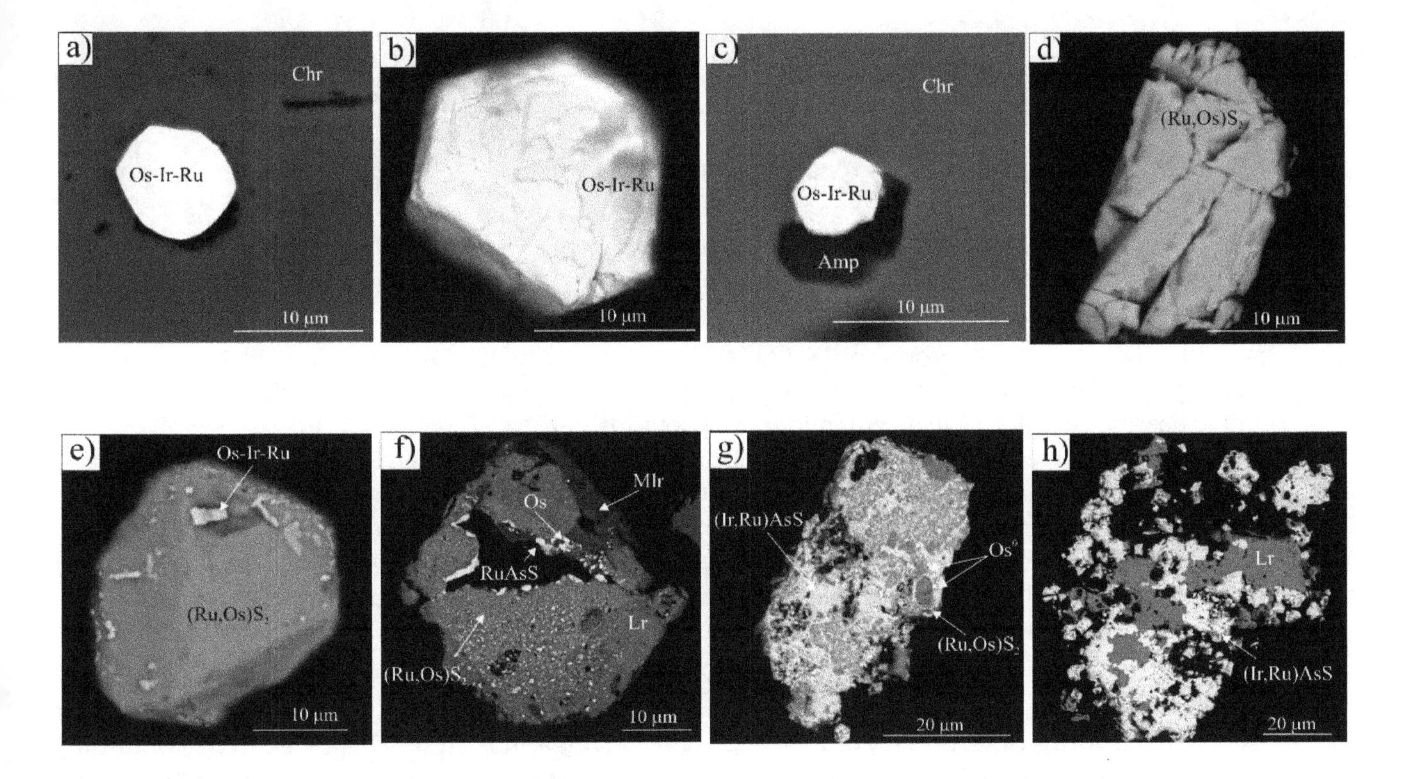

Figure 9. Back Scattered Electron (BSE) images of primary PGM, showing the textural and morphological relations of single and polyphase PGM from the Ulan-Sar'dag chromitites: (**a**) inclusion euhedral Os-Ir-Ru alloy—I in chromite, an. 1; (**b**) individual single-grain Os-Ir-Ru alloys—I, with a dissolution microstructure, an. 2; (**c**) inclusion in Cr-spinel of intergrowth laurite I and hornblende, an. 6; (**d**) individual grain of laurite I, an. 7; (**e**) grain of laurite II (an. 8), with inclusions of micro particle Os-Ir-Ru alloys—II; (**f**) grain laurite II (an. 9), associated with an unnamed phase (Ru,Ni,Os,Ir,Rh)AsS (an. 36) (laurite II is surrounded by Os-poor laurite, which grows with millerite); (**g**) substitution of laurite II by irarsite (an. 32), with remnants of laurite II (an. 10,11) and Os-Ir-Ru alloys—II; (**h**) laurite II (an. 12) surrounded by irarsite (an. 33). Abbreviations: Crsp—chrome spinel; Lr—laurite; Mlr—millerite. Notes: an. No—No analysis, as shown in Table 4.

4.3.2. Secondary Platinum-Group Minerals

The secondary PGMs are presented by $(Os,Ru)S_2$—laurite III, native Os and Ru, Os-Ir-Ru alloys-II of a variable composition, irarsite IrAsS, zaccarinite RhNiAs, and (Ru,Ni,Os,Ir,Rh)(As,S) sulfarsenides of a non-stoichiometric composition. The secondary phases are localized in the chloritized silicate intergranular space of chrome spinels. Some grains (micro particles of osmium and other phases) are very small (less than 2–3 µm), and their chemical composition can be determined only semi-quantitatively. In some cases, micro particles of a variable composition (Ru,Ir,Os,Cu,Te,Ni,Ba,S,O) can be found with secondary PGMs. Further, due to their very small size and partial coincidence with the elemental composition of the host mineral, qualitative analysis can also be hampered.

Os-Ir-Ru Alloys-II

Micro particles of Os-Ir-Ru alloys-II are the common phases and are found in a secondary mineral assemblage. They are localized mainly in laurite II, or they are a part of polyphase aggregates with Ni, Cu sulfides, sulfarsenides of Os, Ir, and Ru, zaccarinite RhNiAs, and laurite III (Figure 10a–d). Their composition varies (Table 4, Figure 8a) from native osmium to native ruthenium. Native osmium is composed of (wt. %): Os (87–92), Ir (0–12), and Ru (3–8). Native Ru is composed of (wt. %): Ru (93), Os (2.5), and Ir (2.1). The composition of the (Os-Ir-Ru) II alloys varies, with a wide range (wt. %): Os (30–74), Ir (6–32), and Ru (8–58).

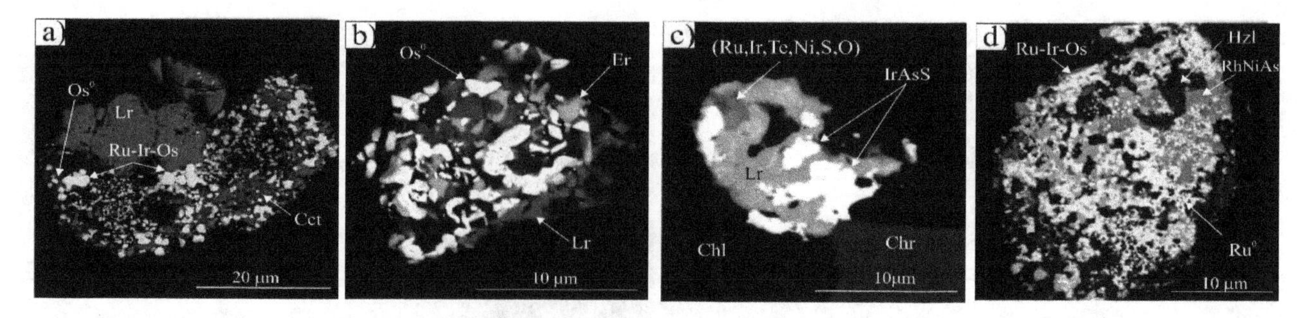

Figure 10. BSE images of secondary PGMs, showing the textural, morphological relations and assemblages of the PGM from the Ulan-Sar'dag chromitites: (**a**) intergrowth of laurite III (an. 26) and PGE bearing chalcocite Cu_2S, with micro inclusions of Os-Ir-Ru alloys- II (an. 18); (**b**) polyphase aggregate of agglomeration PGM particles, consisting of Os-rich laurite III (an.28), native Os^0 (an. 20), and Os-poor laurite III (an. 27); (**c**) composite grain of irarsite (an. 34), laurite III (an. 29), and unnamed phases (Ru, Ir, Te, Ni, S, O) within the interstitial chlorite of chromitites; (**d**) polyphase aggregate, consisting of Os-Ir-Ru alloys-II (an. 21), Ru^0 (an. 22), zaccarinite RhNiAs (an. 35), and heazlewoodite. Abbreviations: Chl—chlorite; Cct—chalcocite; Hzl—heazlewoodite; Irs—irarsite; Lr III—laurite III. Notes: an. No—No analysis, as shown in Table 4.

Laurite III

The occurrence forms of secondary laurite III are very diverse. It is mainly localized in Cr-containing chlorite, where it forms polyphase aggregates with PGE-containing chalcocite (Cu_2S) (Figure 10a), recrystallized PGM aggregates with native Os (Figure 10b), micro particles in association with Ni_3S_2, NiS and phases of a non-stoichiometric composition (Ru,Ir,Os,Cu,Te,Ni,Ba,S,O) (Figure 10c). The following features are characteristic of laurite III: a) A porous structure, with Cr-bearing minerals of the chlorite group in the voids, and b) a very small grain size (micro particles). The chemical composition corresponds to the end member of the laurite-erlichmanite solid solution: RuS_2-OsS_2 (Figure 8b). Laurite III is characterized by very low Os and Ir contents and the absence of Rh. Sometimes it contains Ni impurities. The value of Ru′ varies insignificantly and amounts to 0.93–1 (Table 4); in turn, OsS_2 has low contents of Ru and Ir and is located in a recrystallized aggregate with laurite III and native Os (Figure 10b).

The (Ir,Os,Ru)AsS in the secondary association is in the form of polyphase aggregates. It replaces laurite. It contains Os and Ru in insignificant amounts. Secondary irarsite is presented as very small microparticles (less than 5 μm) in chlorite in association with Ni_3S_2 and laurite III.

(Ru,Ni,Os,Ir,Rh)(As,S) is found in the form of micro particles (7 μm) in a polyphase aggregate, consisting of laurite II and NiS (Figure 9f). There is a deficit of S and As in this phase.

Zaccarinite RhNiAs is found in polyphase aggregates in association with Ni_3S_2, secondary (Os-Ir-Ru) II alloys and native Ru (Figure 10d).

Unknown phase No. 1 (Ir,Ni,Cu,Ru,Os,Cl) is found in the intergrowth with laurite, erlichmanite, and chalcocite. This association is localized in chromite in the cracks filled with Cr-containing chlorite. It is worth noting the presence of Cl in this phase.

Unknown phase No. 2 (Os,Ir,Ru,As,S,O) is found in the millerite cracks in close association with chlorite and chrome-magnetite.

Unknown phase No. 3 (Ru,Ir,Te,Ni,Ba,S,O) is found in chlorite in the intergrowth with laurite and irarsite. The Ba and Te impurities are unusual for the platinum phases of chromitites.

Ni,Fe,Cu sulphides and arsenides. The sulphides of base metals form a dispersed impregnation mainly in serpentine-chlorite aggregate. Heazlewoodite Ni_3S_2 and millerite NiS are the predominant sulfide phases (Figure 5, Table 5). Heazlewoodite is often found in polyphase aggregates with secondary PGE minerals. It has a homogeneous composition and is identical in individual grains and in the intergrowth with platinum metal phases.

Table 5. Chemical composition of Ni, Fe, and Cu sulfides and arsenides (wt. %).

No.	2a-12	6-12	6-13	3-13	6-13	4-12	3-13
Mineral	$(Fe,Ni)_9S_8$	Cu_2S	Ni_3S_2	NiS	Ni_3S_2	Ni_3S_2	Ni_3As_2
Ni	38.9		67.5	64.16	72.82	72.19	64.24
Fe	25.58						
Cu		64.89					
Co	0.87						
Os		4.34					
Ir		1.18					
Ru		8.01	5.29				
Rh			0.42				
S	33.08	19.8	25.81	33.82	27.25	27.2	
As			0.55				36.01
O		1.06					
Total	98.43	99.28	99.57	97.98	100.07	99.39	100.25
	Incl. in olivine	PGE bearing, intergrowth with PGM		PGE-free, individual grains			

5. Discussion

5.1. Chromitite Formation: Composition of Parental Melts

Chromite bodies in ophiolites are formed due to the partial melting of rocks of the upper mantle. The interaction of the melt with mantle peridotites plays an important role in the formation of podiform chromitites. In the channel filled with molten mantle, the ascending olivine-chromite-cotectic melt mixes with the silica-enriched melt, formed through the harzburgite-melt reaction. The formation of surrounding dunites along the host harzburgites is the result of a combination of olivine precipitation from "older" magma and the destruction of orthopyroxene in harzburgite, interacting with the magma channel. Thin chromite schlieren and streaks can be formed by separating the chromite from the cotectic olivine-chromite melt [9,42–46].

In the Pt/Pt*-Pd/Ir diagram (Figure 11), dunites and chromitites are within the partial melting trend. The late metamorphic and hydrothermal processes, during which Pd enrichment occurred, probably cause the high Pd/Ir ratio in some Ulan-Sar'dag chromitites. In the OSMA diagram, most of the ore chrome spinels are in the olivine-spinel equilibrium field, and some of the chrome spinels are beyond this field. This can be explained by the distortion of the magmatic system closedness, changes in the compositions during the tectonic processes and metamorphism of chrome spinels. Chromitites of the USO were formed at the 28–35% degree of partial melting (Figure 12). The data on the dunites, harzburgites, and chromitites of the northern and southern branches of the Ospa-Kitoy ophiolite are presented for comparison (Figure 12). The Mg'(Ol)–Cr'(Sp) ratio in dunites and harzburgites corresponds to the 30–40% degree of partial melting; in ore chrome spinels from the northern branch, it corresponds to 30–40%; in ore chrome spinels of the southern branch, it corresponds to 35%.

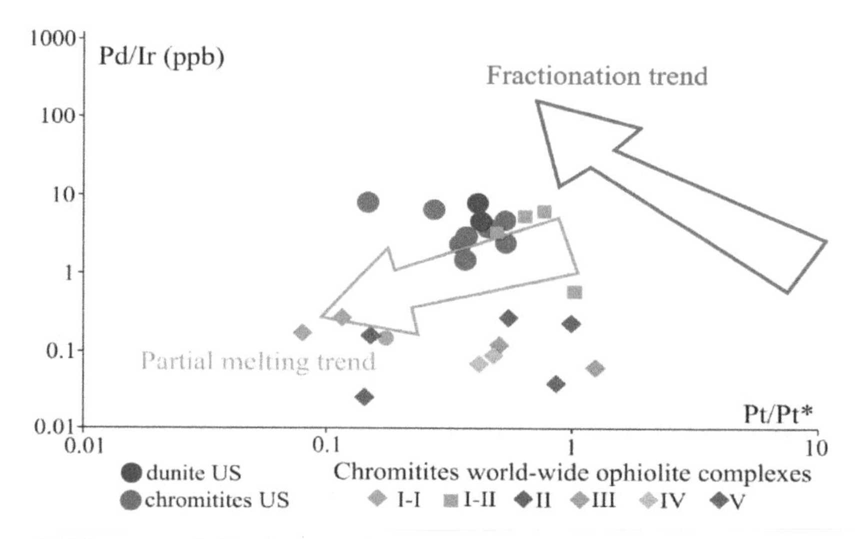

Figure 11. Plot of Pd/Ir versus Pt/Pt* for mantle peridotite and chromitites of the USO. The Pt anomaly is calculated as follows: $Pt/[Pt]^* = (Pt/8.3) \times \sqrt{(Rh/1.6) \times (Pd/4.4)}$ [8]. US—Ulan-Sar'dag ophiolite; ophiolite complexes from around the world: I—Wadi Al Hwanet ophiolite, Saudi Arabia [47]; II—Oman ophiolite, Semail [48]; III—Veria ophiolite, Greece [49]; IV—Shetland Ophiolite Complex, Scotland [50]; V—Ray-Iz ophiolite, Russia [51].

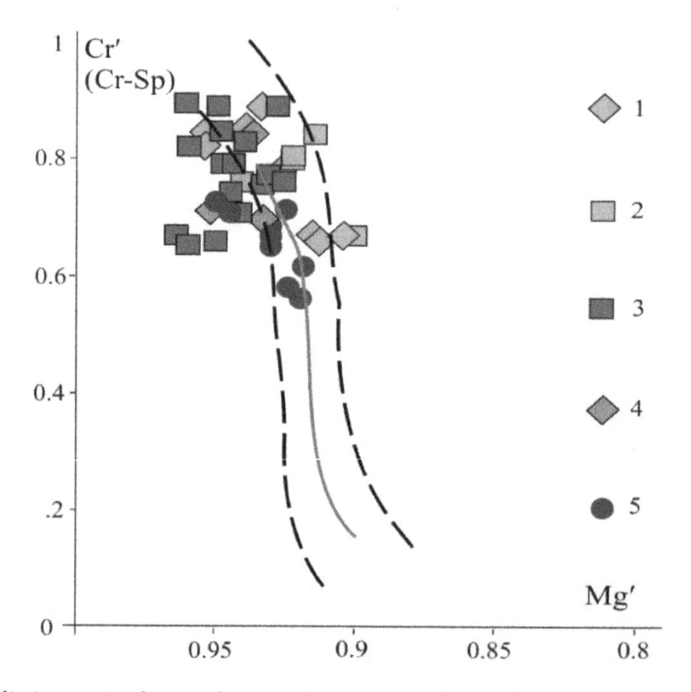

Figure 12. OSMA (olivine-spinel mantle array) is a spinel peridotite restite trend [52]. The chrome spinels are from: 1–4—the Ospa-Kitoy ophiolite: 1—dunites; 2—harzburgites; the chromitites are from: 3—the north-branch Ospa-Kitoy, 4—south-branch Ospa-Kitoy; 5—Ulan-Sar'dag ophiolite. Mg'-Mg/(Mg + Fe) in olivine; Cr'-Cr/(Cr + Al + Fe^{3+}) in chrome spinel.

The composition of the chrome spinels of the USO varies significantly, which may reflect the interaction of mantle peridotites with the melts of various compositions. The joint presence of high-Cr' and high-Al' chrome spinels is common in many ophiolite belts, but as a rule, chrome spinels with different compositions are found in different ophiolite nappes [7,9,52–59]. Their joint occurrence in one ophiolite nappe is less common; in this case, the ophiolite nappe, as a rule, contains peridotites depleted to different degrees [45,60–63]. The composition of chrome spinels of podiform chromitites that includes, as the main components, FeO, MgO, Al_2O_3, and TiO_2 is a function of the composition of the parental melts [9,15,16,18,21,64]. We have calculated the contents of the Al_2O_3, TiO_2 and FeO–MgO ratios in the parental melts, which were in equilibrium with the podiform chromitites (Table 6), according to Equations (1)–(3) [15]. Table 6 shows, for comparison, the values of these parameters for the chrome spinels of ophiolites from around the world.

$$Al_2O_{3Sp}(wt.\ \%) = 0.035 \times (Al_2O_3)_{melt}^{2.42}, \quad (1)$$

$$TiO_{2(melt)}(wt.\ \%) = TiO_{2Sp}^{0.82524} \times e^{0.20203} \quad (2)$$

$$Ln(FeO/MgO)_{Sp} = 0.47 - 1.07 \times Al'_{Sp} + 0.64 \times Fe^{3+\prime} + Ln(FeO/MgO)_{melt}, \text{ where } Al'_{Sp} = Al/(Al + Cr + Fe^{3+}) \text{ and } Fe^{3+\prime}_{Sp} = Fe^{3+}/(Al + Cr + Fe^{3+}) \quad (3)$$

The chemical composition of the chrome spinels I group (Crsp I) and the composition of the parental melt are similar in these parameters for medium-aluminous chrome spinels of the Ospa-Kitoy ophiolite (Table 6). The $(Al_2O_3)_{melt}$ value of Crsp I is 13–18 wt. %, and the Al_2O_{3Sp} / Al_2O_{3melt} ratio corresponds to the spreading trend and abyssal peridotites (Figure 13a). Despite the low TiO_2 content, Crsp I has the TiO_{2Sp}/TiO_{2melt} trend, which is calculated for spinels from the MORB-type peridotites (Figure 13b).

The chemical composition of the chrome spinels II group (Crsp II) and the composition of the parental melt are similar in these parameters for the low-Al′ chrome spinels of the Ospa-Kitoy ophiolite (Table 6). The $(Al_2O_3)_{melt}$ values for the chrome spinels of groups II and III are 10–12 and 10–13 (wt. %), and the (FeO/MgO)melt ratio is 0.4–0.85 and 0.7–1, respectively. The values of the $Al_2O_{3Sp}/Al_2O_{3melt}$ ratios in chrome spinels II and III correspond to the trend of the chrome spinels from the island-arc boninites and chromitites of the Ural-Alaska complexes (Figure 13c). The $(TiO_2)_{melt}$ values for the chrome spinels I, II, and III groups overlap because of the low TiO_2 content (0–0.2 wt. %). In general, the values of $(Al_2O_3)_{melt}$/(FeO/MgO) are similar to the high-Cr′ chrome spinels from the Troodos and Zetford Mine ophiolites (Figure 13d), which were formed in a suprasubduction setting. In the discrimination relationship diagrams, the [$Al_2O_3/Fe^{2+}/Fe^{3+}$], [Mg′/Cr′], and [Al_2O_3/TiO_2] (Figure 14a–c) of the chrome spinels I group are in the field of the MORB–type peridotites. Chromitites were formed during the interaction of harzburgites with primitive MORB-like melts at deep levels of the upper mantle. Through the interaction of MORB-type melts, which are in equilibrium with abyssal dunites [65,66], precipitation of chrome spinels with Cr′ 0.4–0.6, as a product of the melt–peridotite reaction, is possible [43,67]. Among the volcanic rocks of the USO, metabasalts, with enriched mid-ocean ridge basalt (E-MORB) geochemical characteristics, are available [68].

The chrome spinels II group is localized in the boninite field. The TiO_2 and Al_2O_3 content corresponds to the chrome spinels of the suprasubduction peridotites and overlaps with the chrome spinels from the New Caledonia island arc (Figure 14c). The chrome spinels II are formed during the reaction of the peridotites with the island-arc boninite melt in the subduction zone. The chrome spinels III group lies on the boundary of the boninite fields and magmatic complexes of the Ural-Alaskan type. Three mechanisms can be suggested for the formation of the third type of chrome spinels. The first is the interaction with high-iron low-titanium melts [81]. The second mechanism is through plastic deformations under mantle or crust-mantle conditions (low fO_2, Table 1), because of the reactions of the Mg-Fe exchange with olivine. High-iron chrome spinels are found in the structural types of chromite, including chromitites with deformation structures. When compared with chrome spinels of the Alaskan type chromitites, the chrome spinels III group from the chromitites of the USO have low TiO_2 contents, high Cr_2O_3 contents and low (mantle) fO_2 values (Table 1). Variations in TiO_2 contents, from 0 to 0.22 wt. %, may indicate a reaction with TiO_2-containing melts [82]. The third mechanism is through the partial melting of the fluid-metasomatized mantle during the interaction of andesitic melts with rocks of an overlying mantle wedge [71,83,84].

Table 6. Calculated parental melt composition of chrome spinels.

	Cr-Spinels	Al′	Cr#′	Mg′	Parental Melt (wt. %)	References
1	I group	24–60	36–74	45–74	Al_2O_3, 13–18; TiO_2, 0–0.14 FeO/MgO, 0.2–0.7	In this article
2	II group	14–23	74–81	32–48	Al_2O_3, 10–12; TiO_2, 0–0.35 FeO/MgO, 0.4–0.85	In this article
3	III group	13–27	68–81	28–35	Al_2O_3, 10–13; TiO_2, 0.08–0.13 FeO/MgO, 0.7–1	In this article
4	Ospa-Kitoy medium Al′	24—41	59–75	43–70	Al_2O_3, 12–14; TiO_2, 0.01–0.44 FeO/MgO, 0.5–1.1	[38]
5	Ospa-Kitoy low Al′	9–21	77–90	23–59	Al_2O_3, 8–11; TiO_2, 0.01–0.48 FeO/MgO, 0.5–2.4	[38]
6	MORB	35–64	29–57	57–59	Al_2O_3 13–18; TiO_2, 0.3–1.7 FeO/MgO, 0.5–0.7	[18,69];
7	BAB	61	34	75	Al_2O_3, 17.6; TiO_2, 0.4 FeO/MgO, 0.4	[18]
8	OIB	28	61	57	Al_2O_3, 12; TiO_2, 1.6 FeO/MgO, 0.6	[18]
9	IAB	15–29	61–68	58–69	Al_2O_3, 9–12; TiO_2, 0.4–0.7 FeO/MgO, 0.3–0.5	[18]
10	IABon, IAT	5–19	74–89	58–75	Al_2O_3, 6–10; TiO_2, 0.08–0.4 FeO/MgO, 0.2–0.4	[18]
11	LIP	13–35	52–72	35–61	Al_2O_3, 8–13; TiO_2, 0.2–0.5 FeO/MgO, 0.4–1.2	[18]
12	Abissal peridotite	45–77	20–50	64–77	Al_2O_3, 15–19; TiO_2, 0.08–0.1 FeO/MgO, 0.4–0.5	[70]
13	Chromite in ophiolite mantle	40–51	44–52	63–74	Al_2O_3, 14–16; TiO_2, 0.2–0.5 FeO/MgO, 0.3–0.6	[3]
14	Chromite in ophiolite mantle	21–28	67–74	56–68	Al_2O_3, 10–12; TiO_2, 0.1–0.3 FeO/MgO, 0.54–0.6	[3,48]
15	Chromite in Alaskan type	11–14	62–70	45–56	Al_2O_3, 8–9; TiO_2, 0.6–0.9 FeO/MgO, 0.4–0.5	[71]

5.2. Spinel and Olivine Geothermometers and Olivine-Spinel Oxybarometers

Several processes condition the composition of chrome spinels: partial melting, cooling of mantle peridotites and plastic deformations in the upper mantle. Many researchers have shown that, during ultramafite cooling, regardless of their formation (mantle or cumulative), ($Mg \leftrightarrow Fe^{2+}$) exchange reactions take place between coexisting olivines and chrome spinels, as a result of which the coefficient of these elements' distribution increases in favor of chrome spinels, and consequently, the calculated temperatures of the olivine-spinel equilibrium decrease [85–87]. During plastic deformations, exchange processes and rebalancing between olivine and chrome spinels also occur, and schlieren and lenticular segregations of chromitites transform into massive chromite bodies (Figure 4c–f). Based on this, it is assumed that the obtained values for the temperature, pressure and oxygen fugacity correspond not to the formation of ultramafites and chromitites, but to the stages of the formation and transformation of these bodies: the tectonic flow of the upper mantle rocks, the effect of metasomatizing fluids and other processes.

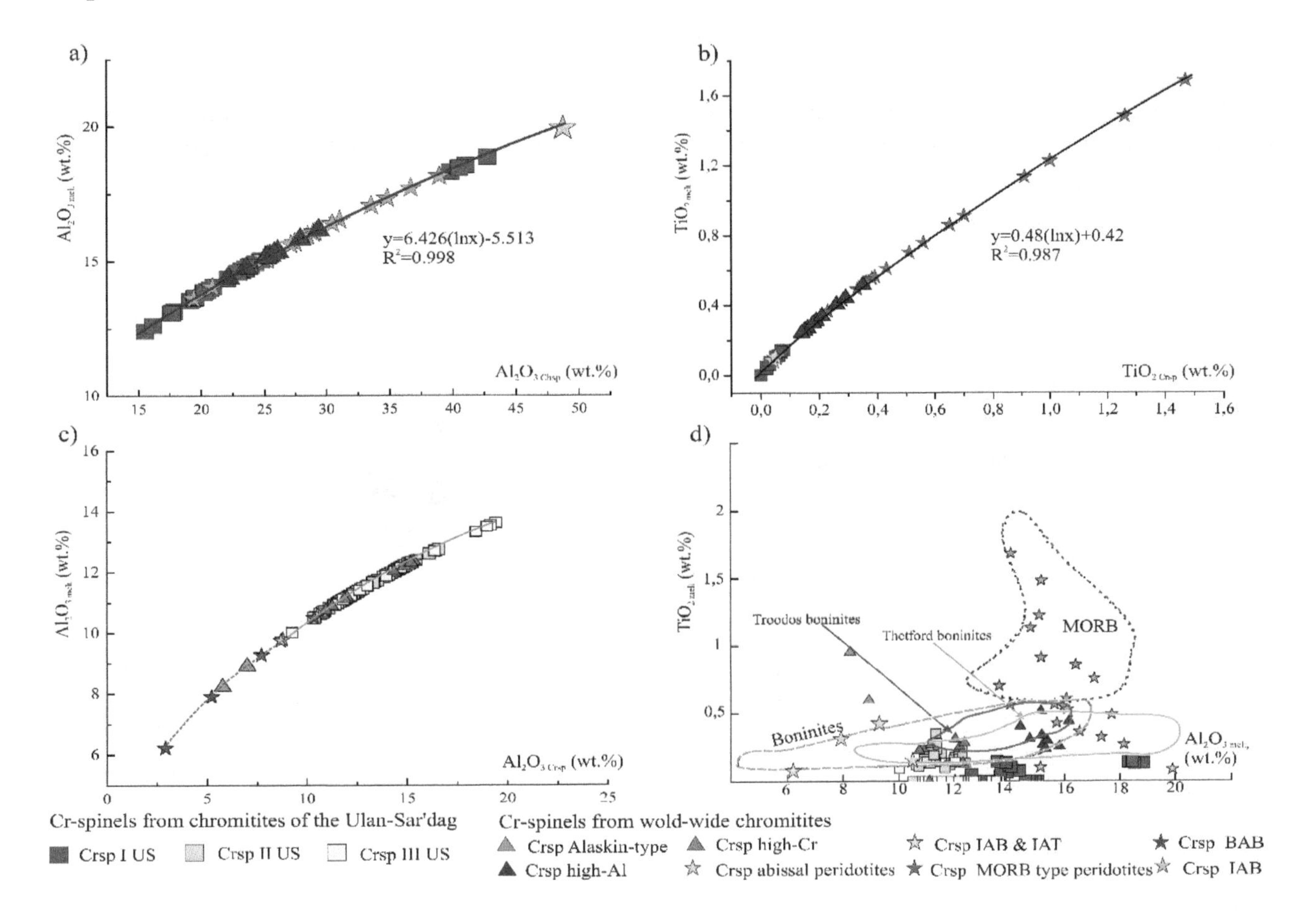

Figure 13. Plots of relations: (**a**) $Al_2O_{3\_Crsp}$–$Al_2O_{3\_melt}$; (**b**) TiO_{2_Crsp}–TiO_{2_melt}; (**c**) $Al_2O_{3\_Crsp}$–$Al_2O_{3\_melt}$; (**d**) calculated abundance of Al_2O_3–TiO_2 in melt, as in the equilibrium with the Cr spinels from the chromitites of the USO and the chrome spinels of chromitites from all over the world. The fields for the chrome spinels of boninites [72–74], Troodos boninites [75,76], Thetford boninites [77] and MORB [78–80] are shown for comparison.

We have calculated and estimated the P-T parameters with the help of an olivine-spinel (Ol-Sp) geothermometer, provided in [88]. The oxygen fugacity was determined using an Ol-Sp oxybarometer, provided in [89], in accordance with oxybarometers [90,91]. For the chrome spinels I group (high Al'), the calculated temperatures of the Ol-Sp equilibrium are $T_{Ol\text{-}Sp}$ = 1020–920 °C, and fO_2 = (−0.7)–(−1.5) (Table 1); for the chrome spinels II group (medium-, low-Al'), the calculated temperatures are $T_{Ol\text{-}Sp}$ = 891–1003 °C, and fO_2 = (−1.1) − (−4.4); for the chrome spinels III group, these values are $T_{Ol\text{-}Sp}$ = 846-738 °C, and fO_2 =(−1.49) − (−3.01). For the chrome spinels containing inclusions of laurite-erlichmanite, $T_{Ol\text{-}Sp}$ = 916–938 °C, and fO_2 = −2.4. For the chrome spinels with the olivine inclusion, $T_{Ol\text{-}Sp}$ = 991.6 °C, and fO_2 = (−1.3). The temperature values overlap, but for the chrome spinels II group, higher temperatures are noted. The values of fO_2 are discrete. It is assumed that a more reducing environment and higher solid-phase reaction temperatures are required for the formation of the chrome spinels II group, in comparison with the chrome spinels groups III and I. The content of impurities (Al_2O_3) in olivine was used as an alternative geothermometer. For our objects, this method was limited by the small amount of preserved olivine. We were interested in the chromitites containing PGE- and PGE-free mineralization. Using the geothermometer of De Hoog and Gall for the Al content in olivine [25], the temperature was estimated by Equation (4). For the chromitites containing PGM, the temperature, according to the olivine thermometer, was 893–1332 °C, and for the PGM-free chromitites, it was 1073–1225 °C (Table 1).

$$T_{ol}(C) = 1087/(7.46 - ln\text{Al}_{ppm}) - 273 \quad (4)$$

An assessment of the temperature using an olivine thermometer and inclusions of primary high-temperature (Os-Ir-Ru)-I alloys gives the values of 1200–1300 °C, which is more consistent with the expected temperatures of the chrome spinel formation from the melt.

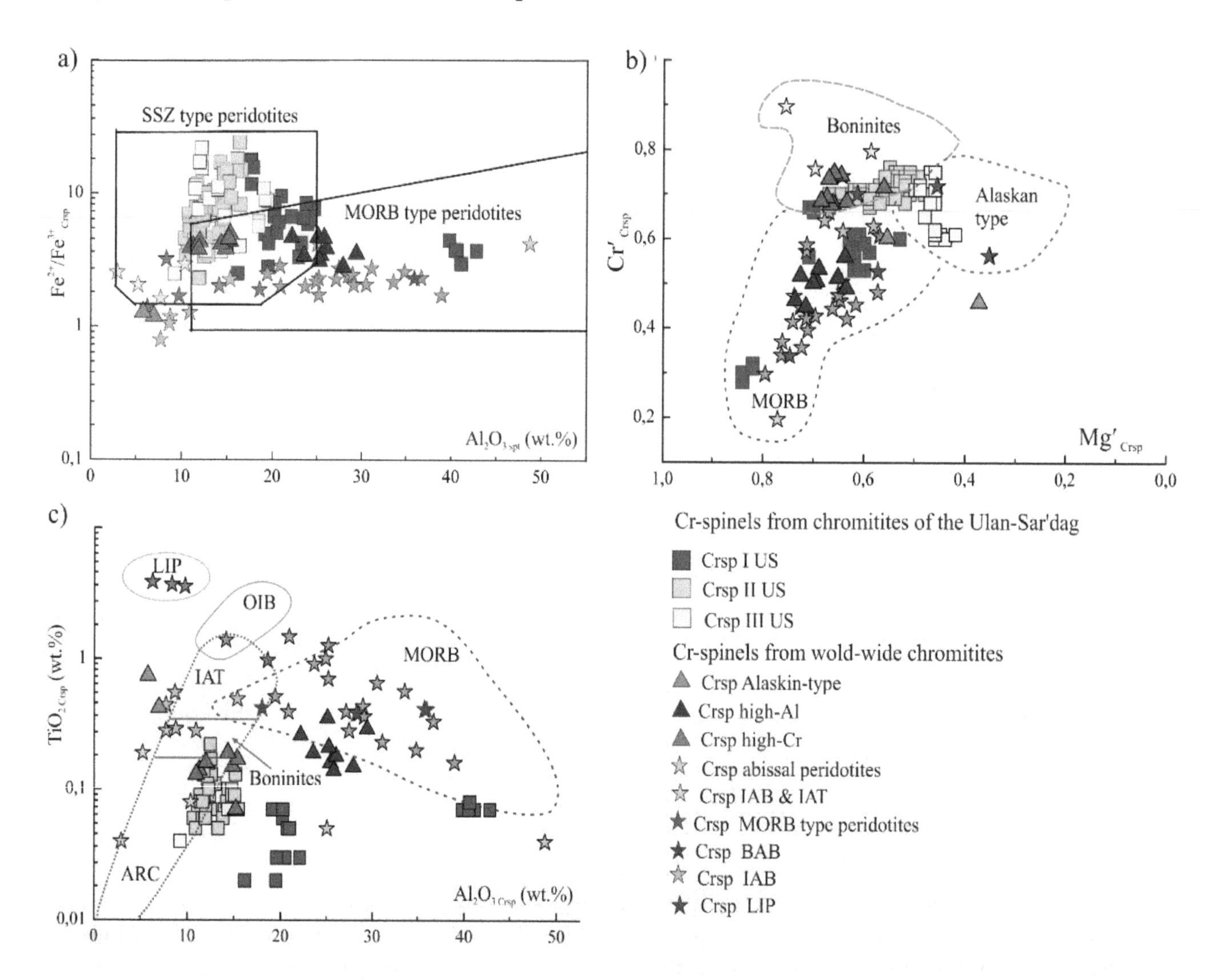

Figure 14. Tectonic discrimination diagrams: (**a**) Al_2O_3–Fe^{2+}/Fe^{3+}; (**b**) Mg'_{sp}–Cr'_{sp}; (**c**) Al_2O_3–TiO_2 from the chromitites of the USO and chrome spinels of chromitites from all over the world. The chrome spinels' composition fields for peridotites from different geodynamic settings are drawn from [18].

5.3. Distribution of PGE in Mantle Peridotites and Chromitites

The form of PGE distribution in the mantle peridotites and chromitites of ophiolites reflects the mantle conditions of chromitite formation and PGE mineralization. In the process of partially melting the mantle source, the PGEs fractionate. The melt is enriched in Rh, Pt, and Pd (PPGE), since PPGE is incompatible with Os, Ir, and Ru (IPGE). The mantle restite will be enriched in IPGE [17,92,93]. This process is confirmed by the form of PGE distribution in the Ulan-Sar'dag peridotites (Figure 15a,b), for which $\sum$IPGE> $\sum$PPGE.

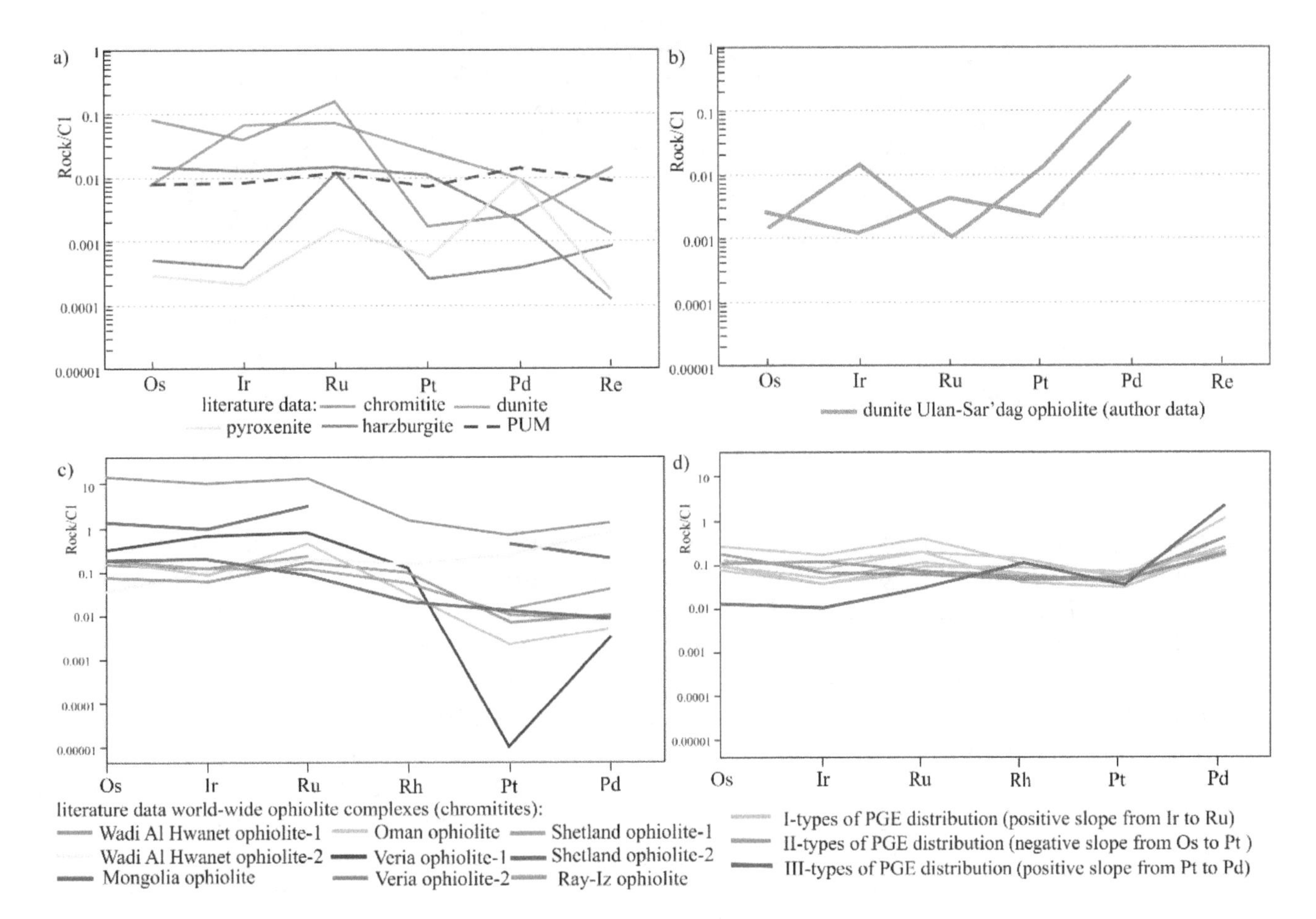

Figure 15. Diagrams: (**a**) PGE distribution in the Ulan-Sardag peridotite literature data [94]; (**b**) PGE distribution in the Ulan-Sar'dag peridotite author data; (**c**) PGE pattern of chromitites of ophiolite complexes from all over the world: The Wadi Al Hwanet ophiolite, Saudi Arabia [47]; Oman ophiolite, Semail [48]; Veria ophiolite, Greece [49]; Shetland Ophiolite Complex, Scotland [50]; Ray-Iz ophiolite, Russia [51]; Mongolian ophiolite [95]; 1—chromitites are enriched in Os-Ir-Ru; 2—chromitites are enriched in Pt-Pd; (**d**) PGE patterns in the Ulan-Sar'dag chromitites.

In the case of a low degree of partial melting, there are no obvious differences between the PPGE and IPGE contents, and the PGE distribution is flat. At a high degree of partial melting, PPGE is depleted in restite mantle rocks, relative to IPGE, and the total PGE contents become lower. In this case, the distributions have a negative slope. The PGE distribution in the mantle peridotite of Ulan-Sar'dag demonstrates: (1) a positive Ru and Pd picks; (2) a flat type of distribution, with a negative slope towards Pd; (3) a positive slope of Ir, Rh, and Pt and a negative slope of Ru (Figure 15b). Podiform chromitites are characterized by a fractionated form of PGE distribution (Figure 15c), which indicates a high degree of partial melting (about 20%–24%) of the mantle source. Three types of PGE distribution are observed in Ulan-Sar'dag chromitites (Figure 15d): (1) a negative slope from Os to Ir and Ru to Pt and a positive slope from Ir to Ru and Pt to Pd; (2) a negative slope of the distribution curve from Os to Pt and a positive slope from Pt to Pd; (3) low contents of Os, Ir and Ru and an enrichment in Rh (Pd is uncharacteristic of chromitites). Chondrite-normalized PGE relations in chromitites of the USO are similar to those of chromitites of ophiolites (Figure 15c,d) formed in a suprasubduction environment from all over the world [7,11,57,63,96–98]. The pronounced positive Ru anomaly is due to the predominant laurite phase in the chromitites of the USO. It is known that Ru has a maximum affinity with S, and the predominance of IPGE sulfides, in turn, indicates a high fugacity of sulfur in the melt, in contrast to the parental melts for chromitites of the northern and southern branches of the Ospa-Kitoy ophiolite (Figure 16a–c).

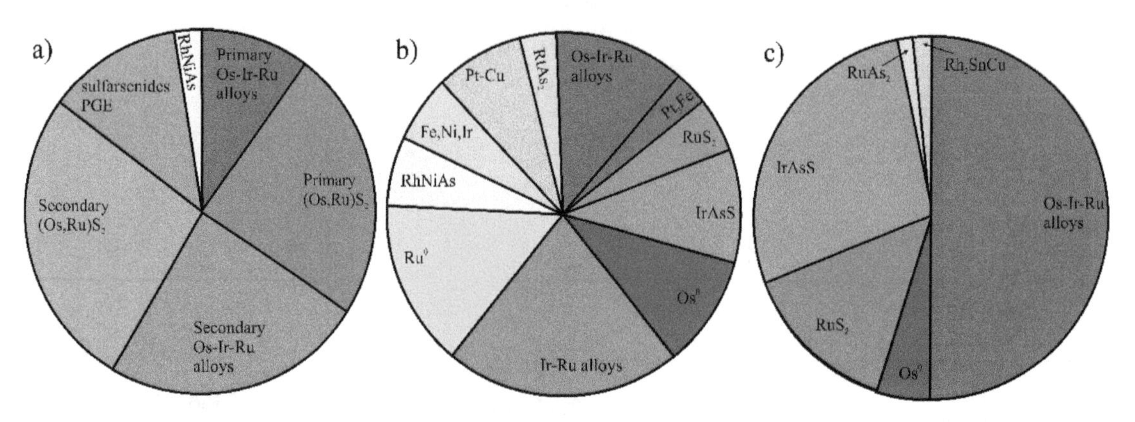

Figure 16. PGE assemblage (**a**) Ulan-Sar'dag chromitites; (**b**) the northern branch of the SEPES ophiolite; (**c**) the southern branch of the SEPES ophiolite [36].

The experimental data show that Os, Ir, and Ru are concentrated by trapping submicroscopic clusters of these elements in the metallic state during chromite crystallization [99–101]. Tsoupas [49] believes that an IPGE enrichment in chromitites can be associated with post-magmatic processes during a long period of deformations, beginning from plastic deformations in the asthenospheric mantle and ending with brittle deformations in the crust. This is confirmed by wide variations in the contents of IPGE, PPGE and the IPGE/PPGE ratio from 0.03 to 2.15 (Table 3) in the studied chromitites. Negative Pt anomalies in chromitites are closely related to the unique properties of Pt itself. The distribution coefficients of the alloy/sulfide liquid at 1000 °C are 1–2 for Pd and more than 1000 for Pt. Thus, the distribution coefficient of the alloy/sulfide melt for Pt is 1000 times greater than it is for Pd [102]. This is confirmed by the distribution diagrams, where a Pt negative anomaly is clearly visible (Figure 15c,d). An extreme Pd enrichment in one of the chromitites is most likely associated with late magmatic processes and exposure to the reduced fluid. Chrome spinels has high FeO contents and low fO_2 values. There are no Pd phases in this chromite. We believe that PGM is concentrated in recrystallized dunite, which requires further study. Detailed studies on the metaperidotites, metagabbros, and metavolcanogenic sedimentary rocks of the USO showed that some rocks have signs of significant exposure to a high-temperature fluid phase, which leads to rock metasomatism [68,103].

5.4. Sequence of the Formation and Transformation of the Platinum-Group Mineral Assemblage

Based on the chemical and textural features of PGM and associations with the magmatic and hydrothermal minerals of chromitites, several stages of PGE mineralization were distinguished (Figure 17).

- I—Magmatic Stage

At the magmatic stage, under the upper mantle conditions, euhedral–subhedral high-temperature Os-Ir-Ru I alloys and laurite I are formed (Figures 8 and 9a–d), which are captured by chromite grains [5,11,22,104–106]. The high osmium content in the primary (Os-Ir-Ru) I alloys is caused by an early crystallization of laurite-erlichmanite (Figure 9c). According to the experimental data, laurite without an Os impurity crystallizes from the melt at a high temperature (T = 1200–1300 °C), P = 5–10 kbar and low log sulfur fugacity (fS_2), from (−0.39) to 0.07 [24,106–109]. A decrease in temperature and increase in fS_2 leads to a replacement of Ru by Os, and as a result, laurite rich in Os is formed. The predominance of laurite-erlichmanite over solid (Os-Ir-Ru) solutions in the chromitites of the USO is a distinctive feature, in comparison with the PGE mineralization of the northern and southern branches of the Ospa-Kitoy ophiolite (Figure 16a–c). In combination with the high and medium Al′ chrome spinels, this indicates the formation of chromitites and PGE mineralization of the USO because of the interaction between the initially S-saturated tholeiite magma and depleted harzburgites. Sulfarsenides and arsenides of Ru and Ir are formed from the residual fluid phase at the late magmatic stage (Figure 10g,h). With magmatic system cooling, volatile components, such as S and As, accumulate with the formation of the residual fluid phase. There is a partial replacement of laurite by irarsite with the formation of laurite II.

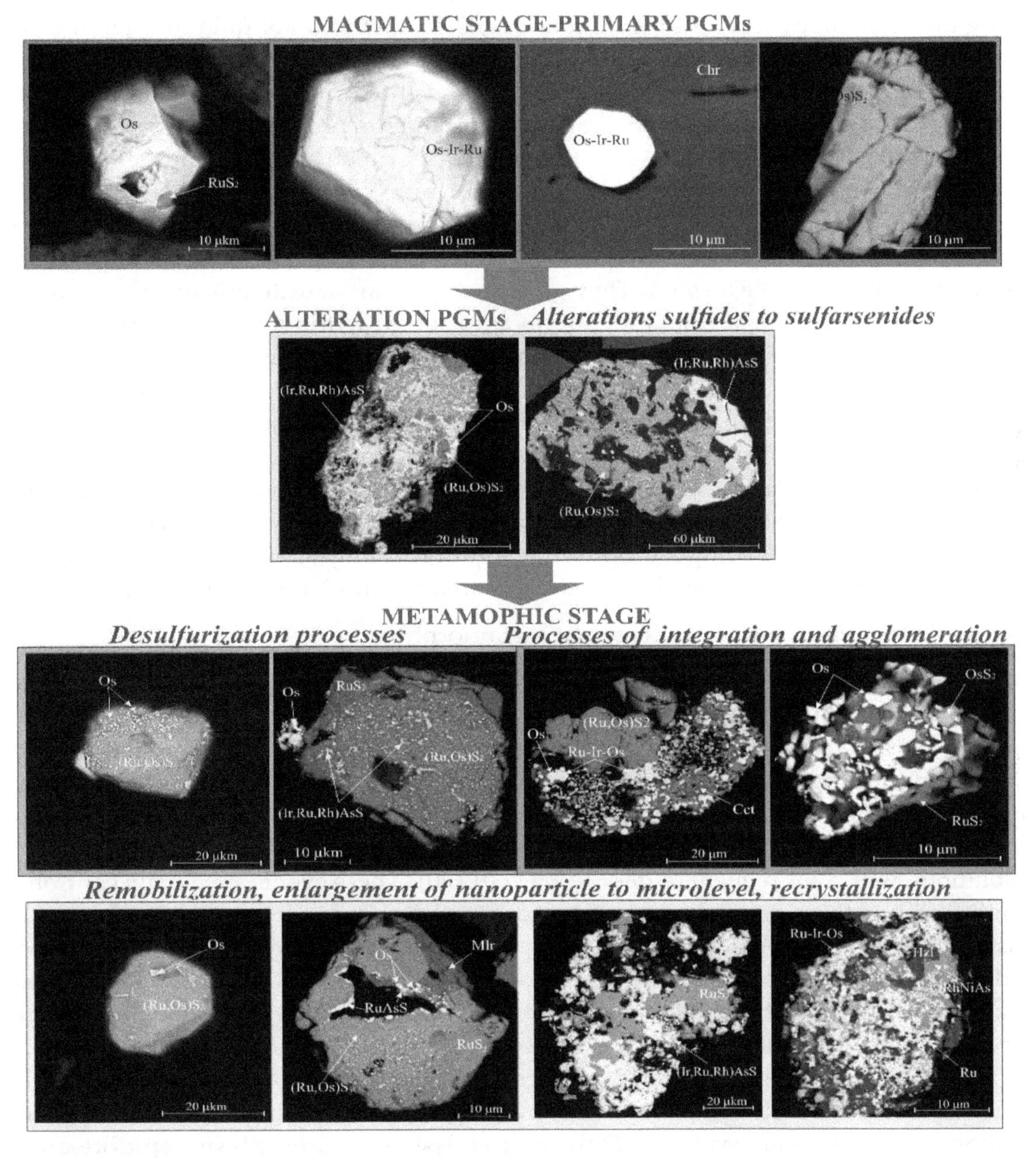

Figure 17. Scheme of the alteration and transformation of platinum-group minerals.

- II—Stage of Serpentinization and Exposure of Fluid

The PGMs in chromitites demonstrate signs of PGE remobilization (Figure 10a–d). The most intense changes of PGMs occur at the stage of the serpentinization of ultramafites. A fluid–rock interaction occurs with the participation of reduced gases (H_2, CH_4) and the H_2O of mantle origin. Dehydrating rocks of the subducting slab, as well as mantle-reduced fluids penetrating along the fault zones in tectonically weakened sectors, serve as the fluid source. At this stage, the following platinum–metal phases are formed: native Os and Ru, (Ir-Ru) alloys, phases of a variable (Os-Ir-Ru) composition, and newly formed laurite III, IrAsS, RhNiAs and RuAs (Figures 9f and 10a–d). Remobilized secondary PGMs form polyphase aggregates in the serpentine-chlorite matrix (Figure 10c) in association with

Ni sulfides, sulfarsenides, and arsenides. The joint occurrence of PGMs with Ni sulfides and the presence of such elements as Ni, Fe, Te, Cu, Co, As, and Sb in the platinum phases indicate PGE mobility in a fluid-saturated medium. The processes of the redistribution and concentration of PGE, including refractory Os, Ir, and Ru, occur at relatively low temperatures, reducing conditions corresponding to the formation of nickel sulfides, sulfarsenides, and arsenides, and low-temperature PGE-bearing intermetallides [110]. At the initial stage, the penetration of fluid through the permeable zones into laurite-erlichmanite led to the desulfurization of sulfides, a deviation from stoichiometry, the appearance of microdefects in the crystal lattice, and the formation of nanopores on the grain surface. These processes led to the formation of microporous structures and the separation of native Os and Ru (Figure 9e,f) [108,109,111,112]. According to the experimental data, congruent $RuS_2 \rightarrow$ native Ru decomposition occurs at T = 300 °C, $\log fS_2 = (-20)$, and P = 0.5 kbar [113–118]. The (Os-Ir-Ru) phases of a variable composition are probably the products of changes in (Os,Ir,Ru)AsS, since Ir has a maximum affinity with As, and irarsite therefore survives for the longest. At the same stage, secondary (newly formed) laurite III can be formed. They grow over primary laurite-erlichmanite or are confined to chloritization zones, and as a rule, they are in association with nickel and copper sulfides and sulfarsenides (Figure 10a). Irarsites IrAsS can be formed during serpentinization. In this case, irarsites are in close association with Ni_3S_2, forming joint aggregates. The microstructural features of such aggregates indicate their simultaneous formation (Figure 10c). The physical–chemical modeling of the forms of PGE transport in the fluid systems indicates the formation of carbonyl, chloride, hydrosulfide, and bisulfide complexes, in the form of which they are transported, and the formation of secondary PGM occurs. When PGEs are transported by bisulfide complexes, and As and Sb appear in the system, PGE solubility decreases, the composition of the solution changes, and the system deviates from equilibrium. Sulfur released from bisulfide complexes reacts with Ni to form Ni_3S_2.

- III—Stage of Ophiolite Obduction. Regional Metamorphism.

As the ophiolite rises to the surface, the rocks undergo repeated processes of serpentinization under the influence of metamorphogenic fluids with an increased activity of O_2, As, and Sb [48]. In chromitites, chrome spinels change into chrome magnetites or magnetites. There are no clear criteria for distinguishing remobilized PGMs under the crustal conditions (under the influence of metamorphogenic fluids). We believe that non-stoichiometric platinum-metal phases, containing Cu, Te, As, Sb, and O, could be formed under the crustal conditions. These elements can be transported by aqueous solutions, with a subsequent re-deposition [119,120]. These events are mainly controlled by the Eh-pH conditions, and these minerals can be formed directly in the supergenic environment. Under the conditions of a changing temperature, varying Eh-pH in the Os-S-O-H system, low $f(S_2)$ and exposure to an oxidizing high-temperature fluid at a temperature of about 500 °C [121], Os becomes more mobile than other PGEs, which leads to the further redistribution and redeposition of osmium. At this stage, the following PGE-containing phases can be formed: (Ru,Ir,Te,Ni,Ba,S,O), (Os,Ir,Ru,As,S,O), and (Ir,Ni,Cu,Ru,Os,Cl). Most often, such phases are localized in the micro voids of early PGMs (laurite, irarsite, etc.). Another process of PGM changing is the enlargement and agglomeration of PGE nanoparticles to the micro level. During progressive metamorphism (epidote-amphibolite and amphibolite facies) and/or a thermal event (introduction of granite intrusion), the changed P-T conditions affect the stability of nanoparticles. As the temperature rises to 590–650 °C, the nanoparticles in the sulfide matrix become unstable [122], which leads to their coalescence (Figure 10b) and enlargement (to micron sizes). The accessory mineralization of the crust-metamorphogenic stage is represented by the products of the changes in Ni, Fe, and Cu sulfides and sulfarsenides, with the appearance of oxygen-containing phases of a non-stoichiometric composition.

6. Conclusions

(1) High- and medium-Al′ chrome spinels were formed through the interaction of mantle peridotites with tholeiite melts in a spreading setting. High-Cr chrome spinels were formed

during the interaction of mantle peridotites with boninite melts in suprasubduction environments. The predominance of PGE sulfides over high-temperature Os-Ir-Ru alloys indicates their formation from S-saturated magma, which is typical of tholeiite melts.

(2) The formation temperatures of magmatic PGM–chromite association are estimated at 1000–1200 °C. The temperatures of olivine–spinel equilibrium, reflecting the formation of chromitites and tectonic deformation processes, range from 1000 to 740 °C, and the *log oxygen* fugacity $f\,(O_2)$ is low, ranging from (−0.76) to (−4.4), which indicates the upper mantle conditions, as well as the effect of reduced mantle fluids.

(3) Platinum-group mineralization in the Ulan-Sar'dag chromitites reflects a long history of formation and transformation, a change in the fluid conditions from magmatic to metamorphic ones. Primary PGMs (Os-Ir-Ru alloys-I, laurite I) were formed under the condition of a high fugacity of sulfur. The physical–chemical conditions were as follows: $T < 1200$–1300 °C, and $\log f\,(S_2) > (-2)/(-1)$. Under the influence of reduced fluids on chromitites, the desulfurization of laurite and the formation of secondary PGMs (native Os and Ru, Os-Ir-Ru alloys-II, laurite III, IrAsS, and RhNiAs) occur in association with serpentine, chlorite, nickel sulfides and arsenides. This association can be formed at

$T = 300$–700 °C, $\log f\,(S_2) = (-20)$, and P = 0.5 kbar.

(4) At the stage of ophiolite obduction and exposure to a metamorphogenic fluid, especially osmium, a further redistribution of PGE occurs. New phases of a non-stoichiometric composition (PGE + Cu, Te, Ba, As, Sb, O, and Cl) are formed. During progressive metamorphism, an enlargement and agglomeration of PGE nanoparticles to the microlevel occur.

Author Contributions: Conceptualization, O.N.K. and E.V.A.; methodology, D.K.B.; software, D.K.B. and E.V.A.; validation, S.M.Z. and E.V.A.; formal analysis, O.N.K.; investigation, O.N.K. and E.V.A.; resources, O.N.K. and S.M.Z.; data curation, O.N.K.; writing—original draft preparation, O.N.K. and E.V.A.; writing—review and editing, O.N.K., S.M.Z. and E.V.A.; visualization, E.V.A. and D.K.B.; supervision, S.M.Z.; project administration, O.N.K.; funding acquisition, O.N.K. All authors have read and agreed to the published version of the manuscript.

Acknowledgments: The work was carried out at the Analytical Center for multi-elemental and isotope research, SB RAS. The authors are grateful of I. Ashchepkov for constructive comments and suggestions when preparing an article.

References

1. Zhou, M.F.; Robinson, P.T. Origin and tectonic environment of podiform chromite deposits. *Econ. Geol.* **1997**, *92*, 259–262. [CrossRef]
2. Arai, S.; Matsukage, K. Petrology of a chromitite micropod from Hess Deep, equatorial Pacific: A comparison between abyssal and alpine-type podiform chromitites. *Lithos* **1998**, *43*, 1–14. [CrossRef]
3. Dönmez, C.; Keskin, S.; Günay, K.; Çolakoğlu, A.O.; Çiftçi, Y.; Uysal, İ.; Türkel, A.; Yıldırım, N. Chromite and PGE geochemistry of the Elekdağ Ophiolite (Kastamonu, Northern Turkey): Implications for deep magmatic processes in a supra-subduction zone setting. *Ore Geol. Rev.* **2014**, *57*, 216–228. [CrossRef]
4. Ahmed, A.H.; Arai, S. Platinum group minerals in podiform chromitites of the Oman ophiolite. *Can. Mineral.* **2003**, *41*, 597–616. [CrossRef]
5. Uysal, I.; Sadiklar, M.B.; Tarkian, M.; Karsli, O.; Aydin, F. Mineralogy and composition of the chromitites and their platinum-group minerals from Ortaca (Mugla-SW Turkey): Evidence for ophiolitic chromitite genesis. *Mineral. Petrol.* **2005**, *83*, 6–13. [CrossRef]
6. Thalhammer, O.A.R.; Prochaska, W.; Mühlhans, H.W. Solid inclusions in chromspinels and platinum group element concentration from the Hochgrössen and Kraubath Ultramafic Massifs (Austria). *Contrib. Mineral. Petrol.* **1990**, *105*, 66–80. [CrossRef]

7. Melcher, F.; Grum, W.; Simon, G.; Thalhammer, T.V.; Stumpfl, E.F. Petrogenesis of the ophiolitic giant chromite deposits of Kempirsai, Kazakhstan: A study of solid and fluid inclusions in chromite. *J. Petrol.* **1997**, *38*, 1419–1458. [CrossRef]
8. Garuti, G.; Fershtater, G.; Bea, F.; Montero, P.G.; Pushkarev, E.V.; Zaccarini, F. Platinum-group element distribution in mafic–ultramafic complexes of central and southern Urals: Preliminary results. *Tectonophysics* **1997**, *276*, 181–194. [CrossRef]
9. Zhou, M.F.; Sun, M.; Keays, R.R.; Kerrich, R.W. Controls on platinum-group elemental distribution of podiform chromitites: A case study of high-Cr and high-Al chromitites from chinese orogenic belts. *Geochimica et Cosmochim. Acta* **1998**, *62*, 677–688. [CrossRef]
10. Gervilla, F.; Proenza, J.A.; Frei, R.; González-Jiménez, J.M.; Garrido, C.J.; Melgarejo, J.C.; Meibom, A.; Díaz-martínez, R.; Lavaut, W. Distribution of platinum-group elements and Os isotopes in chromite ores from Mayarí-Baracoa Ophiolilte Belt (eastern Cuba). *Contrib. Mineral. Petrol.* **2005**, *150*, 589–607. [CrossRef]
11. Uysal, I.; Tarkian, M.; Sadıklar, M.B.; Sen, C. Platinum group-element geochemistry and mineralogy of ophiolitic chromitites from the Kop Mountains, Northeastern Turkey. *Can. Mineral.* **2007**, *45*, 355–377. [CrossRef]
12. Proenza, J.A.; Zaccarini, F.; Escayola, M.; Cábana, C.; Shalamuk, A.; Garuti, G. Composition and textures of chromite and platinum-group minerals in chromitites of the western ophiolitic belt from Córdoba Pampeans Ranges, Argentine. *Ore Geol. Rev.* **2008**, *33*, 32–48. [CrossRef]
13. Zaccarini, F.; Pushkarev, E.; Garuti, G. Platinum-group element mineralogy and geochemistry of chromitite of the Kluchevskoy ophiolite complex, central Urals (Russia). *Ore Geol. Rev.* **2008**, *33*, 20–30. [CrossRef]
14. Barnes, S.J.; Naldrett, A.J.; Gorton, M.P. The origin of the fractionation of platinum-group elements in terrestrial magmas. *Chem. Geol.* **1985**, *53*, 303–323. [CrossRef]
15. Maurel, C.; Maurel, P. Étude expérimentale de la distribution de l'aluminium entre bain silicaté basique et spinelle chromifère. Implications pétrogénétiques: Teneur en chrome des spinelles. *Bull. Minéralogie* **1982**, *105*, 197–202. [CrossRef]
16. Dick, H.J.B.; Bullen, T. Chromium-spinel as a petrogenetic indicator in abyssal and alpine-type peridotites and spatially associated lavas. *Contrib. Mineral. Petrol.* **1984**, *86*, 54–76. [CrossRef]
17. Gueddari, K.; Piboule, M.; Amosee, J. Differentiation of platinum-group elements (PGE) and of gold during partial melting of peridotites in the lherzolitic massifs of the Betico-Rifean range (Ronda and Beni Bousera). *Chem. Geol.* **1996**, *134*, 181–197. [CrossRef]
18. Kamenetsky, V.; Crawford, A.J.; Meffre, S. Factors controlling chemistry of magmatic spinel: An empirical study of associated olivine, Cr-spinel and melt inclusions from primitive rocks. *J. Petrol.* **2001**, *42*, 655–671. [CrossRef]
19. Ahmed, A.H.; Arai, S. Unexpectedly high-PGE chromitite from the deeper mantle section of the northern Oman ophiolite and its tectonic implications. *Contrib. Mineral. Petrol.* **2002**, *143*, 263–278. [CrossRef]
20. Ahmed, A.H. Diversity of platinum-group minerals in podiform chromitites of the late Proterozoic ophiolite, Eastern Desert, Egypt: Genetic implications. *Ore Geol. Rev.* **2007**, *33*, 31–45. [CrossRef]
21. Rollinson, H. The geochemistry of mantle chromitites from the northern part of the Oman ophiolite: Inferred parental melt composition. *Contrib. Mineral. Petrol.* **2008**, *156*, 273–288. [CrossRef]
22. Akmaz, R.M.; Uysal, I.; Saka, S. Compositional variations of chromite and solid inclusions in ophiolitic chromitites from the southeastern Turkey: Implications for chromitite genesis. *Ore Geol. Rev.* **2014**, *58*, 208–224. [CrossRef]
23. Ballhaus, C.; Berry, R.; Green, D. High pressure experimental calibration of the olivine-orthopyroxene-spinel oxygen geobarometer: Implication for the oxidation state of the upper mantle. *Contrib. Mineral. Petrol.* **1991**, *107*, 27–40. [CrossRef]
24. Andrews, D.R.A.; Brenan, J.M. Phase-equilibrium constraints of the magmatic origin of laurite and Os-Ir alloy. *Can. Mineral.* **2002**, *40*, 1705–1716. [CrossRef]
25. De Hoog, J.C.M.; Gall, L. Trace element geochemistry of mantle olivine and its application to geothermometry. *Ofioliti* **2005**, *30*, 182–183.
26. Korolyuk, V.N.; Pokhilenko, L.N. Electron probe determination of trace elements in olivine: Thermometry of depleted peridotites. *Russ. Geol. Geophys.* **2016**, *57*, 1750–1758. [CrossRef]
27. Kuznetsov, A.P.; Kukushkin, Y.N.; Makarov, D.F. The use of nickel matte as a collector of precious metals in the analysis of poor products. *J. Anal. Chem. USSR* **1974**, *29*, 2156–2160. (In Russian)

28. Beklemishev, M.K.; Kuzmin, N.M.; Zolotov, Y.A. Extraction and extraction-kinetic determination of Os using aza analogs of dibenzo-18-crown-6. *J. Anal. Chem. USSR* **1989**, *2*, 356–362. (In Russian)
29. Shlenskaya, V.I.; Khvostova, V.P.; Kadyrova, G.I. Kinetic methods for the determination of osmium and ruthenium (review). *J. Anal. Chem. USSR* **1973**, *28*, 779–784. (In Russian)
30. Rao, N.V.; Ravana, P.V. Kinetic-catalytic determination of osmium. *Mikrochim. Acta* **1981**, *76*, 269–276. [CrossRef]
31. Belichenko, V.G.; Butov, Y.P.; Boos, R.G.; Vratkovskaya, S.V.; Dobretsov, N.L.; Dolmatov, V.A.; Zhmodik, S.M.; Konnikov, E.G.; Kuzmin, M.I.; Medvedev, V.N.; et al. *Geology and Metamorphism of Eastern Sayan*; Nauka: Novosibirsk, Russia, 1988. (In Russian)
32. Dobretsov, N.L.; Konnikov, E.G.; Dobretsov, N.N. Precambrian ophiolitic belts of Southern Siberia (Russia) and their metallogeny. *Precambr. Res.* **1992**, *58*, 427–446. [CrossRef]
33. Zhmodik, S.; Kiseleva, O.; Belyanin, D.; Damdinov, B.; Airiyants, E.; Zhmodik, A. PGE mineralization in ophiolites of the southeast part of the Eastern Sayan (Russia). In Proceedings of the 12th International Platinum Symposium, Abstracts, Russia, 11–14 August 2014; Anikina, E.V., Ariskin, A.A., Barnes, S.-J., Barnes, S.J., Borisov, A.A., Evstigneeva, T.L., Kinnaird, J.A., Latypov, R.M., Li, C., Maier, W.D., et al., Eds.; Institute of Geology and Geochemistry UB RAS: Yekaterinburg, Russia, 2014; pp. 221–225.
34. Kuzmichev, A.B. *The Tectonic History of the Tuva–MongolianMassif: Early Baikalian, late Baikalian, and Early Caledonian Stages*; Probel Publishing House: Moscow, Russia, 2004; 192p. (In Russian)
35. Kuzmichev, A.B.; Larionov, A.N. Neoproterozoic island arcs of East Sayan: Duration of magmatism (from U-Pb zircon dating of volcanic clastics). *Russ. Geol. Geophys.* **2013**, *54*, 34–43. [CrossRef]
36. Kiseleva, O.N.; Zhmodik, S.M.; Damdinov, B.B.; Agafonov, L.V.; Belyanin, D.K. Composition and evolution of PGE mineralization in chromite ores from the Il'chir ophiolite complex (Ospa-Kitoi and Khara-Nur areas, East Sayan). *Russ. Geol. Geophys.* **2014**, *55*, 259–272. [CrossRef]
37. Sklyarov, E.V.; Kovach, V.P.; Kotov, A.B.; Kuzmichev, A.B.; Lavrenchuk, A.V.; Perelyaev, V.I.; Shipansky, A.A. Boninites and ophiolites: Problems of their relations and petrogenesis of boninites. *Geol. Geophys.* **2016**, *57*, 127–140. (In Russian) [CrossRef]
38. Kiseleva, O.; Zhmodik, S. PGE mineralization and melt composition of chromitites in Proterozoic ophiolite complexes of Eastern Sayan, Southern Siberia. *Geosci. Front.* **2017**, *8*, 721–731. [CrossRef]
39. Skopintsev, V.G. *Geological Structure and Mineral Resources of the Upper Rivers Gargan, Urik, Kitoy, Onot*; Results of prospecting works on the site of Kitoy (East Sayan); Report of the Samartin and Kitoy parties; Buryatia Publishing House: Ulan-Ude, Russia, 1995; Book 1; 319p. (In Russian)
40. Pavlov, N.V.; Kravchenko, G.G.; Chuprynina, I.I. *Chromites from the Kempirsai Pluton*; Nauka: Moscow, Russia, 1968. (In Russian)
41. Cabri, L.J. The platinum group minerals. In *The Geology, Geochemistry, Mineralogy and Mineral Beneficiation of Platinum Group Elements*; Published for the Geological Society of CIM; Canadian Institute of Mining, Metallurgy and Petroleum: Montreal, QC, Canada, 2002; Volume 54, pp. 13–131.
42. Kelemen, P.B. Reaction between ultramafic rock and fractionating basaltic magma I. Phase relations, the origin of the calcalkaline magma series, and the formation of discordant dunite. *J. Petrol.* **1990**, *31*, 51–98. [CrossRef]
43. Arai, S.; Yurimoto, H. Podiform chromitites from the Tari-Misaka ultramafic complex, southwestern Japan, as melt-mantle interaction products. *Econ. Geol.* **1994**, *89*, 1279–1288. [CrossRef]
44. Zhou, M.-F.; Robinson, P.T. High-chromium and high-aluminum podiform chromitites, western China: Relationship to partial melting and melt/rock interaction in the upper mantle. *Int. Geol. Rev.* **1994**, *36*, 678–686. [CrossRef]
45. Zhou, M.-F.; Robinson, P.; Malpas, J.; Li, Z. Podiform chromites in the Luobusa Ophiolite (Southern Tibet): Implications for melt-rock interaction and chromite segregation in the upper mantle. *J. Petrol.* **1996**, *37*, 3–21. [CrossRef]
46. Arai, S. Control of wall-rock composition on the formation of podiform chromitites as a result of magma/peridotite interaction. *Resour. Geol.* **1997**, *47*, 177–187.
47. Ahmed, A.H.; Harbi, H.M.; Habtoor, A.M. Compositional variations and tectonic settings of podiform chromitites and associated ultramafic rocks of the Neoproterozoic ophiolite at Wadi Al Hwanet, northwestern Saudi Arabia. *J. Asian Earth Sci.* **2012**, *56*, 118–134. [CrossRef]

48. Prichard, H.M.; Lord, R.A.; Neary, C.R. A model to explain the occurrence of platinum- and palladium- rich ophiolite complexes. *J. Geol. Soc.* **1996**, *153*, 323–328. [CrossRef]
49. Tsoupas, G.; Economou-Eliopoulos, M. High PGE contents and extremaly abundant PGE-minerals hosted in chromitites from Veria ophiolite complex, northern Greece. *Ore Geol. Rev.* **2008**, *33*, 3–19. [CrossRef]
50. O'Driscoll, B.; Day, J.M.D.; Walker, R.J.; Daly, J.S.; McDonough, W.F.; Piccoli, P.M. Chemical heterogeneity in the upper mantle recorded by peridotites and chromitites from the Shetland Ophiolite Complex. Scotland. *Earth Planet. Sci. Lett.* **2012**, *333*, 226–237. [CrossRef]
51. Gurskaya, L.I.; Smelova, L.V.; Kolbantsev, L.R.; Lyakhnitskaya, V.D.; Lyakhnitsky, Y.S.; Shakhova, S.N. *Platinoids of Chromite-Bearing Massifs of the Polar Urals*; Publishing House SPb Card Factory VSEGEI: St. Petersburg, Russia, 2005; 306p. (In Russian)
52. Arai, S. Characterization of spinel peridotites by olivine—Spinel compositional relationships: Review and interpretation. *Chem. Geol.* **1994**, *113*, 191–204. [CrossRef]
53. Leblanc, M.; Violette, J.F. Distribution of Aluminium-rich and Chromium-rich chromite pods in ophiolite peridotites. *Econ. Geol.* **1983**, *78*, 293–301. [CrossRef]
54. Zhou, M.F.; Bai, W.J. Chromite deposits in China and their origin. *Miner. Depos.* **1992**, *27*, 192–199. [CrossRef]
55. Leblanc, M. Chromite and ultramafic rock compositional zoning through a paleotransform fault, Poum, New Caledonia. *Econ. Geol.* **1995**, *90*, 2028–2039. [CrossRef]
56. Graham, I.T.; Franklin, B.J.; Marshall, B. Chemistry and mineralogy of podiform chromitite deposits, southern NSW, Australia: A guide to their origin and evolution. *Mineral. Petrol.* **1996**, *37*, 129–150. [CrossRef]
57. Economou-Ellopoulos, M. Platinum-group element distribution in chromite ores from ophiolite complexes: Implications for their exploration. *Ore Geol. Rev.* **1996**, *11*, 363–381. [CrossRef]
58. Proenza, J.; Gervilla, F.; Melgarejo, J.C.; Bodinier, J.L. Al-and Cr-rich chromitites from the Mayari–Baracoa Ophiolitic Belt (Eastern Cuba): Consequence of interaction between volatile-rich melts and peridotite in suprasubduction mantle. *Econ. Geol.* **1999**, *94*, 547–566. [CrossRef]
59. Ahmed, A.H.; Arai, S.; Attia, A.K. Petrological characteristics of podiform chromitites and associated peridotites of the Pan African Proterozoic ophiolite complexes of Egypt. *Miner. Depos.* **2001**, *36*, 72–84. [CrossRef]
60. Thayer, P.T. Principal features and origin of podiform chromite deposits, and some observations on the Guleman-Soridag district, Turkey. *Econ. Geol.* **1964**, *59*, 1497–1524. [CrossRef]
61. Thayer, T.P. *Chromite Segregations as Petrogenetic Indicators*; 1 (Special Publications); The Geological Society of South Africa: Johannesburg, South Africa, 1970; pp. 380–389.
62. Uysal, I.; Tarkian, M.; Sadiklar, M.B.; Zaccarini, F.; Meisel, T.; Garuti, G.; Heidrich, S. Petrology of Al- and Cr-rich ophiolitic chromitites from the Muğla, SW Turkey: Implications from composition of chromite, solid inclusions of platinum-group mineral, silicate, and base-metal mineral, and Os-isotope geochemistry. *Contrib. Mineral. Petrol.* **2009**, *158*, 659–674. [CrossRef]
63. Xiong, F.; Yang, J.; Liu, Z.; Guo, G.; Chen, S.; Xu, X.; Li, Y.; Liu, F. High-Cr and high-Al chromitite found in western Yarlung-Zangbo suture zone in Tibet. *Acta Petrol. Sin.* **2013**, *29*, 1878–1908.
64. González-Jiménez, J.M.; Proenza, J.A.; Gervilla, F.; Melgarejo, J.C.; Blanco-Moreno, J.A.; Ruiz-Sánchez, R.; Griffin, W.L. High-Cr and high-Al chromitites from the Sagua de Tánamo district, Mayarí-Cristal ophiolitic massif (eastern Cuba): Constraints on their origin from mineralogy and geochemistry of chromian spinel and platinum-group elements. *Lithos* **2011**, *125*, 101–121. [CrossRef]
65. Kelemen, P.B.; Shimizu, N.; Salters, V.J.M. Extraction of mid-ocean-ridge basalt from the upwelling mantle by focused flow of melt in dunite channels. *Nature* **1995**, *375*, 747–753. [CrossRef]
66. Arai, S. Role of dunite in genesis of primitive MORB. *Proc. Jpn. Acad.* **2005**, *B 81*, 14–19. [CrossRef]
67. Arai, S.; Miura, M. Podiform chromitites do form beneath mid-ocean ridges. *Lithos* **2015**, *232*, 143–149. [CrossRef]
68. Kiseleva, O.N.; Airiyants, E.V.; Belyanin, D.K.; Zhmodik, S.M. *Geochemical Features of Peridotites and Volcanogenic-Sedimentary Rocks of the Ultrabasic-Basitic Massif of Ulan-Sar'dag (East Sayan, Russia)*; The Bulletin of Irkutsk State University: Irkutsk, Russia, 2019. (In Russian)
69. Wilson, M. *Igneous Petrogenesis*; Unwin Hyman: London, UK, 1989.
70. Jonson Kevin, T.M.; Dick Henry, J.B. Open System Melting and Temporal and Spatial Variation of Peridotite and Basalt at the Atlantis II Fracture Zone. *J. Geophys. Res.* **1992**, *97*, 9219–9241. [CrossRef]

71. Garuti, G.; Pushkarev, E.V.; Thalhammer, O.A.R.; Zaccarini, F. Chromitites of the Urals (Part 1): Overview of chromite mineral chemistry and geotectonic setting. *Ofioliti* **2012**, *37*, 27–53.
72. Jenner, G.A. Geochemistry of high-Mg andesites from Cape Vogel, Papua New Guinea. *Chem. Geol.* **1981**, *33*, 307–332. [CrossRef]
73. Kamenetsky, V.S.; Sobolev, A.V.; Eggins, S.M.; Crawford, A.J.; Arculus, R.J. Olivine enriched melt inclusions in chromites from low-Ca boninites, Cape Vogel, Papua New Guinea: Evidence for ultramafic primary magma, refractory mantle source and enriched components. *Chem. Geol.* **2002**, *83*, 287–303. [CrossRef]
74. Walker, D.A.; Cameron, W.E. Boninite primary magmas: Evidence from the Cape Vogel Peninsula, PNG. *Contrib. Mineral. Petrol.* **1983**, *83*, 150–158. [CrossRef]
75. Cameron, W.E. Petrology and origin of primitive lavas from the Troodos ophiolite, Cyprus. *Contrib. Mineral. Petrol.* **1985**, *89*, 239–255. [CrossRef]
76. Flower, M.F.J.; Levine, H.M. Petrogenesis of a tholeiite–boninite sequence from Ayios Mamas, Troodos ophiolite: Evidence for splitting of a volcanic arc? *Contrib. Mineral. Petrol.* **1987**, *97*, 509–524. [CrossRef]
77. Page, P.; Barnes, S.J. Using trace elements in chromites to constrain the origin of podiform chromitites in the Thetford Mines Ophiolite, Québec, Canada. *Econ. Geol.* **2009**, *104*, 997–1018. [CrossRef]
78. Shibata, T.; Thompson, G.; Frey, F.A. Tholeiitic and alkali basalts from the mid Atlantic ridge at 43° N. *Contrib. Mineral. Petrol.* **1979**, *70*, 127–141. [CrossRef]
79. Le Roex, A.P.; Dick, H.J.B.; Gulen, L.; Reid, A.M.; Erlank, A.J. Local and regional heterogeneity in MORB from the mid-Atlantic ridge between 54.5° S and 51° S: Evidence for geochemical enrichment. *Geochim. Cosmochim. Acta* **1987**, *51*, 541–555. [CrossRef]
80. Presnall, D.C.; Hoover, J.D. High pressure phase equilibrium constraints on the origin of mid-ocean ridge basalts. *Geochem. Soc. Spec. Pap.* **1987**, *1*, 75–89.
81. Rollinson, H. Chromite in the mantle section of the Oman ophiolite: A new genetic model. *Isl. Arc* **2005**, *14*, 542–550. [CrossRef]
82. Barnes, S.J.; Kunilov, V.Y. Spinels and Mg ilmenites from the Noril'sk and Talnakh intrusions and other mafic rocks of the Siberian flood basalt province. *Econ. Geol.* **2000**, *95*, 1701–1717. [CrossRef]
83. Burns, L.E. The Borger Range ultramafic and mafic complex, south-central Alaska: Cumulative fractionates of island-arc volcanics. *Can. J. Earth Sci.* **1985**, *22*, 1020–1038. [CrossRef]
84. Volchenko, Y.A.; Ivanov, K.S.; Koroteev, V.A.; Auge, T. Structural-substantial evolution of the Urals platiniferous belt's complexes in the time of Uralian type chromite-platinum deposits formation. Part I. *Lithosphere* **2007**, *3*, 3–27.
85. Irvine, T.N. Chromian spinel as a petrogenetic indicator: Part II. Petrologic applications. *Can. J. Earth Sci.* **1967**, *4*, 71–103. [CrossRef]
86. Henry, D.; Medaris, L. Application of pyroxene and olivine-spinel geothermometers to spinel peridotites in South-western Oregon. *Am. J. Sci.* **1980**, *280*, 211–231.
87. Bedard, J.H. A new projection scheme and differentiation index for Cr-spinels. *Lithos* **1997**, *42*, 37–45. [CrossRef]
88. Ashchepkov, I.V. Program of the mantle thermometers and barometers: Usage for reconstructions and calibration of PT methods. *Vestn. Otd. Nauk Zemle* **2011**, *3*, NZ6008. [CrossRef]
89. Ashchepkov, I.V.; Pokhilenko, N.P.; Vladykin, N.V.; Rotman, A.Y.; Afanasiev, V.P.; Logvinova, A.M.; Kostrovitsky, S.I.; Pokhilenko, L.N.; Karpenko, M.A.; Kuligin, S.S.; et al. Reconstruction of mantle sections beneath Yakutian kimberlite pipes using monomineral thermobarometry. *Geol. Soc. Spec. Publ.* **2008**, *293*, 335–352. [CrossRef]
90. O'Neill, H.S.C.; Wal, V.J. The olivine orthopyroxene-spinel oxygen geobarometer, the nickel precipitation curve, and the oxygen fugacity of the Earth's upper mantle. *J. Petrol.* **1987**, *28*, 1169–1191. [CrossRef]
91. Taylor, W.R.; Kammerman, M.; Hamilton, R. New thermometer and oxygen fugacity sensor calibrations for ilmenite and chromium spinel-bearing peridotitic assemblages. In Proceedings of the 7th International Kimberlite Conference, Cape Town, South Africa, 11–17 April 1998; Extended Abstracts. pp. 891–901.
92. Lorand, J.P.; Keays, R.R.; Bodiner, J.R. Copper- and noble metal enrichment across the asthenosphere-lithosphere mantle diapiris: The Lanzo lherzolite massif. *J. Petrol.* **1993**, *34*, 1111–1140. [CrossRef]

93. Brugmann, G.E.; Armdt, N.T.; Hoffmann, A.W.; Tobschall, H.J. Nobel metal abundances in Komatiite suites from Alexo, Ontario and Gorgona Island, Colombia. *Geochim. Cosmochim. Acta* **1987**, *51*, 2159–2169. [CrossRef]
94. Wang, K.-L.; Chu, Z.; Gornova, M.A.; Dril, S.; Belyaev, V.A.; Lin, K.-Y.; O'Reilly, S.Y. Depleted SSZ type mantle peridotites in Proterozoic Eastern Sayan ophiolotes in Siberia. *Geodyn. Tectonophys.* **2017**, *8*, 583–587. [CrossRef]
95. Agafonov, L.V.; Lkhamsuren, J.; Kuzhuget, K.S.; Oidup, C.K.B. *Platinum-Group Element Mineralization of Ultramafic-Mafic Rocks in Mongolia and Tuva*; Tomurtogoo, O., Ed.; Ulaanbaatar Publishing House: Ulaanbaatar, Mongolia, 2005. (In Russian)
96. Page, N.J.; Engin, T.; Singer, D.A.; Haffty, J. Distribution of platinum-group elements in the Bati Kef chromite deposit, Güleman-Elaziğ area, Eastern Turkey. *Econ. Geol.* **1984**, *79*, 177–184. [CrossRef]
97. Yaman, S.; Ohnenstetter, M. Distribution of platinum-group elements of chromite deposits within ultramafic zone of Mersin ophiolite (south Turkey). *Bull. Geol. Congr. Turk.* **1991**, *6*, 253–261.
98. Garuti, G.; Pushkarev, E.V.; Zaccarini, F. Diversity of chromite-PGE mineralization in ultramafic complexes of the Urals. In Proceedings of the Platinum-Group Elements—From Genesis to Beneficiation and Environmental Impact: 10th International Platinum Symposium, Oulu, Finland, 8–11 August 2005; Geological Survey of Finland: Esbo, Finland, 2005.
99. Ballhaus, C.; Sylvester, P. PGE enrichment processes in the Merensky reef. *J. Petrol.* **2000**, *41*, 454–561. [CrossRef]
100. Matveev, S.; Ballhaus, C. Role of water in the origin of podiform chromititedeposits. *Earth Planet. Sci. Lett.* **2002**, *203*, 235–243. [CrossRef]
101. Sattari, P.; Brenan, J.M.; Horn, I.; McDonough, W.F. Experimental constraints in the sulfide-and chromite-silicate melt partitioning behaviour of rhenium and platinum-group elements. *Ecol. Geol.* **2002**, *97*, 385–398. [CrossRef]
102. Fleet, M.E.; Crocket, J.H.; Lin, M.H.; Stone, W.E. Laboratory partitioning of platinum-group elements and gold with application to magmatic sulfide-PGE deposits. *Lithos* **1999**, *47*, 127–142. [CrossRef]
103. Kiseleva, O.N.; Airiyants, E.V.; Belyanin, D.K.; Zhmodik, S.M. Geochemical and mineralogical indicators (Cr-spinelides, Platinum Group Minerals) of the geodynamic settings of formation of maficultramafic Ulan Saridag massif (Eastern Sayan). In Proceedings of the EGU General Assembly Conference Abstracts, Vienna, Austria, 8–13 April 2018.
104. Prichard, H.M.; Tarkian, M. Platinum and palladium minerals from two PGElocalities in the Shetland ophiolite complex. *Can. Mineral.* **1988**, *26*, 979–990.
105. Garuti, G.; Zaccarini, F.; Economou-Eliopoulos, M. Paragenesis and composition of laurite from chromitites of Othrys (Greece): Implications for Os-Ru fractionation in ophiolitic upper mantle of the Balkan peninsula. *Miner. Depos.* **1999**, *34*, 312–319. [CrossRef]
106. Brenan, J.M.; Andrews, D. High-temperature stability of laurite and Ru-Os-Ir alloy and their role in PGE fractionation in mafic magmas. *Can. Mineral.* **2001**, *39*, 341–360. [CrossRef]
107. Ballhaus, C.; Bockrath, C.; Wohlgemuth-Ueberwasser, C.; Laurenz, V.; Berndt, J. Fractionation of the noble metals by physical processes. *Contrib. Mineral. Petrol.* **2006**, *152*, 667–684. [CrossRef]
108. Bockrath, C.; Ballhaus, C.; Holzheid, A. Stabilities of laurite RuS_2 and monosulphide liquid solution atmagmatic temperature. *Chem. Geol.* **2004**, *208*, 265–271. [CrossRef]
109. Finnigan, C.S.; Brenan, J.M.; Mungall, J.E.; McDonough, W.F. Experiments and models bearing on the role of chromite as a collector of platinum group minerals by local reduction. *J. Petrol.* **2008**, *49*, 1647–1665. [CrossRef]
110. Dick, H.J.B. Terrestrial nickel–iron from the josephinite peridotite, its geologic occurrence, associations and origin. *Earth Planet. Sci. Lett.* **1974**, *24*, 291–298. [CrossRef]
111. Zaccarini, F.; Proenza, J.A.; Ortega-Gutiérrez, F.; Garuti, G. Platinum group minerals in ophioliticchromitites from Tehuitzingo (Acatlán complex, southern Mexico): Implications for postmagmatic modification. *Miner. Petrol.* **2005**, *84*, 147–168. [CrossRef]
112. Garuti, G.; Proenza, J.A.; Zaccarini, F. Distribution and mineralogy of platinum-group elements in altered chromitites of the Campo Formoso layered intrusiyn (Bahia State, Brazil): Control by magmatic and hydrothermal processes. *Miner. Petrol.* **2007**, *89*, 159–188. [CrossRef]

113. Stockman, H.W.; Hlava, P.F. Platinum-group minerals in Alpine chromitites from southwestern Oregon. *Econ Geol.* **1984**, *79*, 492–508. [CrossRef]
114. Nilsson, L.P. Platinum-group mineral inclusions in chromitite from Osthammeren ultramafic tectonite body, south central Norway. *Mineral. Petrol.* **1990**, *42*, 249–263. [CrossRef]
115. Bowles, J.F.W.; Gize, A.P.; Vaughan, D.J.; Norris, S.J. Development of platinum-group minerals in laterites—Initial comparison of organic and inorganic controls. *Trans. Inst. Min. Metall. (Sect. B Appl. Earth Sci.)* **1994**, *103*, 53–56.
116. Garuti, G.; Zaccarini, F. In situ alteration of platinum-group minerals at low temperature: Evidence from serpentinized and weathered chromitite of the Vourinos complex, Greece. *Can. Mineral.* **1997**, *35*, 611–626.
117. Bai, W.; Robinson, P.T.; Fang, Q.; Yang, J.; Yan, B.; Zhang, Z.; Hu, X.-F.; Zhou, M.-F.; Malpas, J. The PGE and base-metal alloys in the podiform chromitites of the Luobusa ophiolite, southern Tibet. *Can. Mineral.* **2000**, *38*, 585–598. [CrossRef]
118. Evans, B.V.; Hattori, K.; Barronet, A. Serpentinite: What, Why, Where? *Elements* **2013**, *9*, 99–106. [CrossRef]
119. Bowles, J.F.W. The development of platinum-group minerals in laterites. *Econ. Geol.* **1986**, *81*, 1278–1285. [CrossRef]
120. Bowles, J.F.W.; Lyon, J.C.; Saxton, J.M.; Vaughan, D.J. The origin of Platinum Group Minerals from the Freetown intrusions, Sierra Leone, inferred from osmium isotope systematics. *Econ. Geol.* **2000**, *95*, 539–548. [CrossRef]
121. Xiong, Y.; Wood, A. Experimental quantifycation of hydrothermal solubility of platinum- group elements with special reference to porphyry copper environments. *Mineral. Petrol.* **2000**, *68*, 1–28. [CrossRef]
122. González-Jiménez, J.M.; Reich, M.; Camprubí, T.; Gervilla, F.; Griffin, W.L.; Colás, V.; O'Reilly, S.Y.; Proenza, J.A.; Pearson, N.J.; Centeno-García, E. Thermal metamorphism of mantle chromites and the stability of noble-metal nanoparticles. *Contrib. Mineral. Petrol.* **2015**, *170*, 15. [CrossRef]

11

Grain Size Distribution and Clay Mineral Distinction of Rare Earth Ore through Different Methods

Lingkang Chen [1,2,3,*], Xiongwei Jin [3], Haixia Chen [2], Zhengwei He [1,4,*], Lanrong Qiu [3] and Hurong Duan [5]

1 College of Earth Sciences, Chengdu University of Technology, Chengdu 610059, China
2 College of Sciences, Guangdong University of Petrochemical Technology, Maoming 525000, China; chenhaixia1975@126.com
3 School of Resource and Environmental Engineering, Jiangxi University of Science and Technology, Ganzhou 341000, China; xiongweijjx@126.com (X.J.); lrqiu183@163.com (L.Q.)
4 State Key Laboratory of Geohazard Prevention and Geoenvironment Protection, Chengdu University of Technology, Chengdu 610059, China
5 College of Geomatics, Xi'an University of Science and Technology, Xi'an 710054, China; duanhurong@126.com
* Correspondence: lkchen@jxust.edu.cn (L.C.); hzw@cdut.edu.cn (Z.H.)

Abstract: Although clay mineral content in ion-absorbed rare earth ores is crucial for migrating and releasing rare earth elements, the formation, distribution, and migration of clay minerals in supergene rare earth ores have not been fully understood. Therefore, this study analyzes the characteristics of clay mineral type and content, soil particle size, pH value, leaching solution concentration, and leaching rate. This analysis was performed using different methods, such as regional rare earth mine soil surveys, in situ leaching profile monitoring, and indoor simulated leaching. The results showed that the grain size and volume curve of rare earth ore have unimodal and bimodal shapes, respectively. X-ray diffraction showed the differences in clay mineral types formed by different weathered bedrocks. The principal clay minerals were kaolinite, illite, chlorite, and vermiculite, with their relative abundance varying with parent rock lithology (granite and low-grade metamorphic rocks). In the Ganxian granite weathering profile, the kaolinite content increased from top to bottom. The decomposition of feldspar minerals to kaolinite was enhanced with an increase in the SiO_2 content during weathering. The in situ leaching profile analysis showed that the kaolinite content increased initially and then decreased, whereas the illite/mica content exhibited the opposite trend. Under stable leaching solution concentration and leaching rate, clay mineral formation is favored by lower pH. Low pH, low leaching rate, and highly-concentrated leaching solution (12 wt%) resulted in a slow increase in kaolinite content in the upper part of the profile (30 cm). A lower concentration of the leaching solution (4 wt%) resulted in rapid enrichment of kaolinite after 15 days. Low pH, leaching solution concentration, and leaching rate promoted the formation of distinct kaolinite horizons. We suggest that by disregarding other control factors, rare earth recovery of over 90% can be achieved through leach mining with solutions of 8 wt% and a pH of 5 at a leaching rate of 5 mL/min.

Keywords: clay minerals; grain size characteristics; in situ leaching; simulated leaching; ion-absorbed type rare earth ore

---

1. Introduction

Rare earth elements (REEs) are 16 chemical elements grouped by their atomic number, and classified as light (LREEs), middle (MREEs), and heavy (HREEs). The weathered crust elution-deposited REE ores in southern China have drawn much attention because of their abundance of granitic residuum, their simple extraction processes, and their well-distributed composition [1–3].

The migration and enrichment of REEs are controlled by several factors, such as parent rock lithology, pH value, intensity of weathering, and topography [4–6]. Previous studies have shown that, for chemical index of alteration (CIA) values of 65%–85% in granite, clay minerals increase rapidly with an increasing degree of weathering. There is a positive correlation between the loss on ignition (LOI) of 2%–6% in the weathering crust and REE content [7].

Clay minerals have a controlling effect on the migration and release of REE ore. The completely weathered layer of a weathering crust mainly comprises quartz, feldspar, and clay minerals. The clay mineral content decreases gradually from the weathering crust surface to the lower layer, where clay minerals are converted from hydromica and montmorillonite to halloysite, kaolinite, and gibbsite [8].

The distributions of REEs in the weathering crust are controlled by both the composition of the parent rock and the clay mineral content of the weathering crust. Halloysite, a clay mineral, plays a significant role in the differentiation of cerium [9]. Halloysite has a stronger effect than kaolinite in the adsorption of REEs; however, this adsorption mechanism is not yet fully understood. Previous studies found that the adsorption of REEs is controlled by the properties of the clay minerals rather than the electrolyte solution or dissolved carbon dioxide content [10]. The adsorption capacity of kaolinite increases linearly with increasing pH. A fractionation between HREEs and LREEs due to selective sorption is observed, with HREEs being more sorbed than LREEs at high ionic strengths [10]. For montmorillonite at pHs below 4.5, the REE adsorption capacity is constant, and is modeled by cation exchange [11]. Different clay minerals have different adsorption capacities for REEs. Chi et al. [12] showed that for three common clay minerals, the cation adsorption capacity follows the order: montmorillonite > halloysite > kaolinite. This result shows that different parent rock lithologies will result in different weathering crust structures and clay mineral compositions. Intimate grain-to-grain contacts promote a unique chemical environment at the microscale, bringing about the formation of transient clay mineral phases which quickly disappear in the overlying soil [13]. The bulk of illite in the weathering crust is due to the weathering of mica minerals. A study of unstable soil profiles found that illite is converted into vermiculites or interstratified illite-smectite [14].

Climatic and environmental change is one of the causes of compositional differentiation in clay minerals. Kaolinite and kaolinite interlayer minerals are dominant in strongly leached soil layers [15], while illite and montmorillonite represent a cold and humid climate with weak chemical weathering [16]. Clay minerals of different crystal characteristics differ in physical structure and properties [13]. Clay minerals that host ion-adsorbed REE ores have large specific surface areas and a strong capacity to adsorb REE ions. The clay mineral content thus controls the migration and enrichment of REEs—processes of great significance for REE mineralization.

Although the clay mineralogy of weathered crust elution-deposited REE ores varies, several studies have demonstrated that the clay minerals in these ores commonly comprise halloysite, kaolinite, some illite, and rare montmorillonite [17]. It is widely believed that the horizon enriched in REE generally contains abundant halloysite and kaolinite [18,19], and that clay mineral migration is controlled by soil particle size and specific leaching conditions [20,21]. The metallogenetic mechanism of weathered crust elution REE deposits could involve the weathering of granodiorite and volcanic rocks in warm and humid climates, with the transformation of their parent mineralogy into kaolinite, halloysite, and montmorillonite [22]. In weathering crust elution-deposited REE ores, REEs adsorbed on the clay minerals by ion-exchangeable phases account for more than 80% of the total REE content [3]. However, leaching is controlled by the properties of the REE ore, by the nature and concentration of the leaching reagent, and by the hydrodynamics, kinetics, and mass transfer of the leaching process [22]. We postulate that the weathered crust elution-deposit REE ore is associated with REE ion enrichment, which is dissociated with hydrated or hydroxyl hydrated minerals and adsorbed by clay minerals, which are subsequently deposited, and mineralized in the weathered crust over a long period. In contrast, this is not to say that all REE mineralization can be explained by a single model.

It is important to understand that the clay mineralogy in different environments of REE ore formation varies with different conditions of parent rock, pH values, degrees of weathering [4,5], and mining conditions [9]. This study investigates the types and changing characteristics of clay minerals in several ion-absorbed REE ores in southern Jiangxi Province, China, during weathering and in situ leaching, with an aim of improving the recovery rate. To achieve this, soils on the surface of the weathered crust in a typical rare earth mining area in southern Jiangxi Province were sampled. Then, methods such as in situ leaching profile monitoring and indoor leaching simulation experiments were used to study the characteristics of the clay mineral properties and soil particle size.

2. Background

2.1. Study Site

The study sites are located in the REE mining regions of Longnan County, Anyuan County, Ganxian District, Ganzhou, Jiangxi Province, China (Figure 1). The region is situated in the subtropical monsoon climatic zone, with an average annual precipitation of 1461.2 mm and an annual average temperature of 19.4 °C [23]. The topography is high in the south and east and low in the north and west [23]. The central region consists mostly of basins between hills. REE mines are mainly distributed in these basins.

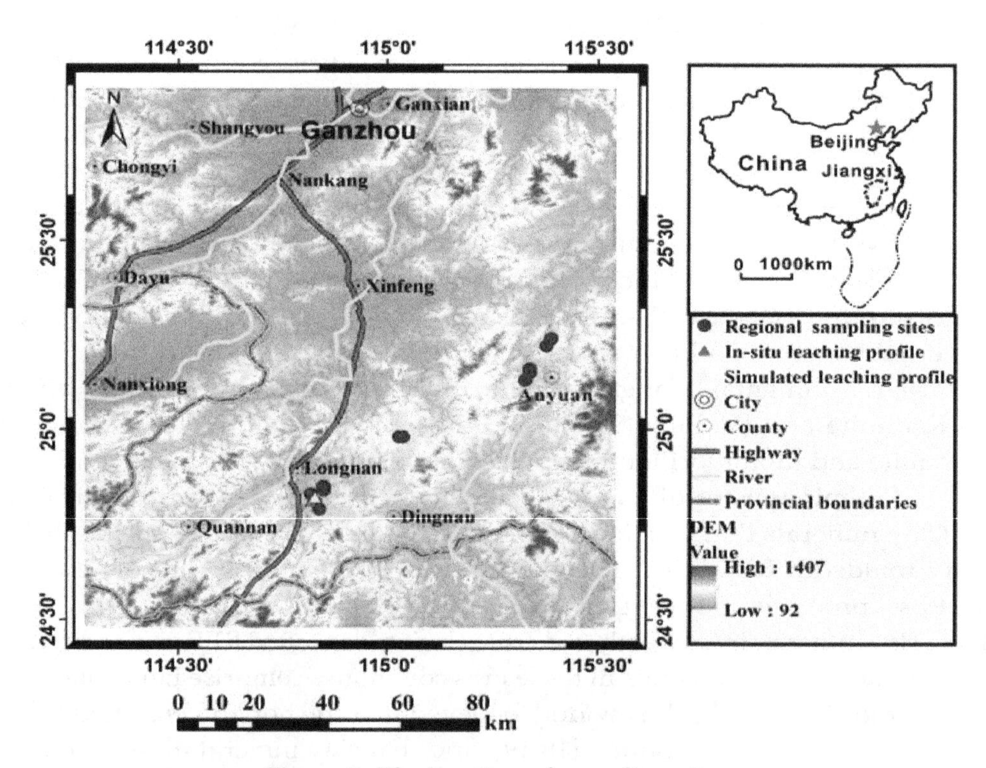

Figure 1. The location of sampling sites.

2.2. Regional Stratigraphy and Lithology

The study area mainly comprises sandstones, slates, phyllites, and carbonates of the Lower Paleozoic (Sinian) to Mesozoic ages, covered by Quaternary sediments, except for the Silurian, Ordovician, and Tertiary. Quaternary sediments consist of loose fluvial deposits in a river terrace with a high ratio [24].

Magmatic activity in the area can be resolved into four cycles: Caledonian, Variscan, Indosinian, and Yanshanian. The main lithology of igneous intrusions is an acidic to medium-acidic rock [25]. These three types of magmatic rock account for 99.2% of all magmatic rocks in the region. This corresponds to the diagenesis times of 461–384, 228–242, and 189–94 of REE ore-bearing rocks, respectively, as given in [26].

3. Materials and Methods

3.1. Sample Collection

A total of 49 surface soil samples were collected from REE mines; among these, 27 were sampled regionally, mainly from Anyuan, Longnan, and Dingnan. Ten profile samples were also collected from Jiangwozi, in Ganxian District [27] and from the Wenlong REE mine in Longnan County (longitude 114°49′31.99″E; latitude 24°49′21.68″N).

The in situ leaching profile of the Wenlong REE mine was sampled on three different occasions, October 2, 2016; January 13, 2017; and April 13, 2017, over 193 days (12 samples in total). These samples were collected at elevations from 255 to 249 m and at depths from 0.5 to 1.5 m from the surface; the location was mapped with GPS. The sampled weathering profile comprises, from the surface to the bedrock, purple clayey-sandy soil, red silty clay, gray yellow clayey- sand. The underlying bedrock lithology comprises siliceous slate with high organic content, fine sandstone, grayish-white medium-crystalline granodiorite, fine-medium biotite granite, and grayish-white medium-crystalline granodiorite.

3.2. Simulated Leaching Experiment

An unexploited weathering crust with soil horizons similar to those of the in situ leaching profile (i.e., completely weathered upper layer; the transitional middle layer; only partly weathered bottom layer) was selected as representative of the textural and structural characteristics of the typical ion-absorbed REE profile. In the laboratory, a custom column device consisting of a liquid injection barrel, soil column tube, and liquid collection basin was used for the leaching experiment.

The soil column tube used a PVC pipe with a height of 120 cm and inner diameter of 11 cm. An array of sampling holes located at 30, 50, 70, 90, and 110 cm from top to bottom were punched through the pipe wall. The tube bottom was sealed by a lid with a floor drain and plug. To prevent leaching of the sample with the liquid discharge, a paper filter of cotton fiber with 11 cm diameter, similar to the PVC pipe, was laid on the floor of the drain, and the bottom lid was perforated and connected to a plastic hose to receive the leaching ore concentrate. The field profile was sampled following its pedostratigraphy, and the samples were placed in the soil column tube to recreate this stratigraphy, from bottom to top layers. Deionized water spraying was used regularly to ensure that the water content and water holding rate in the soil column sample were approximately consistent with those in the field profile. A total of eight soil columns (T1–T8) were made, each tube was filled with 100 cm of soil, and filter paper with a diameter of 11 cm was placed on the top. The leaching solution was made from analytically pure $(NH_4)_2SO_4$ crystals dissolved in deionized water (from AK-RO-UP-500), to simulate leaching and rainfall. HCl was used to adjust the pH value of the solution. The leaching time was 40 days, with samples taken on the 5th, 15th, 23rd, and 40th day. The machine standard was RO < 0.7 mV and UP < 20 MΩ. The specific leaching parameters of each soil column are shown in Table 1·

Table 1. Experimental parameters of simulated leaching.

Soil Column No.	Leaching Solution	pH	Content of $(NH_4)_2SO_4$ (wt%)	Leaching Solution Flow Rate (mL/min)
T1	$(NH_4)_2SO_4$	5	8	3
T2	$(NH_4)_2SO_4$	3	8	3
T3	$(NH_4)_2SO_4$	4	8	3
T4	$(NH_4)_2SO_4$	5	12	3
T5	$(NH_4)_2SO_4$	5	4	3
T6	$(NH_4)_2SO_4$	5	8	1
T7	$(NH_4)_2SO_4$	5	8	5
T8 (Simulated rainfall)	Deionized water	6.8~7.2	0	3

3.3. Particle size Analysis

The particle size analysis was conducted using different analytical testing methods. Regionally collected samples (27) were processed by the wet sieving method [27]. Particles of less than 0.075 mm were tested on an LS908 (A) laser particle size analyzer (Henan Zhengzhou North-south Instrument Equipment Co. LTD, Zhengzhou, China) at the Jiangxi University of Science and Technology School of Resource and Environmental Engineering. The coarser (>0.075 mm) fraction of the in situ leaching profile was also analyzed by wet sieving, while the <0.075 mm fraction was analyzed with a Malvern MasterSizer 2000 laser particle size analyzer at the Peking University School of Urban and Environmental Sciences.

3.4. Clay Mineral Analysis

The XRD analysis of clay minerals followed standard procedures, as described in [28,29]. Bulk samples were pulverized to a fine powder using a planetary ball mill with agate elements. Specimens for XRD analysis were front-loaded using a blade; sieve rotation ensured random grain orientation. The clay fraction was separated in deionized water; the clay suspension was then deposited onto 0.45 μm Whatman filters in vacuum, and transferred to glass slides. Each concentrated clay sample was air-dried before XRD analysis, and then saturated with ethylene glycol for subsequent analysis. Occasionally, heat treatment was necessary; in this case, the slides were heated for one hour at 550 °C before further XRD analysis. Analyses were performed with Rigaku D/max 2550 XRD at the Oil and Gas Laboratory in ALS Houston, USA, and at the Key Laboratory of Nonferrous Metal Materials Science and Engineering of the Ministry of Education, in the Central South University, China. A Bruker Endeavor D4 XRD (Cu radiation, 40 kV, 40 mA) at 0.02564 degree/step/second was used for bulk analysis; clay analysis was performed at 0.02992 degree/step/second.

XRD patterns of separated clay fractions on glass were used for clay mineral identification. Mineral identification was facilitated by JADE (version 9.5). Quantitative analysis of minerals was performed by the Rietveld method [30], and amorphous phases were not accounted. The results were normalized to 100% based on the assumption that the complete mineral content of the sample was accounted for in the XRD patterns. Duplicate samples were analyzed at the two laboratories in the United States and China, and the error was less than 10%.

4. Results

4.1. Particle Size Analysis

A total of 27 regional particle size samples were analyzed. Particle size distribution curves from this sample set show both unimodal and bimodal patterns (Figure 2).

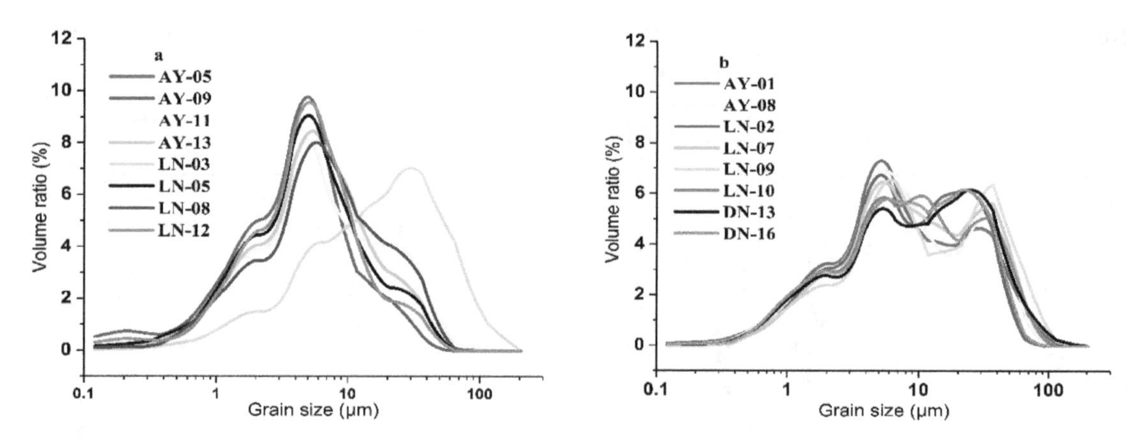

Figure 2. Grain size distribution of 27 regional samples: (**a**) shows unimodal patterns; and (**b**) shows bimodal patterns.

Particle size curves with unimodal patterns presented approximate normal distribution, with the exception of Sample LN-03 (Figure 2a). A significant peak was estimated in the particle size range of 2.1–17.2 μm, with volume ratio from 8.96 to 10.36. The range of grain size ratio was more than 70% and was controlled by fine particles. Significantly, Sample LN-03 showed a unimodal curve with obvious right-hand deviation (Figure 2a), with a grain size peak at 30.5 μm and volume ratio of 7.2%. Therefore, in Sample LN-03, the peak volume ratio was lower and the particle size was coarser than that in other unimodal samples. In samples of bimodal particle size distribution, the first significant peak appeared at 1.7 μm (Figure 2b), with an estimated particle size range of 1.4–2.6 μm and volume ratio of 1.5%–5%. The maximum peak interval was in the particle size range of 2.6–9.7 μm, whereas the volume ratio was in the range of 1.47%–10.07%. The third peak interval was in the particle size range of 14.2–36.9 μm, with a volume ratio of 4% to 7.5%.

Samples with unimodal particle size distribution (Figure 2a; e.g., AY-05 and AY-09) originated in weathering crust comprising red silty clay, indicative of advanced bedrock weathering and pedogenesis. This weathering crust overlies bedrock lithology with siliceous slate and fine sandstone interlayers. In contrast, samples with bimodal particle size distribution originated in soils over grayish-white medium-grained granodiorite with moderate weathering.

In the samples from the weathering crust profile, the grain size increased gradually up-profile, reflecting the degree of weathering in different profile horizons. Sample LN-03, with a coarser particle size overall (Figure 2a), came from a depth of 0.90 m, which is close to party weathering bedrock. The particle volume distribution curves indicate that particle size distribution in the REE mine does not follow a single normal distribution pattern, thus reflecting considerable differences in soil particle size gradation. These analyses show that sample location, degree of weathering, and bedrock lithology affect the distribution of soil particle size.

4.2. *Regional Clay Mineral Analysis*

The clay minerals in the regional samples were mainly kaolinite, followed by illite, chlorite, vermiculite (Table 2). Mineral residues from the parent rock included quartz and potassium feldspar with small quantities of plagioclase in some samples. The kaolinite content varied greatly in the regional samples (maximum: 62.1%; minimum: 8.8%; average: 31.91%; standard deviation of 11.90).

Table 2. Main clay minerals content of regional in soil.

Serial Number	Sample	Clay (%)				Other Minerals (%)		
		Vermiculite	Chlorite	Kaolinite	Illite-Mica	Quartz	Potassium Feldspar	Plagioclase
1	AY-01 *	-	1.8	37	9.4	42.8	8.3	-
2	AY-02	-	-	40.3	10.1	39.8	9.8	-
3	AY-03 *	-	-	37.5	13.5	40.8	6.7	-
4	AY-05 *	-	2.8	37.3	3.4	38	13.4	-
5	AY-06 *	-	2.7	38.1	6.3	30.8	15.9	-
6	AY-07	-	-	62.1	13.9	20.3	3.7	-
7	AY-08	-	-	56	16.4	22.8	4.7	-
8	AY-09 *	-	2.5	31.5	5.9	42.2	10.5	-
9	AY-10	-	-	36.9	9.5	42.2	11.3	-
10	AY-11	-	-	36.2	4.9	55.7	3.2	-
11	AY-12 *	-	3.4	29.2	0	43.8	6	-
12	AY-13 *	6.6	1	45.8	3.1	23.3	23.1	-
13	DN-13 *	-	2.6	28.6	1.8	31.4	31.8	-
14	DN-16 *	1.5	1.5	24.5	1.3	34.1	35.3	-
15	LN-01 *	1.9	0.9	32.8	-	48.7	6.8	-
16	LN-02	3.8	-	25.4	21	35	18.7	-
17	LN-03	-	-	8.8	28.6	54.4	8.2	-
18	LN-04 *	-	-	30.3	2	47.8	18.6	-
19	LN-05	0.1	-	23	19.1	44.1	13.8	-
20	LN-06 *		-	11.9	0.4	31.5	29.3	10

Table 2. *Cont.*

Serial Number	Sample	Clay (%)				Other Minerals (%)		
		Vermiculite	Chlorite	Kaolinite	Illite-Mica	Quartz	Potassium Feldspar	Plagioclase
21	LN-07	1.5	-	12.4	13	59.6	15	-
22	LN-08 *	-	-	30	0.3	47.8	13.2	0.4
23	LN-09 *	1.4	-	33.7	0.6	43.7	16.9	-
24	LN-10 *	0.6	-	37.9	1.3	39.3	15.4	1.2
25	LN-11	0.8	-	13.5	6.4	53.9	26.2	-
26	LN-12 *	-	-	30.1	0.3	38.3	25	1
27	LN-18 *	-	-	30.9	3.2	49.8	6.5	1.5

* Sample analyses were supported by Oil and Gas Laboratory in ALS Houston.

Since the chemical composition of kaolinite is the same as that of halloysite (except for weakly-bound interlayer water) [31], the kaolinite diffraction peak is the same as that of halloysite. Fang et al. [9] studied clay minerals in six REE mining areas in southern Jiangxi Province; the characteristic diffraction peaks of halloysite and kaolinite were 7.30–7.45 Å, 4.5–4.6 Å, 3.58–3.60 Å, and 3.32–3.37 Å. Chi et al. [3] found that the weathered crust leaching type REE ore was mainly composed of clay minerals, i.e., mainly halloysite, illite, kaolinite, and very small amounts of montmorillonite, alongside quartz sand and rock-forming minerals (feldspar). Halloysite is generally formed in the upper layer of a weathering crust from kaolinite interstratified minerals in noncrystalline stage, resulting from weathering and the dissolution of feldspar [13,15,32]. We infer that the kaolinite identified through XRD in our samples may contain kaolinite and kaolinite-interstratified minerals. The coexistence of both minerals in weathering profiles is frequently reported in studies with electron microscopy [32,33].

4.3. Analysis of Clay Minerals in the in Situ Leaching Profile

Samples from the Jiangwozi and Longnan profiles, Ganxian District, mainly comprised kaolinite and illite (i.e., mica), along with quartz and feldspar (Table 3). The kaolinite content ranged from 8.3% to 35.0% (average: 18.88%; standard deviation: 7.82).

Table 3. Main clay minerals content of soil profile (%).

Serial Number	Sample	Kaolinite	Illite-Mica	Quartz	Potassium Feldspar	Depth (cm)
1	GX-01	16.2	19.2	39.7	24.8	20
2	GX-02	18.7	13.5	37	30.8	45
3	GX-03-2	25.9	15.3	33.6	25.2	60
4	GX-03-1	20.3	7.3	39.7	32.7	80
5	GX-04-5	28.2	13.6	31	27.2	105
6	GX-04-4	14.9	15.9	34	35.2	130
7	GX-04-3	16.4	11.3	33.3	39	150
8	GX-04-2	16.5	22.5	32.3	28.6	170
9	GX-04-1	35	17.2	33.1	14.7	190
10	GX-05	9.8	12.4	39.9	37.9	215
11	P01-1-1	12.6	17.2	46.2	31.3	80
12	P01-2-1	12.3	16	40.5	34.1	150
13	P01-3-1	9.5	12.3	44	24.9	195
14	P01-4-1	11.8	10.1	53.2	16.4	280
15	P01-1-2	8.9	13.4	61.3	25.1	80
16	P01-2-2	8.3	13	53.6	18.3	150
17	P01-3-2	33	11.9	36.8	24.2	195
18	P01-4-2	17.7	15	43	20.9	280
19	P01-1-3	22.5	20.7	35.8	30.6	80
20	P01-2-3	22.8	19.9	26.7	21.4	150
21	P01-3-3	31.2	13.8	33.6	34.3	195
22	P01-4-3	22.8	15	27.9	31.3	280

In the Jiangwozi profile, the kaolinite content increases nonlinearly from the surface to the bottom, from 16.2% to 28.2% (Table 2), and then decreases gradually. The kaolinite content reached a maximum of 35.0% at 200 cm from the surface, where the soil texture is coarse and fine sand. At this depth, kaolinite is distributed like a network [27]. The REE ore of the Jiangwozi profile is mainly enriched in the weathering and leaching layer between 90 and 200 cm. The change of clay mineral content is closely related to the chemical weathering rate of the rock: the maximum kaolinite content (35.0%) is in GX-04-1 where potassium feldspar content is at its minimum (14.7%), suggesting that potassium feldspar is strongly weathered. At depths of 0–55 cm, the illite (mica) content showed the opposite trend to that of kaolinite. From 70 to 120 cm, illite (i.e., mica) follows the same trend as kaolinite, but its content is lower. At this depth, the content of coarse particles is relatively increased, probably due to enhancement of chemical weathering and vertical migration of particles. The kaolinite content peaks at about 200 cm, while the illite content peaks earlier, at around 190 cm (at 17.2%). Yang [34] researched the clay mineralogy of the REE weathering crust of Longnan granite, in Jiangxi Province, and found that the crystallization degree of kaolinite gradually increased down-profile, suggesting that the content of kaolinite also increased. This discovery was demonstrated in granite weathering crust profiles where kaolinite was dominant at the bottom of crust [13,35].

The analysis of soil clay minerals and rare earths in the Wenlong mine showed that at the onset of in situ leaching, kaolinite is absent; however, as leaching continues, the kaolinite content increases dramatically at depth of around 200 cm (Table 2). In comparison, with the significant change with depth in the Ganxian District profile, at the Wenlong mine profile, kaolinites formed in the course of weathering increase gradually as in situ leaching progresses. This is the result of the coupling of natural weathering and human activity (profile stripping in the course of REE mining).

4.4. Clay Minerals in the Simulated Leaching Profile

The clay (kaolinite and illite) and other mineral (quartz, potassium feldspar, and plagioclase) content of the eight soil profiles subjected to the simulated leaching experiment are presented in Table 4. Each soil column sample is denoted by a three-digit number: the first digit (T1 to T8) represents the number of the soil column; the second digit (1, 3, and 5) represents the depth of the sample in the soil profile (at 30, 70, and 110 cm, respectively); and the third digit (1, 2, 3, and 5) represents the time of mineral concentration measurement since the onset of the experiment (5th, 15th, 23rd, and 40th day, respectively).

Table 4. Clay and other mineral content of eight soil profiles (T1 to T8) at different times during simulated leaching (%).

Serial Number	Sample	Kaolinite	Illite-Mica	Quartz	Potassium Feldspar	Plagioclase
1	T1-1-1	22.3	7.4	45.7	24.6	-
2	T1-1-2	21.5	7.8	48.1	22.6	-
3	T1-1-3	20.2	5.1	45.2	29.6	-
4	T1-1-5	11.1	12.9	45.9	30.1	-
5	T1-3-1	18.9	8.8	55.6	16.7	-
6	T1-3-2	14.9	9.4	39.4	36.2	-
7	T1-3-3	13.8	13.2	51.7	21.3	-
8	T1-3-5	6.8	14.7	54.6	23.9	-
9	T1-5-1	15.2	14.6	36.6	33.6	-
10	T1-5-2	13.1	8.6	48.1	30.2	-
11	T1-5-3	16.1	12.8	37.2	33.9	-
12	T1-5-5	17.5	8.2	39.2	35.1	-
13	T2-1-1	20.5	11.4	42.5	25.5	-
14	T2-1-2	16.6	12.5	37.7	33.2	-
15	T2-1-3	21.7	7.2	45.3	25.8	-
16	T2-1-5	20.2	10.2	46.3	23.2	-

Table 4. *Cont.*

Serial Number	Sample	Kaolinite	Illite-Mica	Quartz	Potassium Feldspar	Plagioclase
17	T2-3-1	11.2	12.5	44.8	35.1	-
18	T2-3-2	9.2	15.0	43.0	32.8	-
19	T2-3-3	8.8	23.2	35.9	32.1	-
20	T2-3-5	10.5	23.0	43.1	23.5	-
21	T2-5-1	8.9	20.8	44.4	25.8	-
22	T2-5-2	8.5	16.0	42.2	33.4	-
23	T2-5-3	25.3	1.2	45.5	28.0	-
24	T2-5-5	31.5	5.3	31.0	32.2	-
25	T3-1-1	15.3	10.5	55.5	18.7	-
26	T3-1-2	16.2	12.8	40.2	30.8	-
27	T3-1-3	15.7	16.9	39.6	27.8	-
28	T3-1-5	18.4	9.7	48.4	23.5	-
29	T3-3-1	9.2	17.9	43.3	29.6	-
30	T3-3-2	11.2	14.2	45.7	28.9	-
31	T3-3-3	10.5	15.8	43.5	30.3	-
32	T3-3-5	18.6	20.2	38.0	23.2	-
33	T3-5-1	16.5	15.0	28.5	40.1	-
34	T3-5-2	9.2	11.4	44.8	34.6	-
35	T3-5-3	11.8	21.9	28.3	38.1	-
36	T3-5-5	16.2	17.5	25.9	40.5	-
37	T4-1-1	14.1	14.9	43.9	27.2	-
38	T4-1-2	18.1	12.3	48.5	21.1	-
39	T4-1-3	18.2	11.8	42.8	27.2	-
40	T4-1-5	18.5	10.6	34.0	36.9	-
41	T4-3-1	9.9	12.2	46.2	31.7	-
42	T4-3-2	6.9	16.1	41.8	35.2	-
43	T4-3-3	10.4	16.7	29.8	43.1	-
44	T4-3-5	15.7	20.4	37.3	26.6	-
45	T4-5-1	14.3	16.5	34.0	35.2	-
46	T4-5-2	9.1	14.0	42.5	34.4	-
47	T4-5-3	15.3	9.5	50.1	25.1	-
48	T4-5-5	11.2	6.8	51.4	27.4	3.1
49	T5-1-1	14.2	9.6	48.3	27.9	-
50	T5-1-2	16.9	14.0	43.6	25.5	-
51	T5-1-3	16.4	30.6	35.4	17.6	-
52	T5-1-5	26.7	9.6	34.2	29.5	-
53	T5-3-1	16.6	8.9	49.4	25.2	-
54	T5-3-2	12.4	12.1	47.9	27.5	-
55	T5-3-3	16.2	11.1	41.5	31.2	-
56	T5-3-5	12.1	17.1	39.8	31.0	-
57	T5-5-1	15.2	10.2	29.0	37.8	7.9
58	T5-5-2	9.4	8.8	31.8	32.4	17.6
59	T5-5-3	11.4	13.2	36.7	29.7	9.1
60	T5-5-5	9.6	11.5	54.2	24.6	-
61	T6-1-1	10.9	6.9	52.9	23.9	5.4
62	T6-1-2	10.5	8.4	57.7	19	4.4
63	T6-1-3	9.8	8.6	57.7	23.9	-
64	T6-1-5	14.0	12.5	46.9	26.6	-
65	T6-3-1	13.3	13.2	40.6	30.4	2.4
66	T6-3-2	15.1	12.9	49.7	22.3	-
67	T6-3-3	13.7	10.3	47.0	29.0	-
68	T6-3-5	13.3	12.8	60.6	13.2	-
69	T6-5-1	9.8	13.9	47.8	20.8	7.7
70	T6-5-2	7.0	11.6	59.1	22.3	-
71	T6-5-3	15.4	12.9	40.8	26.1	4.8
72	T6-5-5	16.4	12.3	38.0	29.8	-
73	T7-1-1	16.8	10.2	43.0	22.5	7.4
74	T7-1-2	15.7	13.2	40.1	31.0	-
75	T7-1-3	20.6	13.3	39.9	26.2	-
76	T7-1-5	19.1	21.3	38.3	21.3	-

Table 4. *Cont.*

Serial Number	Sample	Kaolinite	Illite-Mica	Quartz	Potassium Feldspar	Plagioclase
77	T7-3-1	7.5	15.5	40.5	30.5	6.1
78	T7-3-2	10.3	13.1	51.3	22.4	2.9
79	T7-3-3	15.3	4.5	38.5	31.1	10.6
80	T7-3-5	11.3	10.9	53.3	19.6	4.9
81	T7-5-1	13.5	8.0	34.1	38	6.4
82	T7-5-2	11.0	14.6	32.8	41.6	-
83	T7-5-3	18.3	12.4	28.0	41.3	-
84	T7-5-5	25.3	17.7	41.5	15.5	-
85	T8-1-1	16.0	9.6	46.4	28.0	-
86	T8-1-5	20.6	19.3	30.7	29.4	-
87	T8-3-1	8.5	5.4	34.7	51.4	-
88	T8-3-5	15.2	11.5	35.7	37.6	-
89	T8-5-1	14.6	11.3	49.9	24.2	-
90	T8-5-5	19.6	17.7	33.8	28.9	-

-: represented that its content was below the detection limit.

The lowest kaolinite content is 6.8%, at T1-3-5, i.e., at sampling port number 3 in the T1 soil column after 40 days of simulated leaching. The highest kaolinite content is 31.5%, at T2-5-5, i.e., sampling port number 5 in the T2 soil column after 40 days of simulated leaching. The average content of kaolinite in the REE ore is 14.67% and the standard deviation is 4.72. In all columns except T1 and T2, the first sampling port (depth: 30 cm) shows that the variation in kaolinite content increases with the longer leaching. This is most obvious in T5, where the kaolinite content after 40 days of simulated leaching (sample No: T5-1-5) is 1.88 times the initial value. In soil columns T1 and T2, the kaolinite content changed little in the first 23 days, but after 40 days, it declined significantly in T1. The highest illite content is 30.6%, at T5-1-3, i.e., sampling port number 1 in the T5 soil column after 23 days of simulated leaching. Overall, the average illite content in the soil column sample set is 12.78%, with a standard deviation of 4.63, slightly less than that of kaolinite. Similarly, the illite-mica content of the first sampling port (depth of 5 cm) increased gradually in T1 and T6–T8 as leaching progressed, while other soil columns showed no obvious variation. However, in the T5 soil column, the illite-mica peaked at 30.6% after 23 days of leaching, and then fell back to the initial level after 40 days of leaching. The second and third sampling ports show that the content of kaolinite, illite and other clay minerals in the RE ore is complex under different leaching conditions (Table 3), which may be controlled by many factors.

The minimum content of potassium feldspar is 13.2%, after 40 days of leaching (sample No: T6-3-5), which is 56.6% lower than the potassium feldspar content at the sampling port at the initial stage of leaching (sample No: T6-3-1: 30.4%). This result indicates that potassium feldspar at the bottom of the soil column may have been weathered and mobilized with the leaching solution after prolonged leaching. The maximum potassium feldspar content is 51.4% (T8-3-1); average value is 28.81%, and the standard deviation is 6.60. Quartz fluctuates from 25.9 to 60.6% in the course of ore leaching, with an average value of 42.6%, and a standard deviation of 7.67, which is greater than that of potassium feldspar. In addition, plagioclase was detected in 15 samples, and its content fluctuated between 2.4% and 17.6%, with an average value of 6.71% and a standard deviation of 3.68. In the other samples, the plagioclase content was below the detection limit.

5. Discussion

5.1. Soil Particle Size and Distribution of Clay Minerals in REE Mining Areas

Ion-absorbed REE ore is mainly formed by advanced weathering of granite. It is a loose, earthy substance comprising quartz, feldspar, and clay minerals [36]. Therefore, this loose soil mantle, formed by surface weathering is closely related to mineral grain size. The cumulative curves of regional particle size distribution are S-shaped (Figure 3), which is consistent with earlier report [36].

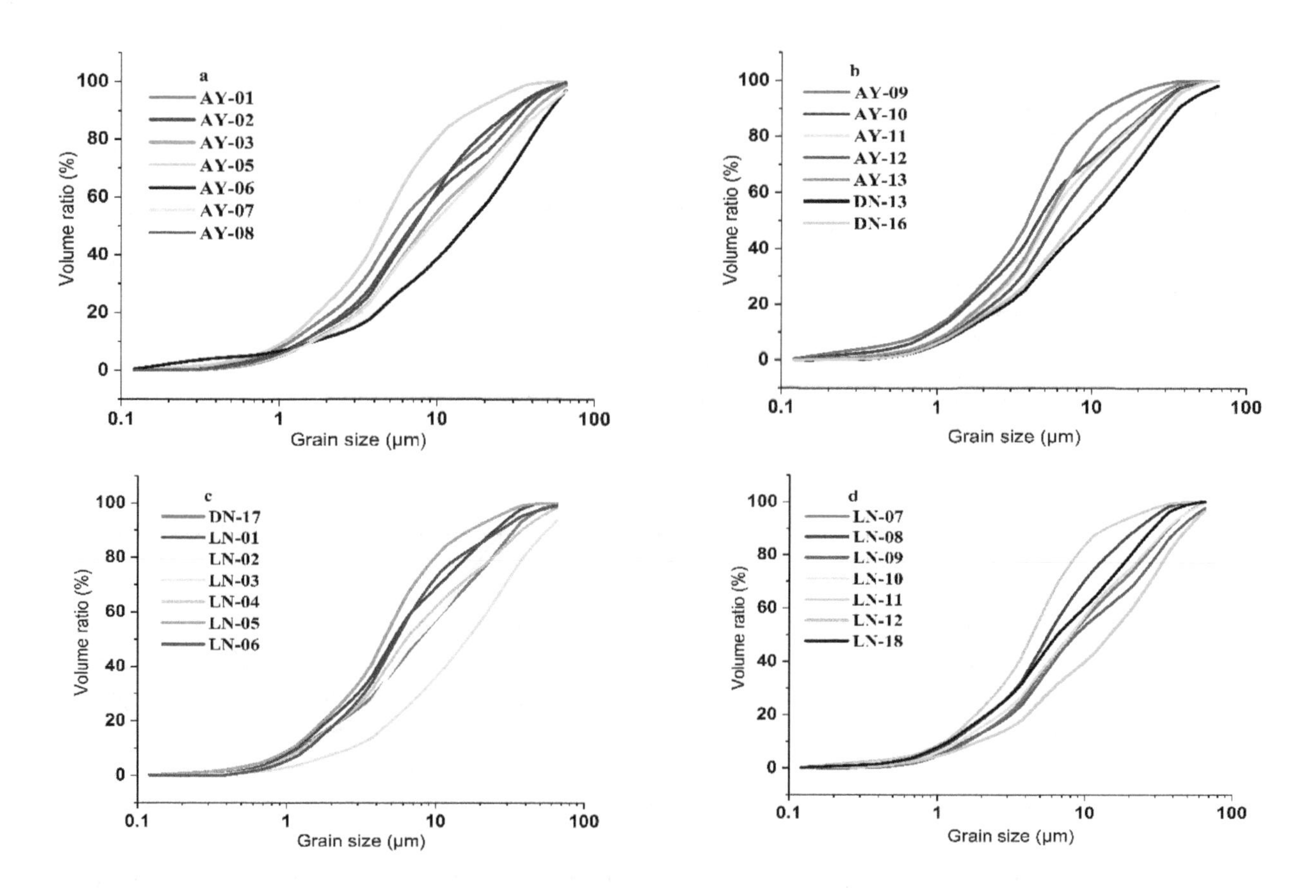

Figure 3. Cumulative particle size distribution curves from ion-absorbed RE in (regional samples). (**a**) shows the curves with AY-01, AY-02, AY-03, AY-05, AY-06, AY-07 and AY-08; (**b**) shows the curves with AY-09, AY-10, AY-11, AY-12, AY-13, DN-13 and DN-16; (**c**) shows the curves with DN-17, LN-01, LN-02,LN-03, LN-04, LN-05 and LN-06; (**d**) shows the curves with LN-07, LN-08, LN-09, LN-10, LN-11, LN-12 and LN-18.

Additionally, the particle distribution curves in Figure 3a–d show that particle size ranges between 3.7 and 30.5 μm. According to Aberg's classification of granular materials [36,37], some of the cumulative curves of the regional sample set are A-shaped (i.e., the left end of the cumulative particle distribution curve is relatively steep, with a concave side downward), indicating that the particle gradation changes significantly with an increase in coarse particles in the soil. Yan et al. [38] used a wet sieving method to classify REE ores into eight types of particle size distribution. In this paper, particle size analysis concentrated on the <0.075 mm size fraction, and the analytical method was different (Malvern-2000 laser particle size analyzer); therefore, our particle size distribution curves are different.

The cumulative particle size distribution curves of AY-05 (Figure 3a), AY-09 (Figure 3b), LN-05 (Figure 3c), and LN-12(Figure 3d), nevertheless, are inclined steeply to left in their upper part (a B-type structure in Arberg's terms [37]), indicating that grain size changes abruptly from fine to coarse. A possible reason for this is that simulated leaching continues to promote rapid decomposition of feldspar (Table 3). Dissolution of feldspar due to leaching releases SiO_2, which migrates downward, where it recrystallizes resulting in particle thickening [36].

Eigenvalue analysis shows that the 10 particle size (the particle size at which the cumulative particle size distribution curve reaches 10% of the volume) was in the range of 0.82–5.03 μm (Table 5). The minimum D10 value (0.82), corresponds to Sample AY-09, a red sand sampled from 1.5 m below- surface, above Sinian feldspar quartz and slate bedrock in the age of Sinian period (Z). Field investigation in this area revealed that the weathering crust is approximately 3.5–3.8-m thick, and that the bedrock is strongly deformed and fractured, and thus particularly susceptible to physical and chemical weathering. D90 (the particle size at which the cumulative particle size distribution curve reaches 90% of the volume) ranges from is 11.99 μm (minimum) to 60.99 μm (maximum), with a standard deviation of 11.4, respectively, thus suggesting that discreteness increases with particle size. The average particle size, Dav and volumetric average particle size, D [4,3] have similar maximum, minimum, and standard deviation values. The standard deviation of the median particle diameter, D50 (the particle size at which the cumulative particle size distribution curve reaches 50% of the volume) is more significant than that of D [3,2] (surface area average particle size), indicating that the median particle diameter is more discrete than the surface area average particle size.

Table 5. Characteristic parameters of particle size distribution; regional weathering crust samples.

Sample No.	D10 (μm)	D50 (μm)	D90 (μm)	Dav (μm)	D [3,2] (μm)	D [4,3] (μm)
AY-01	1.32	6.25	33.88	12.88	3.09	12.88
AY-02	1.43	7.28	29.54	11.65	3.36	11.65
AY-03	1.64	8.47	40.28	15.64	4.09	15.64
AY-05	1.10	4.50	17.10	7.02	2.30	7.02
AY-06	1.79	15.51	50.28	21.57	3.67	21.57
AY-07	2.21	12.76	40.04	17.22	4.75	17.22
AY-08	1.44	6.93	35.50	13.21	3.48	13.21
AY-09	0.82	3.88	11.99	5.40	1.77	5.40
AY-10	0.96	4.57	23.70	8.76	2.21	8.76
AY-11	1.34	5.38	20.86	8.42	2.89	8.42
AY-12	1.13	6.39	28.22	11.06	2.61	11.06
AY-13	1.20	5.02	18.49	7.70	2.58	7.70
DN-13	1.64	12.75	44.85	18.90	4.37	18.90
DN-16	1.37	8.07	30.07	12.42	3.50	12.42
LN-01	1.16	5.14	24.69	9.41	2.84	6.10
LN-02	1.50	7.45	28.71	11.85	3.59	11.85
LN-03	5.03	20.86	61.00	27.80	8.19	27.80
LN-04	1.49	8.39	38.06	14.92	3.67	14.92
LN-05	1.37	6.42	33.75	13.00	3.53	13.00
LN-06	1.41	5.47	26.90	10.20	3.31	10.20
LN-07	1.62	8.17	37.10	14.42	4.07	14.42
LN-08	1.26	5.68	22.95	9.08	2.89	9.08
LN-09	1.55	8.63	43.42	17.11	3.74	17.11
LN-10	1.41	8.08	36.38	13.98	3.58	13.98
LN-11	1.91	15.00	49.19	21.02	4.51	21.02
LN-12	1.12	4.41	13.85	6.45	2.08	6.45
LN-18	1.21	6.68	28.74	11.42	2.79	11.42

With Dav as an independent variable and the other characteristic parameters as dependent variables (Figure 4), the regression coefficient is D90 > D [4,3] > D50 > D [3,2] > D10. This finding illustrates that the increase in surface average particle size has a more significant impact on coarse particles than on fine particles (D10). D [4,3] has the highest correlation with Dav (correlation coefficient: 0.99102), followed by D90, D50, D [3,2], and the lowest correlation with D10 (correlation coefficient: 0.727).

D10 residual analysis (Figure 5a) shows that when the average particle size Dav increases, other residuals decrease (with very few exceptions). This shows that for D10, as the average particle size increases, the volume of particle grain size decreases by less than 10%. Residuals are normally distributed (Figure 5b), and the regression analysis of the dependent variable is similar to that of the independent variable. Residuals present a linear shape at a 99.5% confidence interval (Figure 5c,d).

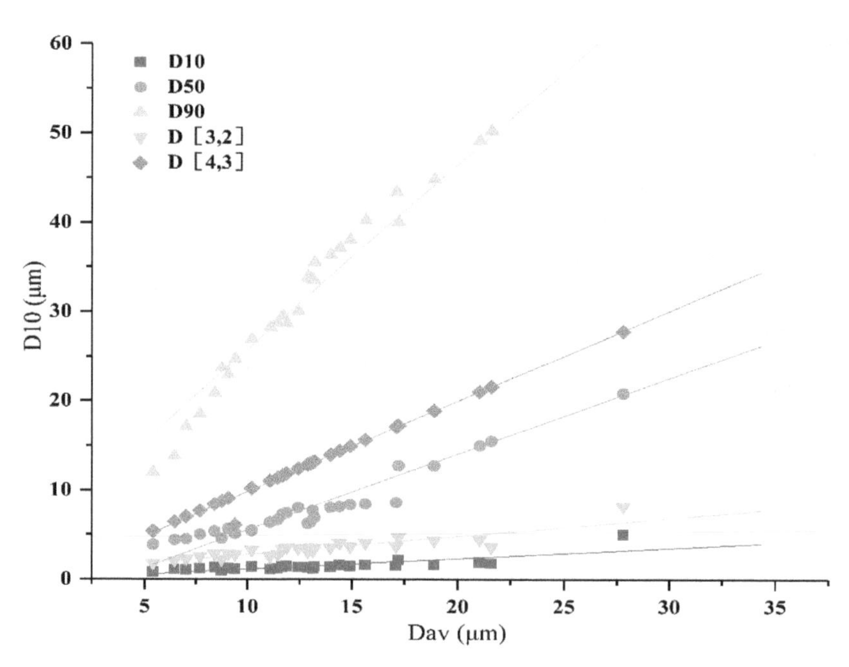

Figure 4. Scatterplot of characteristic parameters of particle size distribution; regional weathering crust samples.

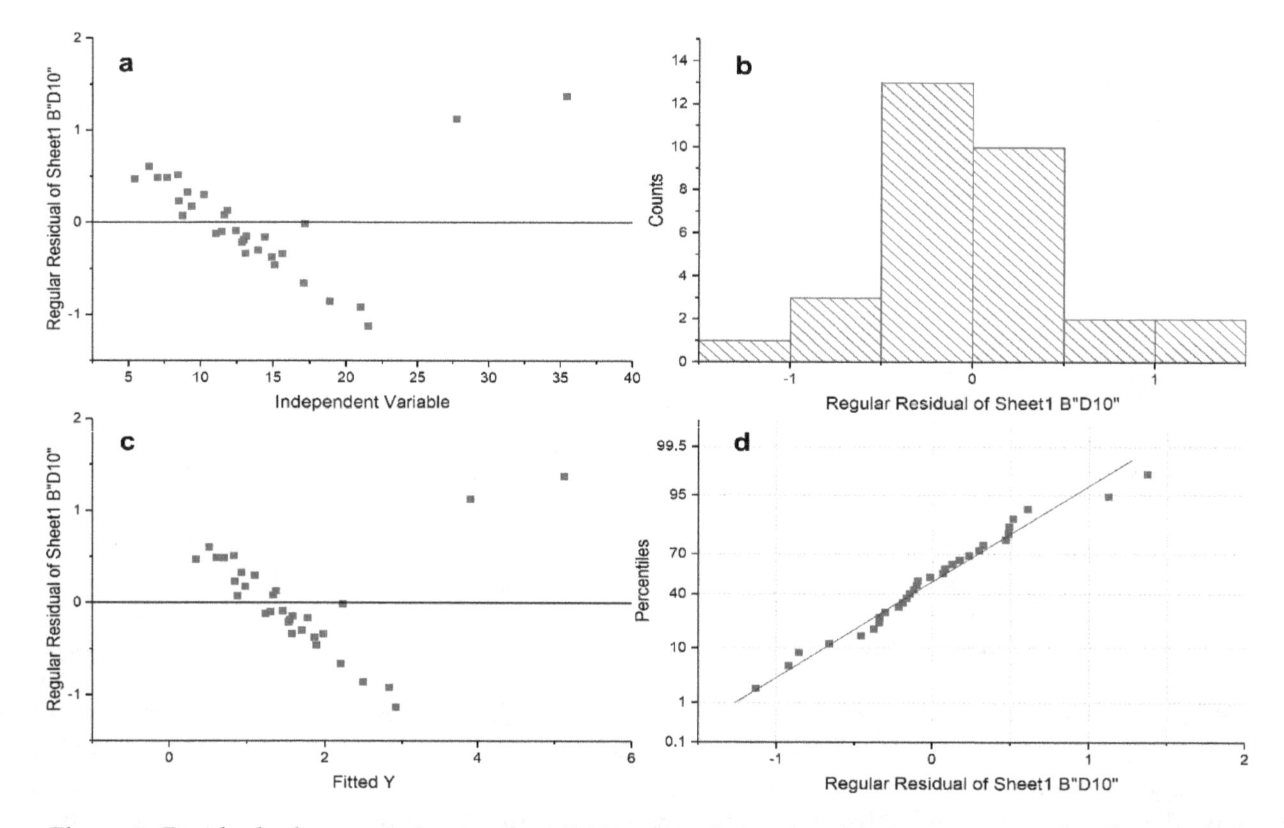

Figure 5. Residuals characteristic parameter D10; regional samples: (**a**) shows scatter plot of residuals; (**b**) shows histogram of regular residual; (**c**) shows scatter plot of residuals with fitted Y; and (**d**) shows scatter plot of residual flatten and significance testing.

The dominant clay minerals in the regional soil samples were kaolinite, followed by illite, and some vermiculite, and chlorite (Table 2). Kaolinite content has a weak correlation with rock-forming minerals, such as potassium feldspar and quartz. This finding indicates a nonlinear process of potassium feldspar alteration into kaolinite during granite weathering. Moreover, clay minerals and quartz were present in the weathering crust over metamorphic sandstone and slate in the Anyuan County area, and their correlation reached 0.68 (Figure 6). This indicates that the conversion ratio of feldspar to clay minerals, after weathering of this metamorphic bedrock, is higher than that of granite in the Longnan County granite.

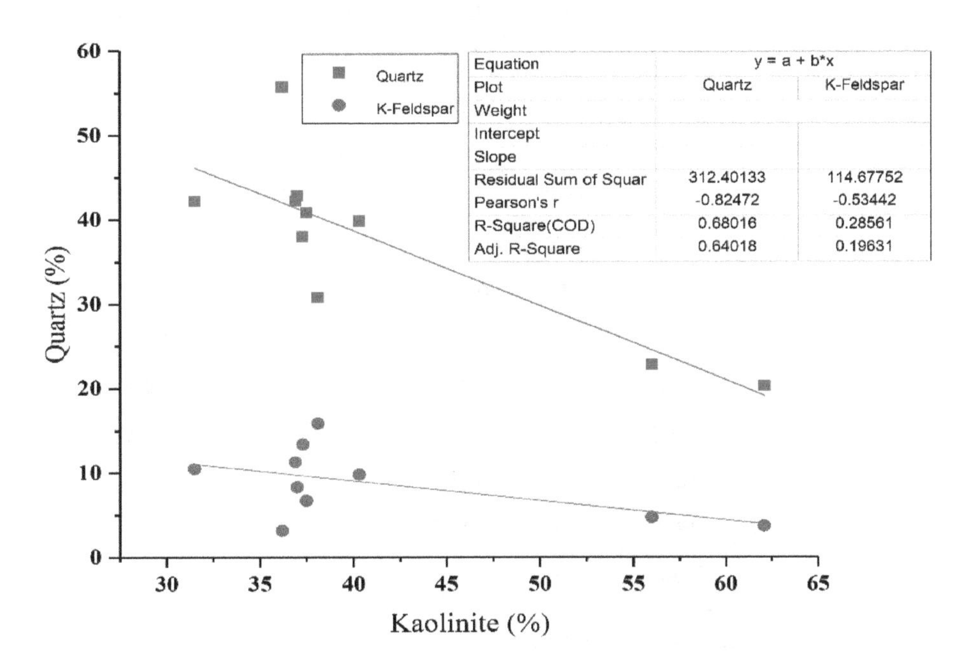

Figure 6. Correlation trend of kaolinite, Potassium feldspar, and quartz in samples from Anyuan County.

The relationship between average particle size (Dav) and kaolinite, quartz, and potassium feldspar indicates that the mineral particle size had only a minor effect on the clay mineral content. Quartz and potassium feldspar are mostly distributed on both sides of kaolinite. Quartz is further away from the *X*-axis, while potassium feldspar is closer to the *X*-axis. This distribution, therefore, further confirms that weathering of potassium feldspar has a significant impact on the formation of kaolinite.

Wang et al. (2018) [39] found that the main minerals of the low-grade metamorphic rocks (e.g., meta-sandstone, meta- siltstone, and slate) in the Anyuan County area of southern Jiangxi are 30–70% quartz, 5–30% feldspar, 3–10% biotite, and 3–12% muscovite. The CIA index is in the range of 68–75%. The Yanshanian granite in Longnan County is mainly compounded of 25–32.7% quartz, 31.1–42.4% potassium feldspar, 17–28.9% plagioclase, 3.4–6% biotite, and 1–3.4% muscovite [40]. According to the analysis of major elements in granite in the Zudong mining area, Longnan County [8,41] (Table 6), CIA is in the range of 61–65%, with an average of 63%, which is significantly lower than that of Anyuan County. This suggests that a higher CIA value reflects more extensive loss of Na^+, K^+, and Ca^{+2} during leaching, enrichment in Al and Si, and more advanced conversion of feldspar to clay minerals [42,43]. This study confirms that the weathering of the parent rock has a significant effect on the formation of soil clay minerals. Weathered feldspar minerals are converted to kaolinites, which are then converted into kaolin minerals under moderate silica and salt-based ion conditions [13]. Furthermore, layered silicate minerals, such as muscovite, biotite, and chlorite, are weathered at varying degrees to form kaolinite minerals [44]. The original rocks of Anyuan County are predominantly metamorphic sandstone, siltstone, and slate, which were found to be relatively broken in the field, and provides favorable conditions for further weathering. As shown by the analysis of kaolinite in Table 1, the average content of kaolinite, illite, and potassium feldspar in Anyuan County (samples AY-01 to AY-12) is 40.66%, 8.03%, and 9.72%, respectively. This indicates that most of the potassium feldspar in the Anyuan County metamorphic bedrock was converted into kaolinite minerals. In Longnan County (samples LN-01 to LN-12), the bedrock is medium-grained granodiorite, with an average kaolinite mineral ratio of 24.67%, i.e., much less than that in Anyuan County.

Table 6. Chemical composition (wt %) of Longnan granites (Zudong mining area).

Sample No.	1	2	3	4	5	6
SiO_2	70.34	75.55	76.14	74.88	72.48	74.58
TiO_2	0.44	0.25	0.03	0.05	0.13	0.15
Al_2O_3	14.59	12.13	12.97	13.44	13.4	13.34
Fe_2O_3	2.03	1.18	0.09	0.21	0.51	0.56
FeO	0.64	1.01	1.07	1.42	1.28	1.25
MnO	0.25	0.07	0.03	0.08	0.11	0.05
MgO	0.57	0.34	0.07	0.19	0.17	0.18
CaO	0.55	0.42	0.62	0.58	1.51	0.73
Na_2O	3.37	2.74	4.25	3.97	3.62	3.65
K_2O	5.7	5.36	4.52	4.61	5.37	4.8
P_2O_5	0.05	0.04	0.02	0.06	0.04	0.05
LOI	0.92	0.7	0.96	0.57	1.62	0.6
Total	100.33	99.85	100.77	100.28	100.24	100.03
CIA	64	63	63	65	61	64

CIA = Chemical Index of Alteration, Data from Bao et al., 2008 [8] and Zhang, 1990 [41].

Differences in kaolinite and illite content between Anyuan and Longnan are not only related to the composition of the parent rock. The sampling depth and topography also have a significant impact on the clay mineral and soil formation [45]. Samples from Anyuan County came from average elevations of between 311 and 468 m, with relatively gentle relief (gradient: 24°–25°) and dense vegetation cover. Owing to the influence of bedrock lithology, geological structure and surface erosion, a concave slope formed in this area, and the weathering profile was deep. Samples from Longnan County, on the other hand, came from average elevations of between 278 and 321 m, from a relief of lightly-weathered residual hills with linear or convex slopes controlled by granite lithology. These nuances of landforms have a crucial influence on parent rock weathering and soil formation.

5.2. Vertical Variation of Clay Minerals in REE Ores

An analysis of the Ganxian District soil profile [27] showed that the main clay minerals are kaolinite and illite (Figure 6). Kaolinite content fluctuates from top to bottom, with the lowest content (9.8%) at 115 cm from the bottom of the section. In this section, the layers are mostly located at the bottom of the semiweathered layer, where the granite structure is visible and the weathering degree of the rock is weakened. The content of rock-forming minerals (i.e., quartz, 39.9%; potassium feldspar, 37.9%) in (Table 2) shows that conversion of potassium feldspar to clay minerals was minor, which is consistent with this deeper level of the weathering profile. The peak of kaolinite (35.0%) appears at a depth of 190 cm; the kaolinite content is relatively low in the 130–190 cm interval (14.9-16.5: Table 3). From 20 to 105 cm below surface, the kaolinite content increases irregularly, reaching 28.2% at 105 cm. This increase is probably due to the rapid conversion of feldspar and mica minerals into kaolinite and other clay minerals.

Furthermore, illite content shows variation similar to that of kaolinite, but with sharper changes. Potassium feldspar and quartz also exhibit different variation characteristics. Two distinct horizons are thus resolved in the Ganxian weathering profile can be divided on the basis of clay mineral distribution, as follows (Figure 7).

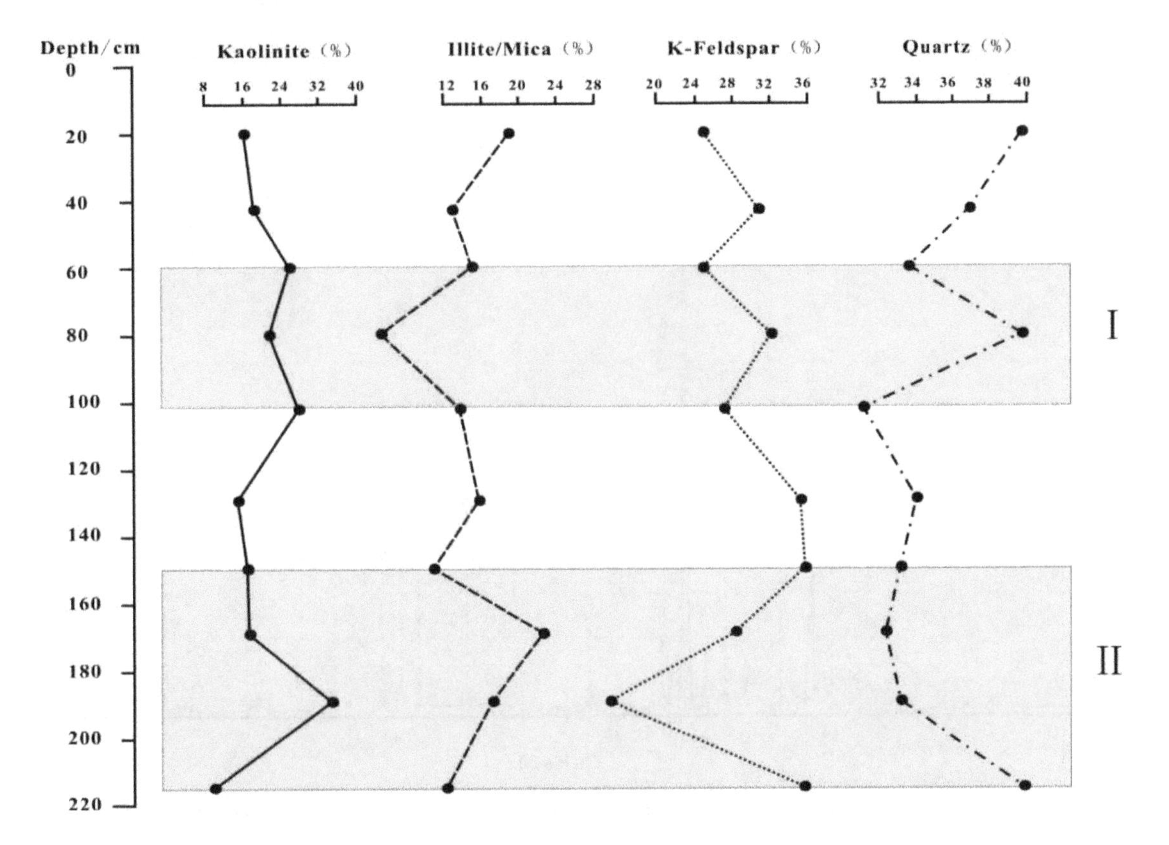

Figure 7. Clay and other mineral content in the Ganxian weathering profile.

Band I (60–105 cm): clay mineral content shows a high-low-high variation trend, while the illite content shows a broader range of variation than that of potassium feldspar. The content of quartz, and feldspar shows the opposite trend to that of clay minerals, and the variation range of the quartz content was broader than that of potassium feldspar. Over the course of weathering, clay minerals are converted into kaolinite minerals due to physical and chemical weathering processes. According to Uzarowicz et al. (2011) [46], the composition of clay minerals in the soil surface follows the acidic soil formation process, i.e., it is strictly controlled by the content of chlorite and mica debris, with subsequent conversion of chlorite and mica to montmorillonite and vermiculite. Our analysis shows that kaolinite and illite were reduced concurrently in the profile, while the quartz and potassium feldspar content increased. It is suggested that large quantities of mica and chlorite minerals were formed after weathering and alteration of the parent rock (i.e., granite), thereby controlling the formation and transformation of kaolinite.

An XRD analysis of Sample GX-04-4, showed seven illite-mica diffraction peaks, between d = 4.47 Å and d = 2.50 Å, with a cumulative peak height of 117.8% (Figure 8). The GX-04-1 sample showed a total of four distinct illite-mica diffraction peaks, with a cumulative peak height of 17.4%. XRD analysis further confirmed that, as a result of surface weathering of the granite bedrock, feldspars have altered to chlorite and mica; other clay minerals are due to the weathering and alteration of mica in the original rock. The formation and conversion of kaolinite were limited.

Band II (150–215 cm): clay mineral shows a low-high-low variation trend, opposite to that of band I. Potassium feldspar and quartz show greater changes in an opposite trend. The content of kaolinite and potassium feldspar shows a particularly evident reversal at 190 cm (increase of kaolinite; decrease of potassium feldspar), indicating that the weathering of potassium feldspar contributes to the formation of kaolinite in the soil. In the supergene weathering realm, the conversion of potassium feldspar into kaolinite can be expressed using the following chemical Equation (1) [47]:

$$4KAlSi_3O_8 + 4H^+ + 2H_2O \rightarrow 4K^+ + Al_4Si_4O_{10}(OH)_8 + 8SiO_2 \quad (1)$$

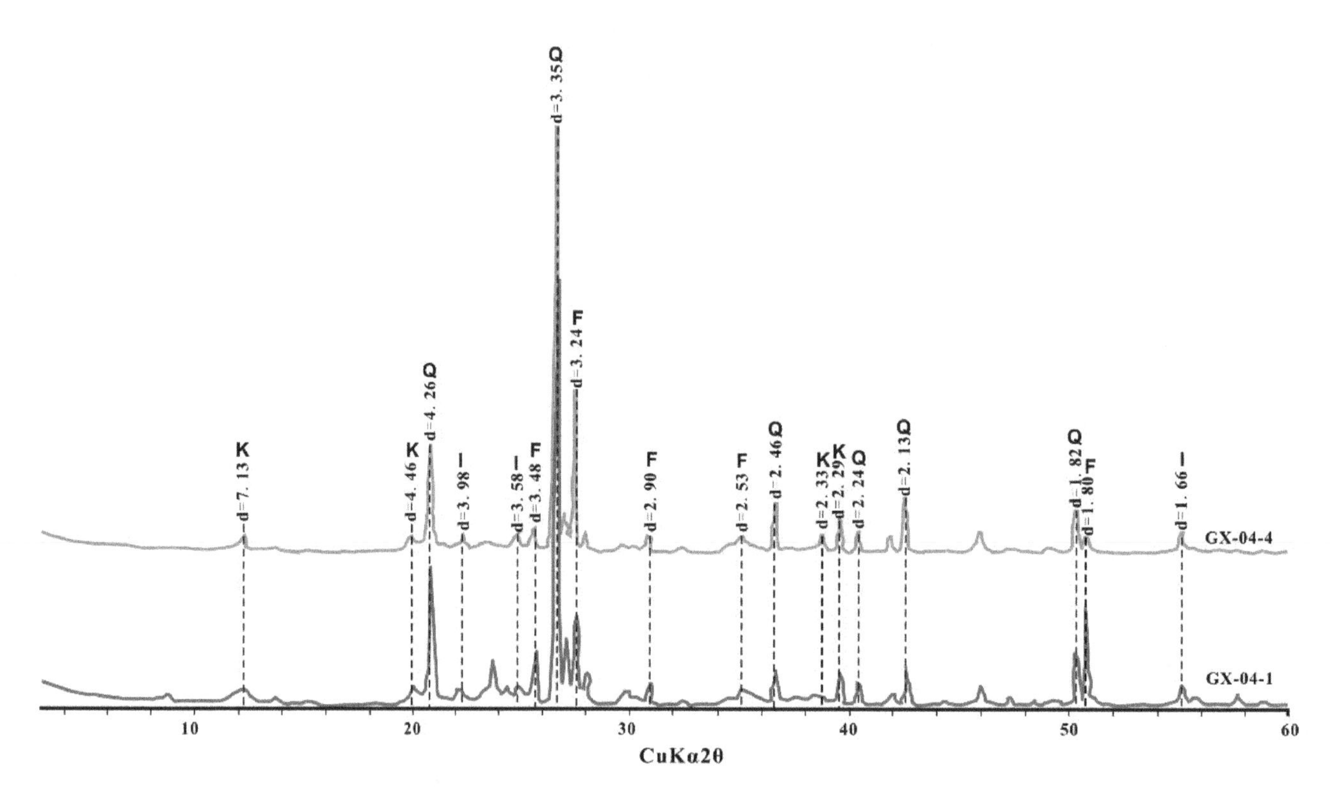

Figure 8. XRD spectra of randomly oriented bulk sample from the Ganxian Profile, I: illite-mica, K: kaolinite, Q: quartz, F: feldspars.

Equation (1) shows that decomposition of potassium feldspar releases a large quantity of free metal cations in the soil. This promotes further enrichment in kaolinite. According to Dixon (1989), the formation of kaolinite in the soil requires an adequate amount of silica and a small number of metal cations [48], while the chemical decomposition of potassium feldspar needs to consume HCO_3^- produced from H^+ and CO_2 [49]. This provides a favorable environment for the formation of kaolinite, forming the distinct volatile characteristic of band II. Moreover, the potassium feldspar diffraction peak in GX-04-1 was significantly reduced as compared to that in GX-04-4 (Figure 9). For quartz, the cumulative diffraction peak height is 195% in Sample GX-04-1, and 79.1% in Sample GX-04-4. This suggests that formation of kaolinite was favored in the SiO_2-rich environment.

An analysis of samples collected at different stages of the in situ leaching profile from Longnan County showed that the clay mineral content fluctuated regularly as leaching progressed (Figure 8). In the early stage of leaching, the kaolinite content in the soil was less than 15%, while the illite-mica content was slightly higher than that of kaolinite, fluctuating from 10.1% to 17.2% (Table 3). The potassium feldspar content fluctuated between 16.4%–34.1%, and the quartz content was relatively high. The total kaolinite content increased, and the total potassium feldspar content decreased as leaching progressed (Figure 8). In the later stage of leaching, from the surface to the bottom of soil column, the kaolinite content increased rapidly from 8.9% to 33.0%, then gradually decreased. The illite-mica content also decreased slightly compared to the previous period, with the exception of the bottom of the Longnan soil column, where it increased weakly (from 10.1% to 15%). The content of quartz increased significantly in the upper layer between 46.2%–61.3% and 40.5%–53.65% (Table 3), with an average increase of 32.5%. In the final stage of leaching, the kaolinite content increased significantly: from 22.5% to 31.2% (average growth: 46.2%; Table 3). The illite-mica content increased by between 15% and 20.7%. The potassium feldspar content increased significantly in the final stage of leaching (average growth: 32.9%), while the content of quartz decreased (27.9% to 35.8%).

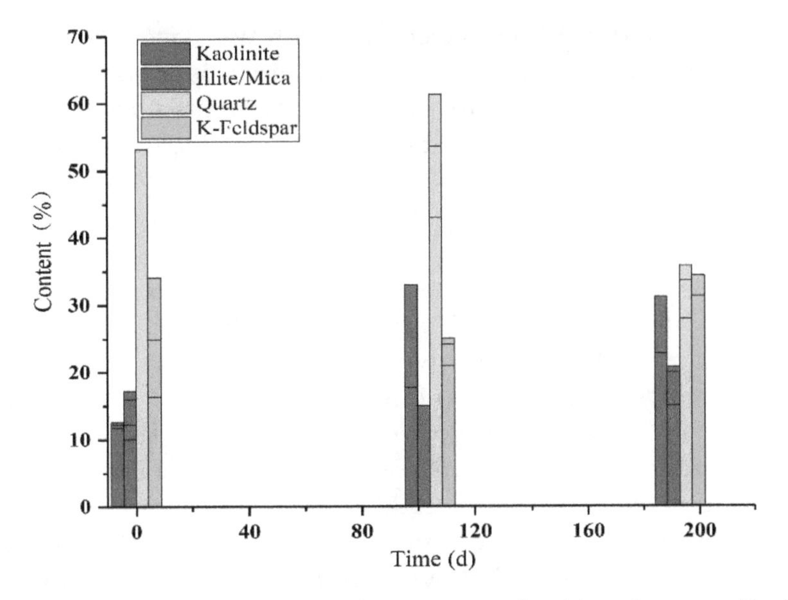

Figure 9. Clay content variation during in situ leaching, Longnan Profile.

Parfitt et al. (1983) [20] found that the Si concentration in the soil solution decreased as rainfall increased, reflecting increasing leaching of the soil. Wu et al. (2016) [36] suggested that granite with compact structure has higher strength, a low degree of weathering, and a higher content of residual feldspar. In conditions of sustained weathering by acid rain, cations increased in the leached upper soil, and residual feldspar decomposed rapidly, demonstrating a process of feldspar decomposition, as described by Equation (1). Fine particles of SiO_2 gradually migrated from the upper to the lower soil horizons, leading to silica enrichment in the latter.

An analysis of the major clay minerals in different stages of in situ leaching (Figure 10) showed that the vertical migration of clay minerals was significant. Kaolinite did not change significantly in the initial stage of leaching; its content fluctuated between 9.5% and 12.6%, with a standard deviation of 1.22. In the course of leaching, potassium feldspar was consistently weathered and converted to kaolinite [36]. The kaolinite content peaked at a depth of 200 cm, reaching a maximum value of 33%, which is 3.47 times the initial value. During the late leaching stage, the kaolinite content fluctuated slightly. Although it increased slightly from 200 cm, it rapidly reduced afterward. Field investigation in the Longnan profile revealed that about 50 m from the northeast end of the profile, there were rows of 150–180 cm-deep injecting holes along the hill slope (i.e., along 330°–150° direction).

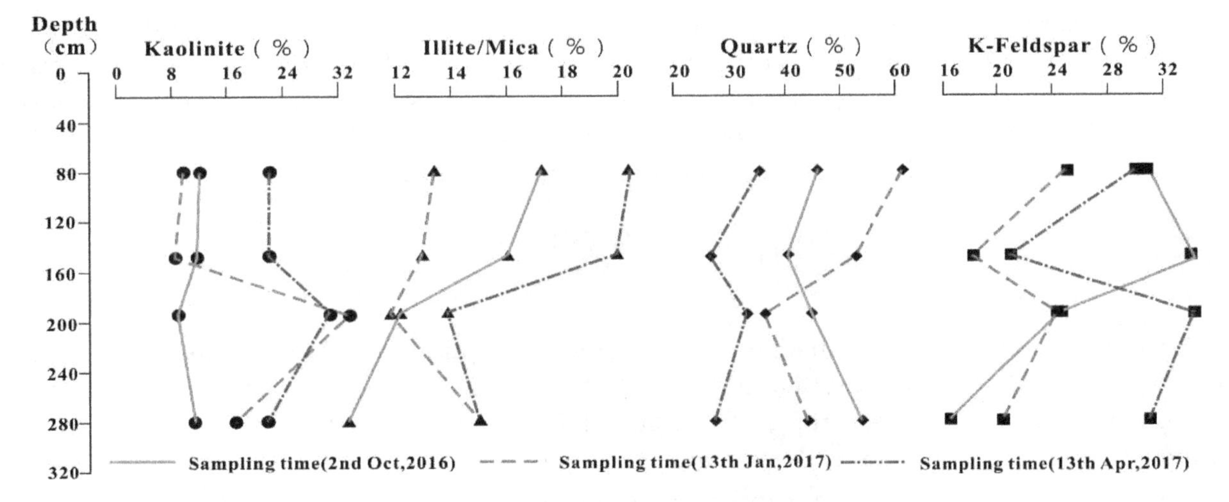

Figure 10. Clay mineral, quartz, and potassium feldspar content on the Longnan soil column at different stages of the leaching experiment.

These injection holes received $(NH_4)_2SO_4$ electrolyte solutions over a long period [50]. Due to this, soil in the lower part of the profile was saturated, and its acidity was enhanced, which further enhanced kaolinization of potassium feldspars. In the initial stage of simulated leaching, the illite-mica content was high near the surface of the sediment column (17.2%), and decreased downwards to only 10.1% at the bottom of column. As leaching progressed, the illite-mica content continued to decrease with increasing depth in the 0–200 cm interval; from 200 cm downwards, however, it increased significantly in the later stage of leaching. The range of illite-mica contents was similar to that of the leaching process, but higher than that in the leaching process. Different clay minerals have different geochemical behavior, and may have different physical and chemical responses to factors such as pH, salinity, and blocking cations [51]. Previous studies have shown that illite forms from potassium feldspar alteration in two different types of microsystems [14]: a) in the early stages of the weathering process, along crystal joints of orthoclase with muscovite or biotite; and b) in the final stages of weathering, where the original structure of the parent rock is destroyed. In both cases, illite forms in association with other clay minerals, i.e., smectite in the early weathering stage, and kaolinite in the late weathering stage. With the exception of its interlayer charge and consequent potassium content, illite is, in many ways, similar to phengite mica [14]. As indicated above, this depth presented an increase in the illite-mica contents due to potassium feldspar weathering and dissolution, silicon release, and hydrated layers mineral formation.

Comparing kaolinite with illite-mica, no correlation between the two was evident in the initial stage of in situ leaching, while a significant negative correlation was present in the later stage of leaching. This trend indicates that, besides the decomposition of potassium feldspar into kaolinite, a large quantity of interlayer silicates such as illite-mica are converted to kaolinite as leaching progresses.

The change in quartz and potassium feldspar content became complex with increasing depth in the soil column (Figure 10). In the early leaching stage, the quartz content initially decreased, then increased, while potassium feldspar content showed the opposite trend. For both minerals, the inflection points were at a depth of 150 cm. Between 80 and 150 cm in the soil column, the content of both quartz and potassium feldspar decreased as leaching progressed. However, quartz inherited the characteristics of the initially decreasing trend, i.e., downwards from 195 cm, it turned into an increasing trend. In contrast, potassium feldspar showed the opposite trend downwards from 150 cm. In the later leaching stage, both quartz and potassium feldspar showed a decreasing-increasing-decreasing trend with depth, but the variation range of potassium feldspar was broader than that of quartz.

Although the content of vermiculite was not measured in the simulated leaching experiment, regional sampling of REE ore revealed low vermiculite content (Table 2). Vermiculite formation occurs in two stages: a) in the early stage, the common mica weathering products are dioctahedral vermiculites whose layer charge is lower than that of the parent mica; b) in the second stage, mica dissolution advances further, and corroded zones of polyphase assemblage of dioctahedral hydroxy-vermiculite appear within mica crystals [14]. Vermiculite has good ion-adsorption properties; its adsorption capacity of REE ions is nearly 0.2 mmol/g. Vermiculites adsorbing REE ions can be regenerated by cation ion-exchange reagents according to the following reaction [52]:

$$\begin{aligned}&\left\{(\mathrm{Mg,Fe,Al})_6[(\mathrm{Si,Al})_8\mathrm{O}_{20}](\mathrm{OH})_4\right\}_m\cdot \mathrm{nRE}^{3+}\cdot \mathrm{eH_2O} + 3\mathrm{nM}^+\\ &\quad = \left\{(\mathrm{Mg,Fe,Al_6})[(\mathrm{Si,Al})_8\mathrm{O}_{20}](\mathrm{OH})_4\right\}_m\cdot 3\mathrm{nM}^+\cdot \mathrm{eH_2O} + \mathrm{nRE}^{3+}\end{aligned} \quad (2)$$

Chemical Equation (2) shows that the decomposition of potassium feldspar releases large quantities of free Al, Fe and Mg cations, and Si in the soil. This favors the formation of vermiculite (Figure 10). However, in the in situ leaching profile, the vermiculite content was low, probably due to the flow of leaching liquid flow and surface water elution.

Previous studies indicated that ion-absorbed REE ores mainly contain halloysite, illite, and kaolinite, and less smectite [1–3,9]. The factors that strongly favor the formation of smectite include low-lying topography, poor drainage, and base-rich parent material, leading to chemical conditions of high pH, high silica activity, and an abundance of basic cations [13]. In leaching conditions with lower pH, as in our leaching experiment, it was impossible to form abundant smectite. With the exception of vermiculite transformed to smectite, the original REE ores in our area of study contained less than 1% smectite [3].

5.3. *Simulating Migration of Clay Minerals during Leaching*

A total of eight simulated soil columns were subjected to different experimental conditions of pH, immersion concentration, and leaching rate. A high-acidity leaching mining solution was used to further decompose the remaining feldspar in the soil to clay minerals [17]. The simulated leaching experiment showed that kaolinite was further enriched in the soil column. The leaching solution concentration and leaching rate also had an effect the rate of decomposition of silicate minerals in the soil columns [53].

In the T1 soil column (Figure 11), the kaolinite content was initially high; subsequently, it decreased and then increased slowly as leaching progressed. The most prominent diffraction peak of kaolinite (d = 7.20 Å in Figure 10) is relatively weak in the middle part of the soil column, compared with the upper and lower parts. Similar results have been reported from the weathering profiles of other Mesozoic granites [54]. Other relatively prominent kaolinite diffraction peaks were at d = 2.33 and 1.99 Å. As shown in Table 3, the variation was also weaker in the middle of the soil column compared with the upper and lower parts. Kaolinite and illite-mica had similar diffraction peak characteristics across the soil column. For illite-mica, the most prominent initial diffraction peak corresponded to d = 10.01 Å, and the (002) crystal planes showed significantly high diffraction. Other evident peaks were at d = 5.0 Å and d = 4.46 Å. As the diffraction angle increased, strong diffraction peaks appeared at d = 2.44 Å, and d = 1.99 Å. Quartz showed a prominent diffraction peak for d = 4.26 Å (peak height: 2615; diffraction intensity: 21.9%). High quartz diffraction peaks at d = 3.34 Å were present in all samples. Potassium feldspar presented the first evident diffraction peak for d between 6.6 and 6.45 Å (corresponding to a diffraction angle (2-Theta) at between 13.4° and 13.7°. This finding reflects the different diffraction intensities of different crystal faces. Significant diffraction peaks were also present at d = 3.24, 2.28, and 1.98 Å.

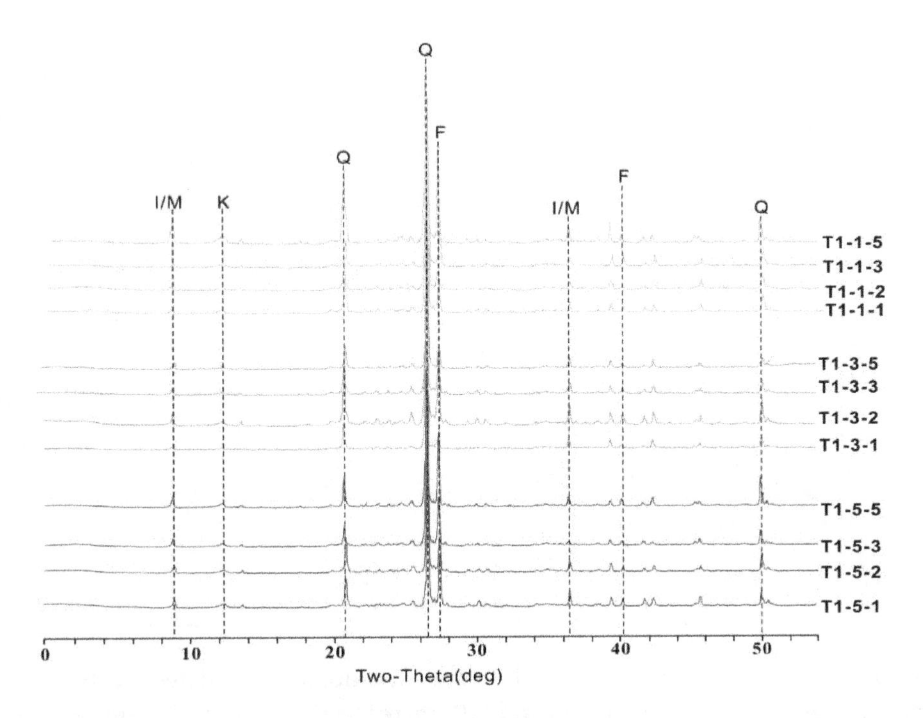

Figure 11. XRD spectra of randomly oriented bulk samples from Soil Column T1 (simulated leaching). I/M: illite-mica; K: kaolinite; Q: quartz; F: potassium feldspars.

Here, we discuss clay mineral content under three key conditions of simulated leaching.

① Same concentration of leaching solution and leaching rate; different pH values (soil columns T1–T3 in Table 3, Figure 12):

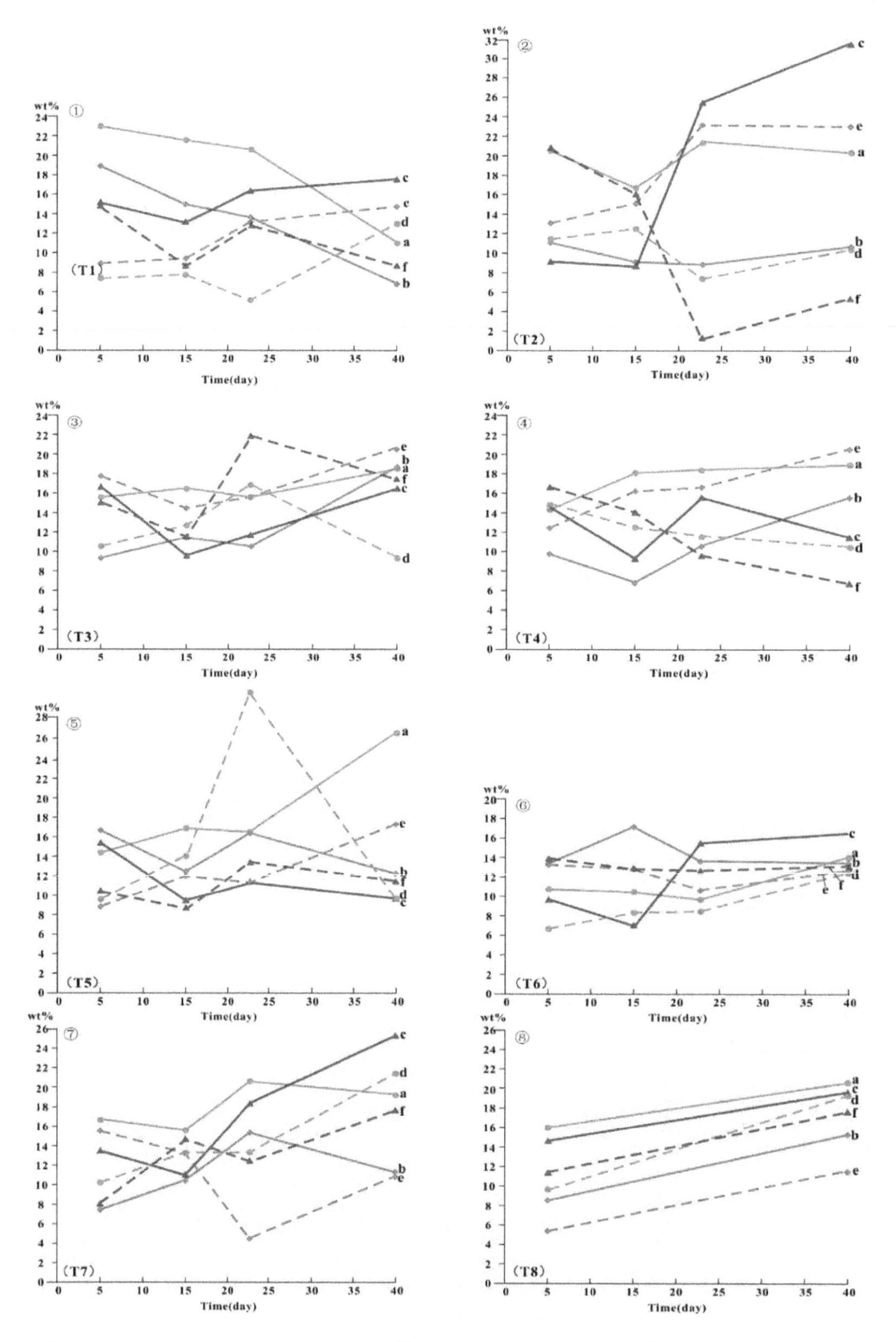

Figure 12. Changing clay mineral content in different conditions of simulated leaching. (**a–c**) kaolinite content sample at sample locations of 30, 70, and 110 cm, respectively; (**d–f**) illite-mica content at sample locations of 30, 70, and 110 cm, respectively.

With the concentration of leaching solution and rate of leaching being stable, the migration and enrichment of clay minerals is controlled by pH. Soil column T1 shows that as leaching progresses, the kaolinite content decreases at depths of 30 and 70 cm in the soil column. After two weeks of leaching, the soil column was gradually enriched in the clay minerals, and the content of kaolinite content in the bottom layer increased (① in Figure 12); the lower the pH, the more favorable the soil conditions for the decomposition of parent rock, particularly for the hydrolysis of feldspars and the formation of clay minerals. As indicated above, the soil column T2 was leached with solution of pH = 3, the lowest pH value in this group. After 15 days of leaching, the kaolinite content in soil Column T2 increased markedly, reaching a maximum of 31.5% (② in Figure 12). In the final leaching stage, the kaolinite content increased significantly at a depth of 110 cm, indicating that REEs were adsorbed and released due to the high recovery rate. Other studies have demonstrated that the higher the pH of the leaching solution, the higher the adsorption capacity of clay minerals for rare earth ions. In a weak acid environment (pH = 4), the kaolinite content in different layers of soil Column T3 soil increased slowly (③ in Figure 12).

② Same pH and leaching rate; different concentration of the leaching solution (soil Columns T4, T5):

In soil Columns T4 and T5, the kaolinite content increased gradually in the near-surface layer (30 cm) as leaching progressed (③ in Figure 12 and ④ in Figure 12. The higher the concentration of the leaching solution, the flatter the change in clay mineral composition. After 15 days of leaching, the concentration of the leaching solution in soil Column T5 decreased to 4% $(NH_4)_2SO_4$, while the clay mineral content near the surface increased rapidly, eventually reaching 26.7%. At a depth of 70 cm, Soil Columns T4 and T5 showed different responses: with the lower-concentration leaching solution of T5, the kaolinite content in this layer decreased slowly, with an average decrease of about 25%; with the more concentrated leaching solution, i.e., T4, the kaolinite content showed increasing volatility. At a depth of 110 cm, different leaching concentrations produced similar effects on the kaolinite content.

At the same time, illite-mica minerals exhibited different responses under different leaching conditions. The high-concentration leaching solution in soil Column T4 gradually reduced the illite-mica content. On the 23rd day, with the 4% $(NH_4)_2SO_4$ solution, the illite-mica content increased abruptly, further verifying the enhanced adsorption capacity of illite in a weakly acidic environment. As indicated above, at the upper level of the soil columns, clay minerals were concentrated by the high-concentration leaching solution, but at the bottom of the soil columns, this was not observed (⑤ in Figure 12). Our experiment suggested that the effective concentration of leaching solution was 8%.

③ Same pH and leaching concentration; different leaching rates:

Soil Columns T6 and T7 showed that the leaching rate had an impact on the content of kaolinite (⑥ in Figure 12, ⑦ in Figure 12). At 30 cm, the content of kaolinite in soil Column T6 increased slowly as leaching progressed. After 15 days, the kaolinite content increased significantly in soil column T7; in soil column T6, at 70 cm, it increased slightly, and gradually decreased as leaching progressed. In soil column T7, after 20 days of leaching, the total kaolinite content increased, as it did in the upper layer (30 cm), demonstrating that clay minerals have similar structures in REE ore. At a depth of 110 cm, the kaolinite content decreased after 15 days in columns T6 and T7, and then increased again. This change probably reflects the translocation of fine-grained kaolin minerals from the upper and middle to the lower parts of the profile as leaching progresses. The longer the leaching time, the higher the content of fine-grained clay minerals at the bottom of profile; some of these even clog the porosity, which reduces the flushing out of rare-earth ions and hampers leaching [36]. In the leaching conditions of this sample group, with a leaching rate of 5 mL/min, enrichment in clay minerals was at a rate of over 1 mL/min. This indicated that a high recovery rate is not possible at a slower rate of leaching.

A comparison between different simulated leaching conditions showed that the content of kaolinite and other clay minerals tended to increase from the initial stage until the completion of leaching. The fluctuation of clay mineral content is the result of the combination of different pHs, leaching concentrations, and leaching rates (⑧ in Figure 12). However, it cannot be assumed that a certain leaching condition determines the outcome of the leaching mining process. Leaching mining is a complex chemical process, and the variation of clay mineral content only reflects one aspect of it. It is impossible to adequately simulate ore leaching conditions in sediment columns, in view of the boundary restrictions of a soil column, the horizontal flow of ore leaching solution, and the ore texture and structure. For these reasons, our experimental results were not as expected, although the leaching conditions in our soil column experiment were controlled. However, comparing the results experimental leaching of soil Columns T2, T4, T6, and T7 at 110 cm after 23 days of leaching (Figure 11) with the results of in situ leaching in the Longnan section (Figure 9), in both cases, the kaolinite content increased while the illite-mica content decreased. We expect that simulated leaching experiments applying many different leaching conditions will permit us to explore how various factors influence clay mineral fluctuation during leaching.

The REE distribution on the kaolinite-water interface is considered to be the result of the adsorption of REE ions by kaolinite, and is strongly controlled by pH [55]. Tian et al. (date) [52] found that the REE recovery was up to over 96% with a NH_4^+ concentration in the raffinate solution of 0.2 g/L and a pH of 2. Other research suggested if the pH of the leaching agent is either too high or too low, recovery of REE is reduced. The optimal pH values were between 4 and 8. The maximum leaching efficiency of REE was 91% [56]. In our experimental study, the pH values were between 3 and 5. Disregarding other external factors, we suggest that REE recovery of over 90% can be achieved through leach mining with a leaching solution of 8 wt% concentration and a pH of 5, at a leaching rate of 5 mL/min.

6. Conclusions

Analyses of REE ore clay mineral properties and grain size, and the distribution and variation of clay minerals, quartz, and feldspars in in situ leaching and simulated leach mining led to the following conclusions:

(1) In the surveyed REE mine areas, the soil particle size (i.e., volume frequency) curve showed unimodal and bimodal distribution. Many cumulative particle size distribution curves had a "B" shape, with particle sizes of 3.74–30.46 μm. Other curves showed an "A" shape, which indicates that the increase of coarse particles in the soil affects particle gradation. An analysis of particle size eigenvalues showed that D10 was 0.82–5.03 μm, while D90 ranged from 11.99 to 60.99 μm, with a standard deviation of 11.44. These values reflect discreteness among the coarse particle fraction. Taking the Dav parameter as the independent variable and other characteristic parameters as the dependent variables of the regression analysis, the regression coefficient order was D90 > D [3,4] > D50 > D [2,3] > D10. This finding revealed that the increase of surface average particle size in soils has a greater effect on coarse particles. A D10 residual analysis showed that with an increase of the average particle size of Dav content, the volume fraction below 10% was reduced. The residual error showed a linear relationship with a 99.5% confidence interval.

(2) A regional clay mineral analysis showed that ion-absorbed REE ores formed on different bedrock lithologies have similar clay mineral contents. The main clay minerals were kaolinite, illite, chlorite, and vermiculite. The kaolinite content ranged from a minimum of 8.8% to a maximum of 62.1%. In the granite weathering area of Longnan County, the kaolinite content had weak correlation with the content of potassium feldspar, quartz, and other rock-forming minerals. The chemical weathering index (CIA) was in the range of 61–65, and the average kaolinite content was low. In the metamorphic terrain of Anyuan County, the kaolinite content was strongly correlated with quartz, and the CIA ranged from 68 to 75. This reveals a higher degree of feldspar weathering and conversion to clay minerals in metamorphic bedrock.

(3) Studying the natural weathering profile of granite in the Ganxian District showed that the content of kaolinite was relatively low at 60–105 cm below surface. There, conversion of feldspars to chlorite and mica in the course of weathering, and the increased content of mica in the parent rock, limited the formation of kaolinite. At 150–215 cm below the surface, the kaolinite content increased.

This was attributed to the desiliconization of overlying feldspar minerals, which increased the SiO_2 content in the weathering mantle and promoted the decomposition of feldspars into kaolinite.

(4) An analysis of in situ leaching mining profile showed that in the early stage of leaching, the content of kaolinite in the soil is relatively low, i.e., lower than that of illite-mica. As the leaching progressed and potassium feldspar continued to convert to kaolinite, significant kaolinite peaks appeared in the profile, and the illite-mica content became slightly reduced. This indicates that interlayer silicate minerals such as illite-mica were progressively converted to kaolinite. In a later stage of leaching, the kaolinite content became slightly reduced in comparison with that of the earlier leaching stage, and the content of illite-mica and potassium feldspar increased.

(5) Simulated leaching mining reveals that under the same conditions of leaching solution concentration and leaching rate, the migration and enrichment of clay minerals were controlled by pH; the lower the pH value, the more favorable the conditions for clay mineral formation. With the same pH and leaching rate, but different concentrations of the leaching solution, a high-concentration leaching solution (i.e., 12% $(NH_4)_2SO_4$)) resulted in a slow increase in the kaolinite content in the upper part of the soil profile (30 cm). Under stable pH and immersion concentration conditions, the variation of the leaching rate influences soil formation and clay mineralization processes in different horizons within the soil profile. Based on these results, and disregarding other external factors, we suggest that REE recovery of over 90% can be achieved through leach mining with a leaching solution of 8% concentration and a pH of 5, at a leaching rate of 5 mL/min.

Author Contributions: Conceptualization and methodology, L.C., and H.C.; investigation, L.C., X.J., L.Q., and H.C.; experimental analysis, X.J.; writing-review and editing, L.C. and Z.H.; partial writing, X.J., and L.Q.; plotting, H.C.; software, H.D. All authors have read and agreed to the published version of the manuscript.

Acknowledgments: We are indebted to Kaixing Wu and Tao Sun for their helpful discussions and suggestions. We are also sincerely thanks for Houston–Oil and gas laboratory and the Key Laboratory of Nonferrous Metal Materials Science and Engineering of the Ministry of Education. The authors are grateful to two anonymous reviewers for their helpful suggestions and comments. We want to thank Editage for the English language editing.

References

1. Yang, Y.Q.; Hu, C.S.; Luo, Z.M. Geological characteristic of mineralization of rare earth deposit of the ion-absorption type and their prospecting direction. *Bull. Chin. Acad. Geol. Sci.* **1981**, *2*, 102–118. (In Chinese with English abstract)
2. Chen, D.Q.; Wu, J.S. The mineralization mechanism of ion-adsorbed REE deposit. *J. Chin. Rare Earth Soc.* **1990**, *8*, 175–179. (In Chinese with English abstract)
3. Chi, R.A.; Tian, J. Review of weathered crust rare earth ore. *J. Chin. Rare Earth Soc.* **2007**, *25*, 641–650. (In Chinese with English abstract)
4. Harlavan, Y.; Erel, Y. The release of Pb and REE from granitoids by the dissolution of accessory phases. *Geochim. Cosmochim. Acta.* **2002**, *66*, 837–848. [CrossRef]
5. Galan, E.; Fernandez-Caliani, J.C.; Miras, A.; Aparicio, P.; Marquez, M.G. Residence and fractionation of rare earth elements during kaolinization of alkaline peraluminous granites in NW Spain. *Clay Miner.* **2007**, *42*, 341–352. [CrossRef]
6. Sanematsu, K.; Murakami, H.; Watanabe, Y.; Duangsurigna, S.; Vilayhack, S. Enrichment of rare earth elements (REE) in granitic rocksand their weathered crusts in central and southern Laos. *Bull. Geol. Surv. Jpn.* **2009**, *60*, 527–558. [CrossRef]
7. Zhao, Z.; Wang, D.H.; Chen, Z.H.; Chen, Z.Y. Progress of Research on Metallogenic Regularity of Ion-adsorption Type REE Deposit in the Nanling Range. *Acta Geol. Sin.* **2017**, *91*, 2814–2827.
8. Bao, Z.W.; Zhao, Z.H. Geochemistry of Mineralization with Exchangeable REY in the Weathering Crusts of Granitic Rocks in South China. *Ore Geol. Rev.* **2008**, *33*, 519–535. [CrossRef]

9. Fan, C.Z.; Zhang, Y.; Chen, Z.H.; Zhu, Y.; Fan, X.T. The study of clay minerals from weathered crust rare earth ores in southern Jiangxi Province. *Acta Petrol Mineral.* **2015**, *34*, 803–810. (In Chinese with English abstract)
10. Coppin, F.; Berger, G.; Bauer, A.; Castet, S.; Loubet, M. Sorption of lanthanides on smectite and kaolinite. *Chem. Geol.* **2002**, *182*, 57–68. [CrossRef]
11. Brasbury, M.H.; Baeyens, B. Sorption of Eu on Na- and Ca-montmorillonites: Experimental investigations and modelling with cation exchange and surface complexation. *Geochim. Cosmochim. Acta* **2002**, *66*, 2325–2334. [CrossRef]
12. Chi, R.A.; Tian, J.; Luo, X.P.; Xu, Z.G.; He, Z.Y. The basic research on the weathered crust elution-deposited rare earth ores. *Nonferrous Met. Sci. Eng.* **2012**, *3*, 1–13. (In Chinese with English abstract)
13. Wilson, M.J. The origin and formation of clay minerals in soils pat, present and future perspectives. *Clay Miner.* **1999**, *34*, 7–25. [CrossRef]
14. Meunier, A.; Velde, B. *The Geology of Illite*; Verlag Berlin/Heidelberg, Springer: Berlin/Heidelberg, Germany, 2004; pp. 63–143.
15. Zhang, Z.Y.; Huang, L.; Liu, F.; Wang, M.K.; Fu, Q.L.; Zhu, J. Characteristics of clay minerals in soil particles of two Alfisols in China. *Appl. Clay Sci.* **2016**, *120*, 51–60. [CrossRef]
16. Cudahy, T.; Caccett, M.; Thomas, M.; Hewson, R.; Abrams, M.; Kato, M.; Kashimura, O.; Ninomiya, Y.; Yamaguchi, Y.; Collings, S.; et al. Satellite-derived mineral mapping and monitoring of weathering, deposition and erosion. *Sci. Rep.* **2016**, *6*, 23702. [CrossRef] [PubMed]
17. Moldoveanu, G.A.; Papangelakis, V.G. An overview of rare-earth recovery by ion-exchange leaching from ion-adsorption clays of various origins. *Mineral. Mag.* **2016**, *80*, 63–76. [CrossRef]
18. Ishihara, S.; Hua, R.; Hoshino, M.; Murakami, H. REE abundance and REE minerals in granitic rocks in the Nanling range, Jiangxi province, Southern China, and generation of the REE-rich weathered crust deposits. *Resour. Geol.* **2008**, *58*, 355–372. [CrossRef]
19. Wang, X.; Lei, Y.; Ge, J.; Wu, S. Production forecast of China's rare earths based on the generalized Weng model and policy recommendations. *Resour. Pol.* **2015**, *43*, 11–18. [CrossRef]
20. Parfitt, R.L.; Russell, M.; Orbell, G.E. Weathering sequence of soils from volcanic ash involving allophane and halloysite, New Zealand. *Geoderma* **1983**, *29*, 41–57. [CrossRef]
21. Botero, Y.L.; López-Rendón, J.E.; Ramírez, D.; Zapata, D.M.; Jaramillo, F. From Clay Minerals to Al_2O_3 Nanoparticles: Synthesis and Colloidal Stabilization for Optoelectronic Applications. *Minerals* **2020**, *10*, 118. [CrossRef]
22. Tian, J.; Yin, J.Q.; Chi, R.A.; Rao, G.H.; Jiang, M.T.; Ouyang, K.X. Kinetics on leaching rare earth from the weathered crust elution-deposited rare earth ores with ammonium sulfate solution. *Hydrometallurgy* **2010**, *101*, 166–170.
23. Wu, D.R.; Wu, A.Q.; He, X.Y.; Li, M. Trends of Extreme Climate in South Jiangxi from 1956 to 2013. *J. Yangtze River Sci. Res. Inst.* **2017**, *34*, 24–29.
24. Jiangxi Bureau of Geological and Mineral Resources. *Regional Geology of Jiangxi Province*; Geological Publishing House: Beijing, China, 1984; pp. 1–308.
25. Jiangxi Bureau of Geological and Mineral Exploration and Development. *Drawing Instructions of Digital Geological Map of Jiangxi Province*; Geological Publishing House: Beijing, China, 1998; pp. 1–80.
26. Zhao, Z.; Wang, D.H.; Chen, Z.Y.; Chen, Z.H.; Zheng, G.D.; Liu, X.X. Zircon U-Pb Age, Endogenic Mineralization and Petrogenesis of Rare Earth Ore-bearing Granite in Longnan, Jiangxi Province. *Acta Geosci. Sin.* **2014**, *35*, 719–725. (In Chinese with English abstract)
27. Liu, C.Y.; Liu, J.H.; Zhong, L.X.; Xu, S.; Chen, L.K. Soil particle size distribution characteristics of ionic rare earth: A case study in rare earth mine of Jiangwozi in Gan County. *Nonferrous Met. Sci. Eng.* **2017**, *8*, 125–130. (In Chinese with English abstract)
28. Hu, X.P.; Shi, Y.X.; Dai, X.R.; Wang, J.T.; Liu, B.; Wang, N.Y. Characteristics and source of city wall's earth in the Neolithic Liangzhu City based on XRD analyses of clay minerals. *Acta. Petrol. Mineral.* **2013**, *32*, 373–382.
29. Mukasa-Tebandeke, I.Z.; Ssebuwufu, P.J.M.; Nyanzi, S.A.; Schumann, A.; Nyakairu, G.W.A.; Ntale, M.; Lugolobi, F. The Elemental, Mineralogical, IR, DTA and XRD Analyses Characterized Clays and Clay Minerals of Central and Eastern Uganda. *Adv. Mater. Phys. Chem.* **2015**, *5*, 67–86. [CrossRef]
30. Rietveld, H.M. A Profile Refinement Method for Nuclear and Magnetic Structures. *J. Appl. Cryst.* **1969**, *2*, 65–71. [CrossRef]

31. Gi, Y.J. Formation of Vermicular Kaolinite from Halloysite Aggregates in the Weathering of Plagioclase. *Clays Clay Miner.* **1998**, *46*, 270–279.
32. Banfield, J.F. The mineralogy and chemistry of granite weathering. Master's Thesis, Australian National University, Canberra, Australia, 1985; 130p.
33. Banfield, J.E.; Eggleton, R.A. Analytical transmission electron microscope studies of plagiodase, muscovite, and Potassium feldspar weathering. *Clays Clay Miner.* **1990**, *38*, 77–89. [CrossRef]
34. Yang, Z.M. A study on clay minerals from the REE-rich weathered crust developed on the Longnan granite in Jiangxi. *Sci. Geo. Sin.* **1987**, *1*, 70–81. (In Chinese with English abstract)
35. Righi, D.; Meunier, A. *Origin of Clays by Rock Weathering and Soil Formation*; Springer: Berlin/Heidelberg, Germany, 1995; pp. 1–119.
36. Wu, K.X.; Zang, L.; Zhu, P.; Chen, L.K.; Tian, Z.F.; Xin, X.Y. Research on Particle Size Distribution and Its Variation of Ion-adsorption Type Rare Earth Ore. *Chin. Rare Earths.* **2016**, *37*, 67–74. (In Chinese with English abstract)
37. Aberg, B. Void ratio of noncohesive soils and similar materials. *J. Geotech Eng.* **1992**, *118*, 1315–1334. [CrossRef]
38. Yan, J.B.; Wu, K.X.; Liu, H.; Liao, C.L.; Jing, X.W.; Zhu, P.; Ouyang, H. Patterns of particle size distribution and genesis of Ion-adsorption REE Ore-exemplified by Dabu REE Ore Deposit, Ganxian County, Southern Jiangxi Province. *Chin. Rare Earths* **2018**, *36*, 372–384. (In Chinese with English abstract)
39. Wang, Z.; Zhao, Z.; Zou, X.Y.; Chen, Z.Y.; Tu, X.J. Petrogeochemical Characteristics and Metallogenetic Potential of Epimetamorphic Rocks in South Jiangxi Province. *Rock Miner. Anal.* **2018**, *37*, 96–107. (In Chinese with English abstract)
40. Zhang, L.; Wu, K.X.; Chen, L.K.; Zhu, P.; Ouyang, H. Overview of Metallogenic Features of Ion-adsorption Type REE Deposits in Southern Jiangxi Province. *J. Chin. Soc. Rare Earth.* **2015**, *33*, 10–17. (In Chinese with English abstract)
41. Zhang, Z.H. A Study on weathering crust ion adsorption type REE deposits, south China. *Contrib. Geol. Miner. Resour. Res.* **1990**, *5*, 57–71. (In Chinese with English abstract)
42. Goldberg, K.; Humayun, M. The applicability of the Chemical Index of Alteration as a paleoclimatic indicator: An example from the Permian of the Paraná Basin, Brazil. *Palaeogeogr. Palaeoclimatol. Palaeoecol.* **2010**, *293*, 175–183. [CrossRef]
43. Nesbitt, H.W.; Young, G.M. Early Proterozoic climates and plate motions inferred from major element chemistry of lutites. *Nature* **1982**, *299*, 715–717. [CrossRef]
44. Samotoin, N.D.; Bortnikov, N.S. Formation of Kaolinite Nanoand Microcrystals by Weathering of Phyllosilicates. *Geol. Ore Depos.* **2011**, *53*, 340–352. [CrossRef]
45. Oliveira, D.P.; Sartor, L.R.; Souza Júnior, V.S.; Corrêa, M.M.; Romero, R.E.; Andrade, G.R.P.; Ferreira, T.O. Weathering and clay formation in semi-arid calcareous soils from Northeastern Brazil. *Catena* **2018**, *162*, 325–332. [CrossRef]
46. Uzarowicz, Ł.; Skiba, S.; Skiba, M.; Segvic, B. Clay-mineral Formation in Soils Developed in the Weathering Zone of Pyrite-Bearing Schists: A Case Study from the Abandoned Pyrite Mine in Wieściszowice, Lower Silesia, SW Poland. *Clays Clay Miner.* **2011**, *6*, 581–594. [CrossRef]
47. Liu, S.K.; Han, C.; Liu, J.M.; Li, H. Hydrothermal decomposition of potassium feldspar under alkaline conditions. *RSC Adv.* **2015**, *5*, 93301–93309. [CrossRef]
48. Dixon, J.B.; Weed, S.B.; Parpitt, R.L. Minerals in Soil Environments. *Soil Sci.* **1989**, *150*, 283–304.
49. Yuan, G.H.; Cao, Y.C.; Schulz, H.M.; Hao, F.; Gluyas, J.; Liu, K.; Yang, T.; Wang, Y.Z.; Xi, K.L.; Li, F.L. A review of feldspar alteration and its geological significance in sedimentary basins: From shallow aquifers to deep hydrocarbon reservoirs. *Earth-Sci. Rev.* **2019**, *191*, 114–140. [CrossRef]
50. Liu, J.H.; Chen, L.K.; Liu, C.Y.; Qiu, L.R.; He, S. Pb speciation in rare earth minerals and use of entropy and fuzzy clustering methods to assess the migration capacity of Pb during mining activities. *Ecotox. Environ. Safe.* **2018**, *165*, 334–342.
51. Cuevas, J.; Leguey, S.; Garralón, A.; Rastrero, M.R.; Procopio, J.R.; Sevilla, M.T.; Jiménez, N.S.; Abad, R.R.; Garrido, A. Behavior of kaolinite and illite-based clays as landfill barriers. *Appl. Clay Sci.* **2009**, *42*, 497–509. [CrossRef]

52. Tian, J.; Yin, J.Q.; Chen, K.H.; Rao, G.H.; Jiang, M.T.; Chi, R.A. Extraction of rare earths from the leach liquor of the weathered crust elution-deposited rare earth ore with non-precipitation. *Int. J. Miner. Process.* **2011**, *98*, 125–131.
53. Xiao, Y.; Feng, Z.; Huang, X.; Huang, L.; Chen, Y.; Wang, L.; Long, Z. Recovery of rare earths from weathered crust elution-deposited rare earth ore without ammonia-nitrogen pollution: I. Leaching with magnesium sulfate. *Hydrometallurgy* **2015**, *153*, 58–65.
54. Tan, P.; Oberhardt, N.; Dypvik, H.; Riber, L. Weathering profiles and clay mineralogical developments, Bornholm. Denmark. *Mar. Pet. Geol.* **2017**, *80*, 32–48. [CrossRef]
55. Wan, Y.X.; Liu, C.Q. Study on Adsorption of Rare Earth Elements by Kaolinite. *J. Rare Earth* **2005**, *23*, 337–341.
56. He, Z.Y.; Zhang, Z.Y.; Yu, J.X.; Zhou, F.; Xu, Y.L.; Xu, Z.G.; Chen, Z.; Chi, R.A. Kinetics of column leaching of rare earth and aluminum from weathered crust elution-deposited rare earth ore with ammonium salt solutions. *Hydrometallurgy* **2016**, *163*, 33–39.

12

Mineralogical and Geochemical Constraints on the Origin of Mafic–Ultramafic-Hosted Sulphides: The Pindos Ophiolite Complex

Demetrios G. Eliopoulos [1], Maria Economou-Eliopoulos [2,*], George Economou [1] and Vassilis Skounakis [2]

[1] Institute of Geology and Mineral Exploration (IGME), Sp. Loui 1, Olympic Village, GR-13677 Acharnai, Greece; eliopoulos@igme.gr (D.G.E.); Georgeoik7@gmail.com (G.E.)
[2] Department of Geology and Geoenvironment, University of Athens, 15784 Athens, Greece; vskoun@geol.uoa.gr
* Correspondence: econom@geol.uoa.gr

Abstract: Sulphide ores hosted in deeper parts of ophiolite complexes may be related to either primary magmatic processes or links to hydrothermal alteration and metal remobilization into hydrothermal systems. The Pindos ophiolite complex was selected for the present study because it hosts both Cyprus-type sulphides (Kondro Hill) and Fe–Cu–Co–Zn sulphides associated with magnetite (Perivoli-Tsoumes) within gabbro, close to its tectonic contact with serpentinized harzburgite, and thus offers the opportunity to delineate constraints controlling their origin. Massive Cyprus-type sulphides characterized by relatively high Zn, Se, Au, Mo, Hg, and Sb content are composed of pyrite, chalcopyrite, bornite, and in lesser amounts covellite, siegenite, sphalerite, selenide-clausthalite, telluride-melonite, and occasionally tennantite–tetrahedrite. Massive Fe–Cu–Co–Zn-type sulphides associated with magnetite occur in a matrix of calcite and an unknown (Fe,Mg) silicate, resembling Mg–hisingerite within a deformed/metamorphosed ophiolite zone. The texture and mineralogical characteristics of this sulphide-magnetite ore suggest formation during a multistage evolution of the ophiolite complex. Sulphides (pyrrhotite, chalcopyrite, bornite, and sphalerite) associated with magnetite, at deeper parts of the Pindos (Tsoumes), exhibit relatively high Cu/(Cu + Ni) and Pt/(Pt + Pd), and low Ni/Co ratios, suggesting either no magmatic origin or a complete transformation of a preexisting magmatic assemblages. Differences recorded in the geochemical characteristics, such as higher Zn, Se, Mo, Au, Ag, Hg, and Sb and lower Ni contents in the Pindos compared to the Othrys sulphides, may reflect inheritance of a primary magmatic signature.

Keywords: sulphides; ophiolites; ultramafic; selenium; gold; Pindos

---

1. Introduction

Traditionally, the sulphide mineralization associated with ophiolite complexes is that of Cyprus-type volcanogenic massive sulphide (VMS) deposits. They may be derived from the interaction of evolved seawater with mafic country rocks, under greenschist facies metamorphic conditions and subsequent precipitation on and near the seafloor, when ore-forming fluids are mixed with cold seawater [1–5]. They are associated with basaltic volcanic rocks and are important sources of base and trace metals (Co, Sn, Se, Mn, Cd, In, Bi, Te, Ga, and Ge) [1]. Massive sulphide deposits have been described in the Main Uralian Fault Zone (Ivanovka and Ishkinino deposits), southern Urals; they are mafic–ultramafic-hosted VMS deposits and show mineralogical, compositional, and textural analogies with present-day counterparts on ultramafic-rich substrates [6]. Recently, an unusual association of magnetite with sulphides of Cyprus-type VMS deposit was described in Ortaklar, hosted in the

Koçali Complex, Turkey, which is part of the Tethyan Metallogenetic Belt [7]. Additionally, the largest magnetite deposit in a series of apatite and sulphide-free magnetite orebodies hosted in serpentinites of Cogne ophiolites, in the Western Alps, Italy, is characterized by typical hydrothermal compositions [8].

Although the Fe–Cu–Ni–Co-sulphide mineralization was initially considered an unusual type in ophiolite complexes, several occurrences have been located, like those in pyroxenite cumulates of the Oregon ophiolite [9], in dunites associated with chromitites of the Acoje ophiolite, Philippines [10], in layered gabbros of the Oman ophiolite [11], in dunites of the upper mantle–crust transition zone of the Bulqiza (Ceruja, Krasta), Albania ophiolite [12], Shetland (Unst), UK ophiolite [13], and the Moa-Baracoa ophiolitic massif (Cuba) [14]. On the basis of magmatic texture features and steep positive chondrite-normalized Platinum-Group Elements (PGE) patterns, sulphide mineralization of that type has been interpreted as reflecting the immiscible segregation of sulphide melts [9,13,14]. Moreover, the occurrence of Fe–Ni–Cu±Zn-sulphide mineralization (with dominant minerals pyrrhotite, chalcopyrite, and minor pentlandite) in mantle serpentinized peridotites and mafic to ultramafic rocks of ophiolite complexes of Limassol, Cyprus, Othrys (Eretria) in Greece, Pindos (Tsoumes) ophiolite and elsewhere has been the topic of research for extensive studies [15–20]. On the other hand, texture and geochemical characteristics, including PGE contents, and a very low partition coefficient for Ni and Fe between olivine and sulphides are inconsistent with sulphides having an equilibrium with Ni-rich host rocks at magmatic temperature [16,18]. Although the initial magmatic origin is not precluded, present characteristics of the highly transformed ore at the Eretria (Othrys) area may indicate that the magmatic features have been lost or that metals were released from the host rocks by a low-level hydrothermal circulation process [16]. Fe–Cu–Zn–Co–Ni mineralization is also reported in seafloor VMS deposits from modern oceans (as well as in their possible analogues on several ophiolites on land, e.g., Urals) indicating that these deposits can be formed by purely hydrothermal processes [19].

Despite the extensive literature data on a diverse array of sulphide mineralizations, sulphide ores hosted in mafic–ultramafic ophiolitic rocks are characterized by structure as well as mineralogical and geochemical features, suggesting either magmatic origin or links to serpentinization processes and metal remobilization from primary minerals into hydrothermal systems. The present study is focused on some new SEM/EDS and geochemical data on Cyprus-type and Fe–Cu–Co–Zn-type sulphides hosted in deeper parts of the Pindos ophiolite complex, aiming to improve our understanding of the factors controlling trace element incorporation into sulphide minerals and their origin.

2. Materials and Methods

2.1. Mineral Chemistry

Polished sections (20 samples) from sulphide ores were carbon-coated and examined by a scanning electron microscope (SEM) using energy-dispersive spectroscopy (EDS). The SEM images and EDS analyses were carried out at the University of Athens (NKUA, Athens, Greece), using a JEOL JSM 5600 scanning electron microscope (Tokyo, Japan), equipped with the ISIS 300 OXFORD automated energy-dispersive analysis system (Oxford, UK) under the following operating conditions: accelerating voltage 20 kV, beam current 0.5 nA, time of measurement (dead time) 50 s, and beam diameter 1–2 μm. The following X-ray lines were used: FeKα, NiKα, CoKα, CuKα, CrKα, AlKα, TiKα, CaKα, SiKα, MnKα, and MgKα. Standards used were pure metals for the elements Cr, Mn, Mo, Ni, Co, Zn, V, and Ti, as well as Si and MgO for Mg and Al_2O_3 for Al.

2.2. Whole Rock Analysis

The studied sulphide samples were massive and disseminated mineralizations, derived from large (weighing approximately 2 kg) samples, which is necessary to obtain statistical representative trace element distribution in sulphide ores. They were crushed and pulverized in an agate mortar. Major and minor/trace elements were determined at the SGS Global—Minerals Division Geochemistry Services Analytical Laboratories Ltd., Vancouver, BC, Canada. The samples were dissolved using sodium peroxide fusion, combined Inductively Coupled Plasma and Atomic Emission Spectrometry, ICP-AES and Mass Spectrometry, ICP-MS (Package GE_ICP91A50). On the basis of the quality control report provided by Analytical Labs, the results of the reference material analysis in comparison to expected values, and the results from the multistage analysis of certain samples, showed an accuracy and a precision of the method in good agreement with the international standard (<10%).

3. A Brief Outline of Characteristics for the Studied Sulphides

The Pindos ophiolite complex, of Middle to Upper Jurassic age, is located in the northern-western part of Greece (49° N, 21° E), lies tectonically over Eocene flysch of the Pindos zone, and contains a spectrum of lavas from Mid-Ocean Ridge, MOR basalts through island arc tholeiites (IATs) to boninite series volcanics (BSVs) [21–23]. Two tectonically distinct ophiolitic units can be distinguished: (a) the upper unit (Dramala Complex), including mantle harzburgites, and (b) a lower unit, including volcanic and subvolcanic sequences at the Aspropotamos Complex (Figure 1). This complex consists of a structurally dismembered sequence of ultramafic and mafic cumulate ophiolitic rocks, including gabbros, which is locally underlain by sheets of serpentinite [21,22]. Sulphide mineralization in the Pindos ophiolite complex is located near the Aspropotamos dismembered ophiolite unit, belonging to the lower ophiolitic unit of the complex and includes the following.

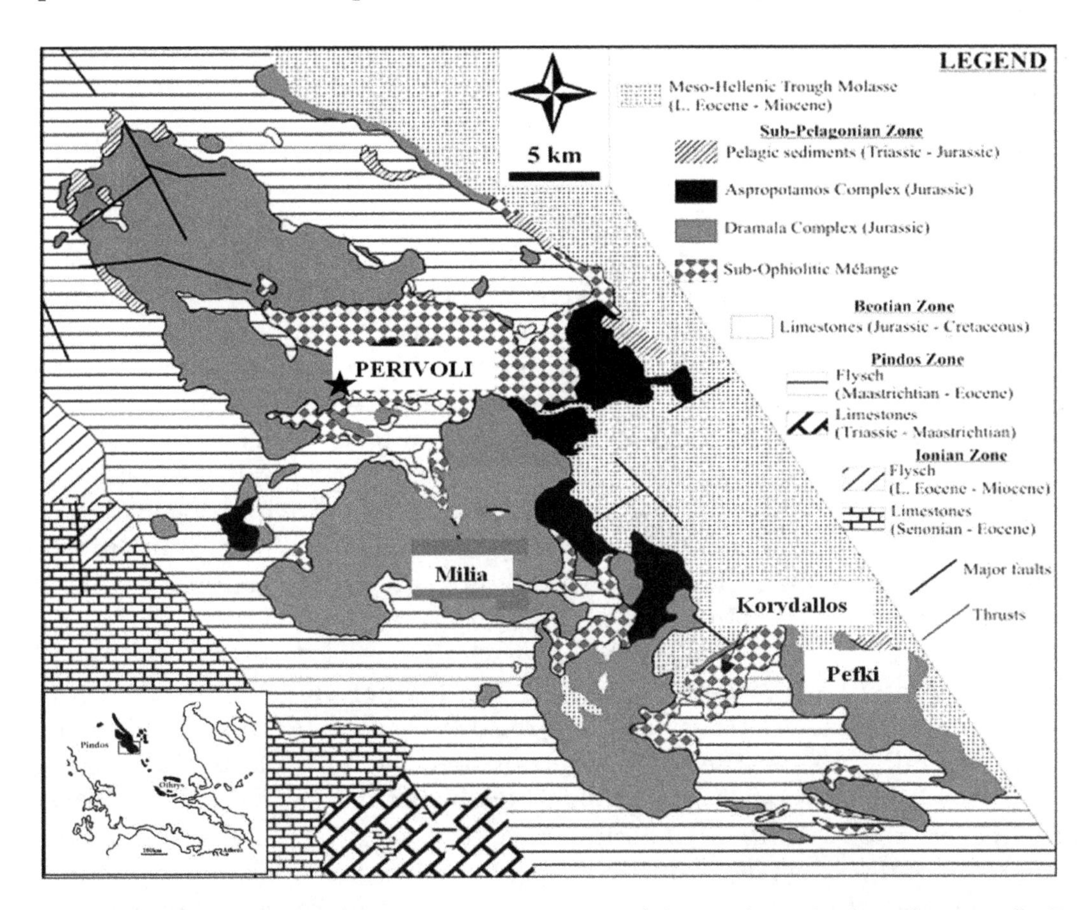

Figure 1. Simplified geological map of the southern part of the Pindos ophiolite, showing the Perivoli (Tsoumes–Kondro) area (modified after [22,24]).

A volcanic and subvolcanic sequence composed mainly of basalts and basaltic andesite pillow lavas ranging from high to low Ti affinity [20–24]. The different magmatic groups may have been derived from different mantle sources and/or various degrees of partial melting [20–24]. Massive Cyprus-type sulphide ore occurrences in the form of small lenses (maximum 4 × 40 m), are located in an abandoned mine, at the Kondro Hill, very close to the village of Perivoli (Figure 1). The estimated ore potential is about 10,000 tons with an average 6.6 wt.% Cu and 9.4 wt.% Zn [25]. They occur on the top of diabase (massive or pillow lavas) and are directly overlain by metalliferous (Fe-Mn-oxide-bearing sediments). Due to the tectonic disruption of the Aspropotamos unit, the spatial association between massive and stockwork disseminated mineralization is unclear.

Several sub-vertical veins of quartz with veinlets and a brecciated pipe-shaped diabase dike (stockwork ore zone) have been described in the Neropriona area of the Aspropotamos unit, Kondro Hill, with disseminations of pyrite + chalcopyrite, and dominant mineral-altered plagioclase and clinopyroxene, penninite, kaolinite, quartz, epidote, and calcite [20,25–27]. Small irregular to lens-like occurrences (4 × 1.5 m) of massive Fe–Cu–Zn–Co-type sulphide mineralization associated with magnetite are exposed at the Perivoli (Tsoumes) Hill (Figure 1). These are hosted within gabbro, close to its contact with serpentinized harzburgite [20,25], consisting of pyrrhotite, pyrite, chalcopyrite, and sphalerite associated with magnetite. The contact between ore and hosting rock is not sharp, appearing as irregular nets of veinlets. Rounded fragments of highly altered rock and massive fragments of sulphide ore are broadly parallel to the shear plane of a thrust fault.

4. Mineralogical Features

4.1. Cyprus-Type Sulphides

The massive ore is mainly composed of pyrite, chalcopyrite, bornite, and in lesser amounts covellite, siegenite, sphalerite, and clausthalite, while pyrrhotite is lacking (Figure 2). Chalcopyrite, bornite, and sphalerite occur in at least two different generations. Pyrite grains vary from euhedral to subhedral and rarely framboidal. Textural relationships indicate that early pyrite, commonly occurring as large crystals but often exhibiting dissolution, is extensively penetrated and replaced by fine-grained chalcopyrite, bornite, and sphalerite in a matrix of quartz (Figure 2). Copper-bearing sphalerite, with up to 3.6 wt.% Fe, 4.2 wt.% Cu, and 1.7 wt.% Bi, occurs within pyrite crystals and/or cements minor chalcopyrite and pyrite (Figure 2b,e,f; Table 1). Pyrite is extensively replaced by intergrowths between chalcopyrite or bornite and Fe-poor sphalerite (Figure 2b,d; Figure 3) and occasionally contains Co (Table 1). Fine-grained intergrowths of framboidal or colloform pyrite-bornite, occurs in a matrix of quartz (Figure 2g,h). Fine-grained chalcopyrite or bornite are often found in cross-cutting veins, hosting selenides (mainly clausthalite, PbSe) (Table 1), the telluride mineral melonite ($NiTe_2$), gold, galena, and barite [26].

Furthermore, present investigation reveals the formation of aggregates of secondary minerals, occurring as characteristic crusts on bornite surfaces (Figure 2i–l). These minerals are present-day grown minerals, on the surface of polished sections of sulphide ore, exposed to air, under room conditions (20–25 °C) and moderate air humidity (atmospheric water). Gold, as inclusions in chalcopyrite reaching a maximum size of 20 μm with Ag contents of up to 9 wt.%, is a rare component of the ores [26]. Additionally, we observed the presence of submicroscopic gold, i.e., <1 μm and thus invisible under an optical microscope, in grains of pyrite, chalcopyrite, and bornite, that increases with decreasing crystal size, reaching contents up to 7.7 ppm Au in pyrite, 8.8 ppm Au in very fine intergrowths between pyrite and sphalerite, and 17.3 ppm Au in fine intergrowths between pyrite and bornite [20].

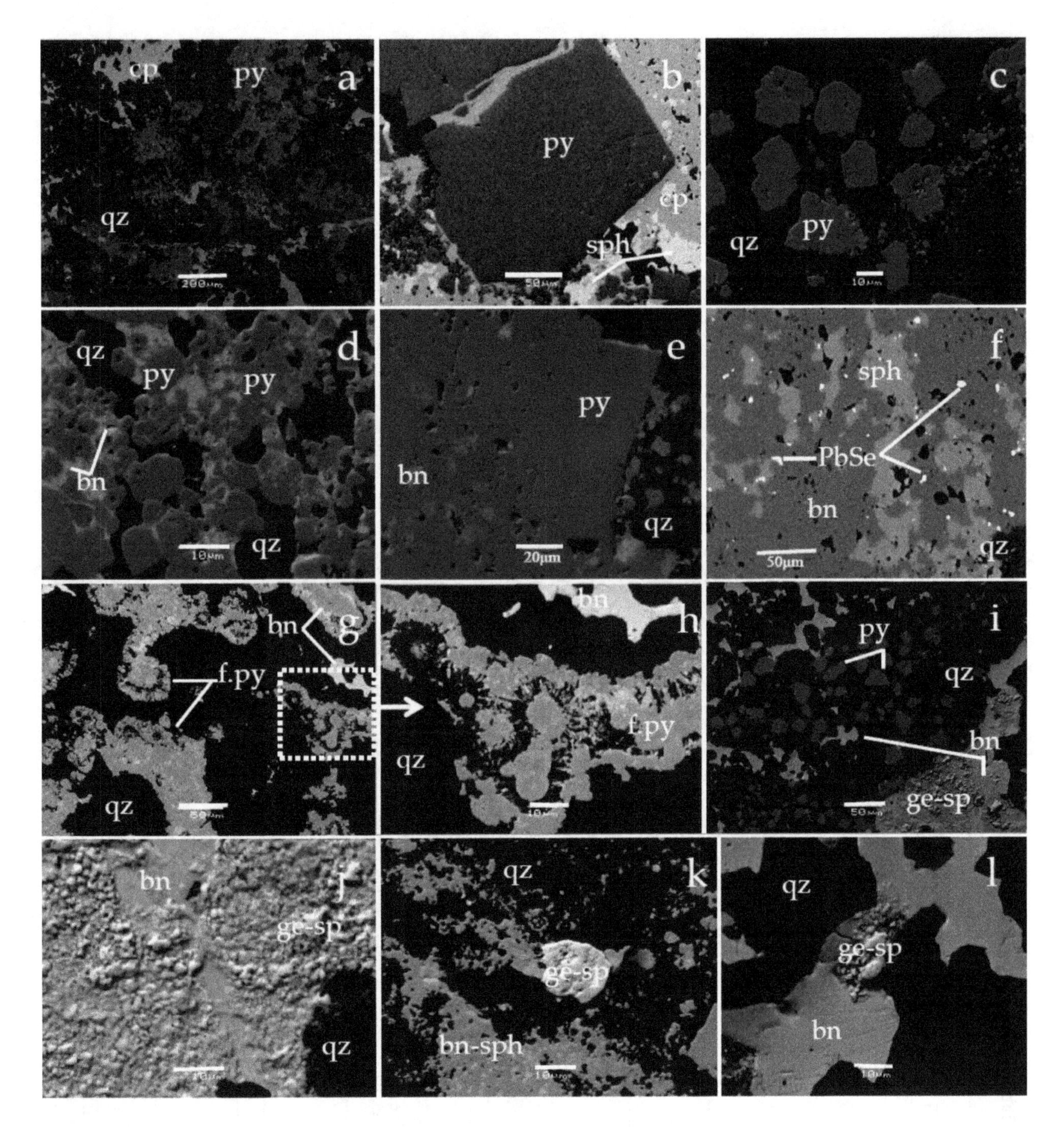

Figure 2. Backscattered electron (BSE) images representative of various morphological forms and textures of pyrite associated with chalcopyrite and sphalerite in a matrix of quartz (**a–d**); euhedral pyrite displaying erosion and replacement by chalcopyrite–sphalerite intergrowths (b); a close intergrowth between pyrite and bornite (**d,e**); selenides (clausthalite) as inclusions within bornite (**f**); fine-grained intergrowths of framboidal or colloform pyrite-bornite, in a matrix of quartz (**g,h**); replacement of bornite by neo-formed intergrowths of Cu minerals (**i–l**), which are Cu-enriched and Fe-depleted (Table 1). Scale bar: 200 μm (**a**); 50 μm (**b,f,g,i**); 20 μm (**e**); 10 μm (**c,d,h,j,k,l**). Abbreviations: py = pyrite; cp = chalcopyrite; sph = sphalerite; bn = bornite; PbSe = clausthalite; qz = quartz; f.py = framboidal forms of pyrite with tiny inclusions of Cu minerals; ge-sp = intergrowths of neo-formed Cu minerals with composition corresponding to geerite and spionkopite.

Table 1. Representative SEM/EDS analyses of minerals from the Pindos sulphide ores. (n.d.: Lower than detection limit).

	Sulphides of Cyprus Type									Kondro		
Mineral	Pyrite			Sphalerite			Chalcopyrite			Bornite		
wt%	1	2	3	4	5	6	7	8	9	10	11	12
Fe	44.6	43.7	42.8	3.6	1.9	1.8	28.7	29.9	29.2	11	11.7	11.9
Cu	n.d.	0.5	2.1	4.2	1.2	0.4	33.6	34.0	35.3	60.6	60	62.2
Zn	n.d.	n.d.	n.d.	57.7	62.7	63.3	1.6	n.d.	n.d.	n.d.	n.d.	n.d.
Co	n.d.	1.2	n.d.	n.d.	n.d.	n.d.	n.d.	n.d.	n.d.	n.d.	n.d.	n.d.
Bi	1.2	n.d.	n.d.	1.7	1.4	1.5	0.8	1.2	n.d.	1.2	n.d.	n.d.
Se	n.d.	n.d.	n.d.	n.d.	n.d.	n.d.	n.d.	n.d.	1.1	n.d.	n.d.	0.6
S	54.1	53.9	53.9	32.8	33.3	32.9	35.3	35.4	32.9	27.1	27.8	25.3
Total	99.9	99.3	99.3	100	100.5	99.9	100	100.4	99.8	99.9	99.5	100

Mineral	Epigenetic Cu-minerals (Figures 2 and 3)							Selenides-claousthalite			Tellurides-Melonite	
wt%	13	14	15	16	17	18	19	20	21	22	23	24
Fe	3.8	3.2	2.2	0.8	3.2	1.4	0.6	3.2	2.2	2.1	0.9	1.2
Cu	70.8	70.6	71.1	82.4	73.2	73.2	76.2	4.6	1.9	2.3	1.5	0.9
Zn	n.d.	4.6	1.8	n.d.	n.d.	n.d.	0.9	n.d.	n.d.	n.d.	n.d.	n.d.
Se	n.d.	n.d.	n.d.	n.d.	n.d.	n.d.	n.d.	21.6	25.1	24.8	1.4	0.5
Pb	n.d.	n.d.	n.d.	n.d.	n.d.	n.d.	n.d.	70.5	71	69.8	n.d.	n.d.
Co	0.3	n.d.	n.d.	n.d.	n.d.	n.d.	n.d.	n.d.	n.d.	n.d.	1.9	3.8
Ni	n.d.	n.d.	n.d.	n.d.	n.d.	n.d.	n.d.	n.d.	n.d.	n.d.	13.8	12.4
Te	n.d.	n.d.	n.d.	n.d.	n.d.	n.d.	n.d.	n.d.	n.d.	n.d.	80.4	80.9
S	24.5	21.6	23.9	26.2	24.2	25	23	n.d.	n.d.	n.d.	n.d.	n.d.
Total	99.4	100	99	99.4	100.6	99.6	100.7	99.9	100.2	99	99.9	99.7

Sulphides associated with Magnetite Tsoumes

	Pyrrhotite			Pyrite		Chalcopyrite			Sphalerite	
	25	26	27	28	29	30	31	32	33	34
Fe	61.0	58.9	59.7	44.9	44.5	32.0	30.6	30.4	7.5	6.4
Cu	n.d.	n.d.	n.d.	n.d.	n.d.	31.4	33.8	33.6	n.d.	n.d.
Zn	n.d.	0.8	n.d.	n.d.	n.d.	2.3	n.d.	n.d.	57.8	58.6
Bi	n.d.	1.4	n.d.	n.d.	n.d.	n.d.	n.d.	n.d.	1.4	1.9
S	39.4	39.8	40	55.3	54.2	34.3	35.9	10	33.1	33.2
Total	100.4	100.6	99.7	100.2	99.7	100	100.3	99.6	99.8	100.1

1 = $Fe_{3.2}Bi_{0.02}Cu_{0.05}S_{6.8}$
2 = $Fe_{3.1}Co_{0.1}Cu_{0.03}S_{6.7}$
3 = $Fe_{3.0}Cu_{0.14}S_{6.8}$
4 = $Zn_{4,3}Fe_{0.3}Bi_{0.04}Cu_{0.3}S_{5.0}$
5 = $Zn_{4,7}Fe_{0.2}Bi_{0.03}Cu_{0.09}S_{5.0}$
6 = $Zn_{4.8}Fe_{0.2}Bi_{0.04}S_{5.0}$
7 = $Cu_{2.4}Fe_{2.4}Zn_{0.02}S_{5.1}$
8 = $Cu_{2.5}Fe_{2.4}Bi_{0.02}S_{5.1}$
9 = $Cu_{2.5}Fe_{2.4}Bi_{0.04}S_{5.1}$
10 = $Cu_{4.8}Fe_{1.0}Bi_{0.03}S_{4.3}$
11 = $Cu_{4.8}Fe_{1.0}S_{4.2}$
12 = $Cu_{4.9}Fe_{1.1}Se_{0.04}S_{4.0}$
13 = $Cu_{5.7}Fe_{0.4}S_{4.1}$
14 = $Cu_{5.8}Fe_{0.3}Zn_{0.4}S_{3.5}$
15 = $Cu_{5.8}Fe_{0.2}Zn_{0.2}S_{3.9}$
16 = $Cu_{6.1}Fe_{0.0}S_{3.8}$
17 = $Cu_{5.9}Fe_{0.3}Co_{0.3}S_{3.9}$
18 = $Cu_{5.9}Fe_{0.12}S_{4.0}$
19 = $Cu_{6.2}Fe_{0.06}Zn_{0.07}S_{3.7}$
20 = $Pb4.6Se_{3.7}Fe_{0.8}Cu_{1.0}$
21 = $Pb4.7Se_{4.4}Fe_{0.5}Cu_{0.4}$
22 = $Pb4.6Se_{4.3}Fe_{0.5}Cu_{0.5}$
23 = $Te6.97Ni_{2.6}Fe_{0.2}Cu_{0.2}Se_{0.2}$
24 = $Te7,17Ni_{2.4}Fe_{0.3}Cu_{0.6}Se_{0.07}$
25 = $Fe_{4.7}S_{5.1}$
26 = $Fe_{4.6Zn0.05Bi3.03}S_{5.3}$
27 = $Fe_{4,6}S_{5.4}$
28 = $Fe_{3.2}S_{6.8}$
29 = $Fe_{3.2}S_{6.8}$
30 = $Fe_{2.6}Cu_{2.3}Zn_{1.6}S_{4.9}$
31 = $Fe_{2.5}Cu_{2.4}S_{5.1}$
32 = $Fe_{2.5}Cu_{2.4}S_{5.1}$
33 = $Zn_{4.3}Fe_{0.6}Bi0.03S_{5.0}$
34 = $Zn_{4.4}Fe_{0.6}Bi_{0.04}S_{5.0}$

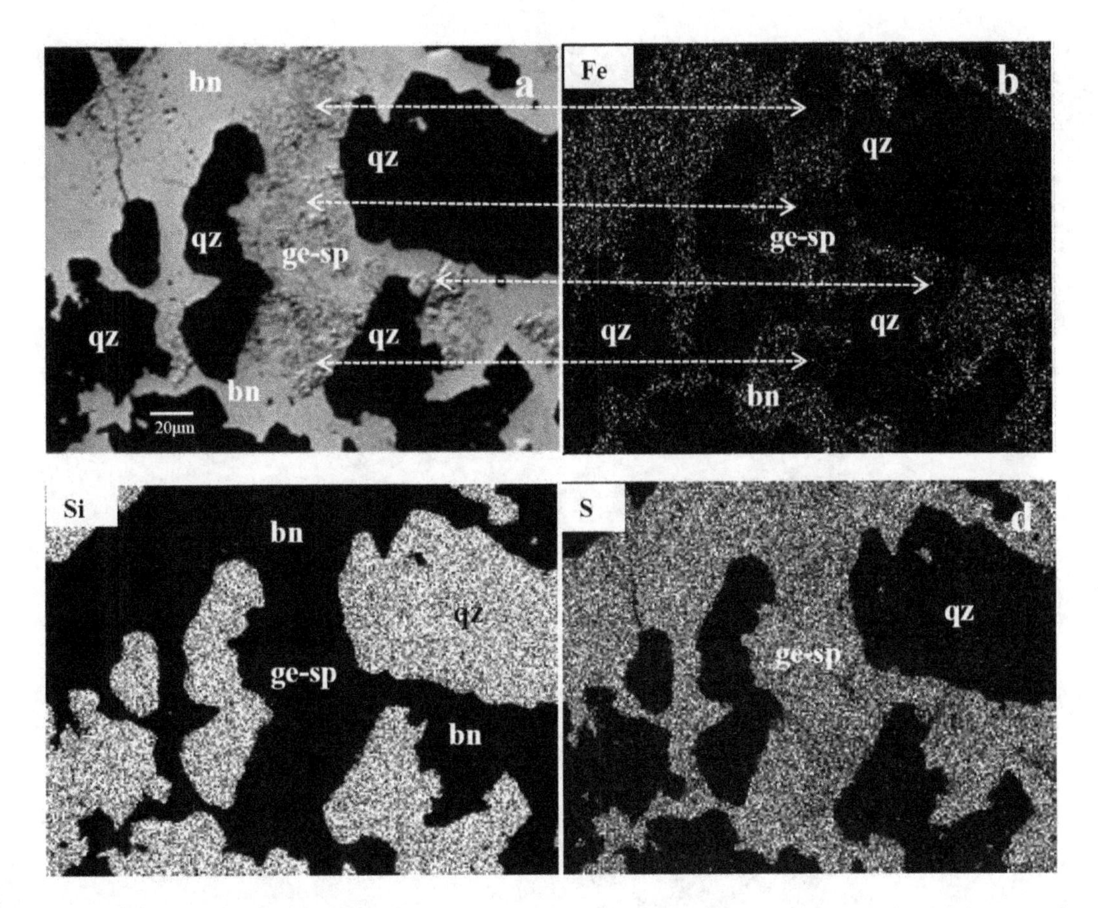

Figure 3. Backscattered electron (BSE)images showing bornite, partially replaced by neo-formed intergrowths of epigenetic minerals resembling geerite or spionkopite, in a matrix of quartz (**a**) and the single scanning for Fe (**b**), Si (**c**), and S (**d**). White arrows indicate the Fe depletion in the present-day formed intergrowths of high Cu minerals. Symbols, as in Figure 2. Scale bar for b, c, d as in 3a (20 μm)

4.2. *Breccia Pipe*

Disseminated pyrite, minor chalcopyrite, and sphalerite occur mostly in vesicles filled by quartz, kaolinite, chlorite, and epidote, within brecciated pipe-form diabase (a discharge pathway) underlying the Kondro massive ore [27]. Samples (n = 10) of pyrite separates from the diabase breccia have shown a limited range for $\delta^{34}S$ values from +1.0 to +1.5‰ [27].

4.3. *Massive Fe–Cu–Co–Zn-Type Sulphides Associated with Magnetite*

The sulphide ore is mainly composed of pyrrhotite, while pyrite, chalcopyrite, sphalerite, malachite, and azurite are present in lesser amounts. Pure magnetite, often forming a network texture, is associated with sulphides, either as massive ore with inclusions of sulphides (chalcopyrite, pyrite, and pyrrhotite), or as individual grains dispersed within sulphide ore (Figure 4). A characteristic feature of magnetite is its elongated and curved form and its textural relationship with the sulphides. Pyrrhotite, which is the most abundant sulphide, is followed by chalcopyrite and sphalerite, all showing an irregular contact with the magnetite, that often occurs surrounding sulphides (Figure 4a–c). Additionally, a salient feature is the occurrence of an (Fe/Mg) phyllosilicate associated with sphalerite, chalcopyrite, and magnetite (Figure 4d–f; Figure 5).

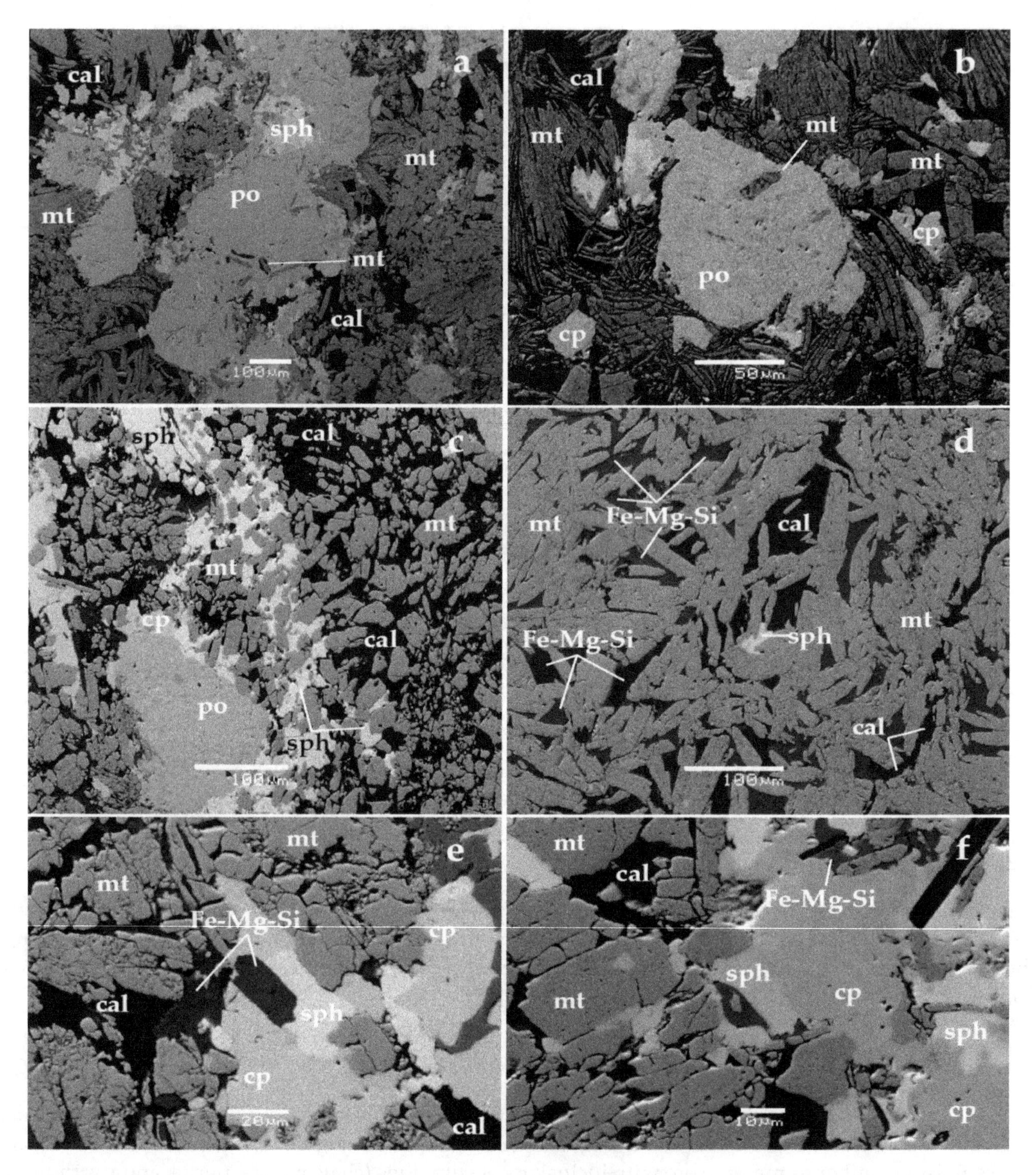

Figure 4. Backscattered electron (BSE) images showing intergrowths between pyrrhotite, sphalerite, and magnetite in a matrix of calcite (**a**); elongated curved crystals of magnetite and inclusions within pyrrhotite (**b**); intergrowths between pyrrhotite, sphalerite, chalcopyrite, and magnetite in a matrix of calcite (**c**); magnetite associated with sphalerite in a matrix of calcite and (Fe,Mg) silicate (**d**); sphalerite adjacent to chalcopyrite with inclusions of (Fe,Mg) silicate (**e**); transitional contact between chalcopyrite and sphalerite and their intergrowths with magnetite (**f**) Scale bar: 100 µm (**a**,**c**,**d**); 50 µm (**b**); 20 µm (**e**); 10 µm (**f**). Symbols: cal = calcite; Fe-Mg-Si = (Fe,Mg) silicate, and as in Figure 2.

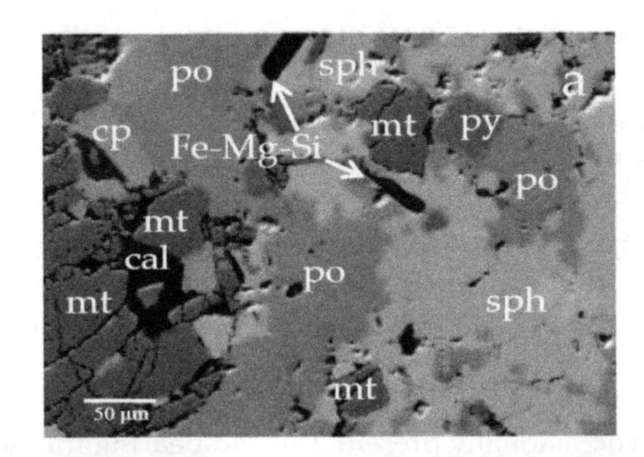

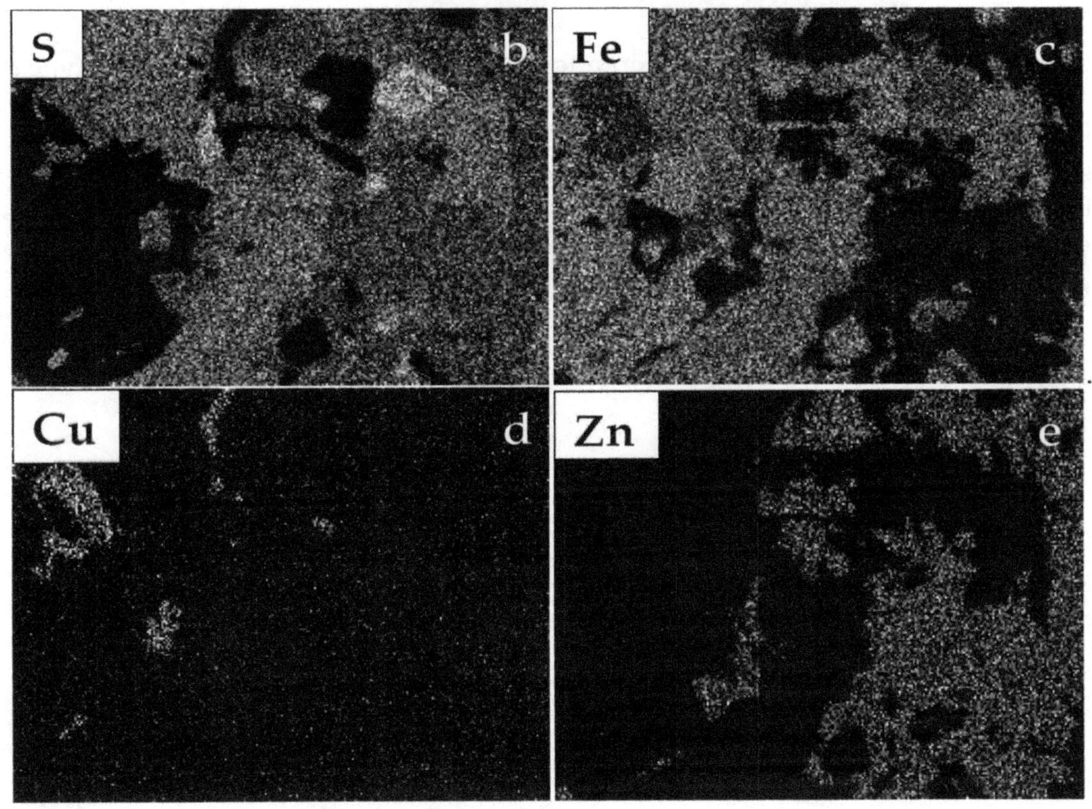

Figure 5. Backscattered electron (BSE) images showing intergrowths of pyrrhotite, pyrite, chalcopyrite, and sphalerite in a matrix of calcite and an unknown (Fe–Mg) silicate (**a**), Table 2; single scanning for S (**b**), Fe (**c**), Cu(**d**), and Zn (**e**). Scale bar for b, c, d as in 5a (50 µm). Symbols, as in Figures 2 and 4.

Table 2. Representative SEM/EDS analyses of (Fe,Mg) silicates from the Tsoumes massive sulphides.

wt %	Tsoumes (Fe,Mg)-Silicate	Laramie Complex [28] Hisingerite		
SiO_2	35	33.2	42.7	37.8
Al_2O_3	0.9	1.1	1.2	0.03
Cr_2O_3	n.d.	n.d.	0.03	0.04
Fe_2O_{3t}	42.3	43.2	35.83	46.94
MnO	n.d.	n.d.	0.15	0.49
MgO	11.9	12.1	7.46	2.15
NiO	n.d.	n.d.	0.05	0.01
CaO	n.d.	n.d.	0.81	0.39
Total	90.1	89.6	88.22	87.85

5. Geochemical Characteristics of Sulphides

Massive sulphide ores from the Kondro Hill exhibit uncommon high contents in Au (up to 3.6 ppm), Ag (up to 56 ppm), Se (up to 1900 ppm), Co (up to 2200 ppm), Mo (up to 370 ppm), Hg (up to 280 ppm), Sb (up to 10 ppm), and As up to 150 ppm, which are much higher than those in the Fe–Cu–Zn–Co-type sulphide hosted in ultramafic parts of the complex, along a shearing zone, close to a contact with gabbros (Table 3), as well as within brecciated pipe-form diabase (a discharge pathway) underlying the Kondro massive ore [20]. Major and minor elements, such as Fe, Cu, and Zn, are hosted in sulphides (pyrrhotite and pyrite, chalcopyrite, bornite, and epigenetic high Cu minerals, and sphalerite, respectively). Magnetite and Fe silicates (Tables 1 and 2), selenides, tellurides, gold, galena, and barite are occasionally present, but Mo-bearing minerals were not identified.

Major and trace elements in massive sulphide ores of Cyprus- and Fe–Cu–Co–Zn-type sulphides from the Pindos ophiolite complex, along with those from comparable ophiolites such as the Othrys and Troodos ophiolite complexes [29–33], are plotted in Figure 6. Although there are overlapping fields, it seems likely that the Pindos sulphides can be distinguished by their higher Zn (Figure 6a,c) and Co (Figure 6b) contents (Figure 6). The highest Se contents were recorded in the Pindos and the Apliki ores (Cyprus) accompanied by Cu and Au contents (Figure 6d,e). A positive trend is also clear between Au and As (Figure 6f).

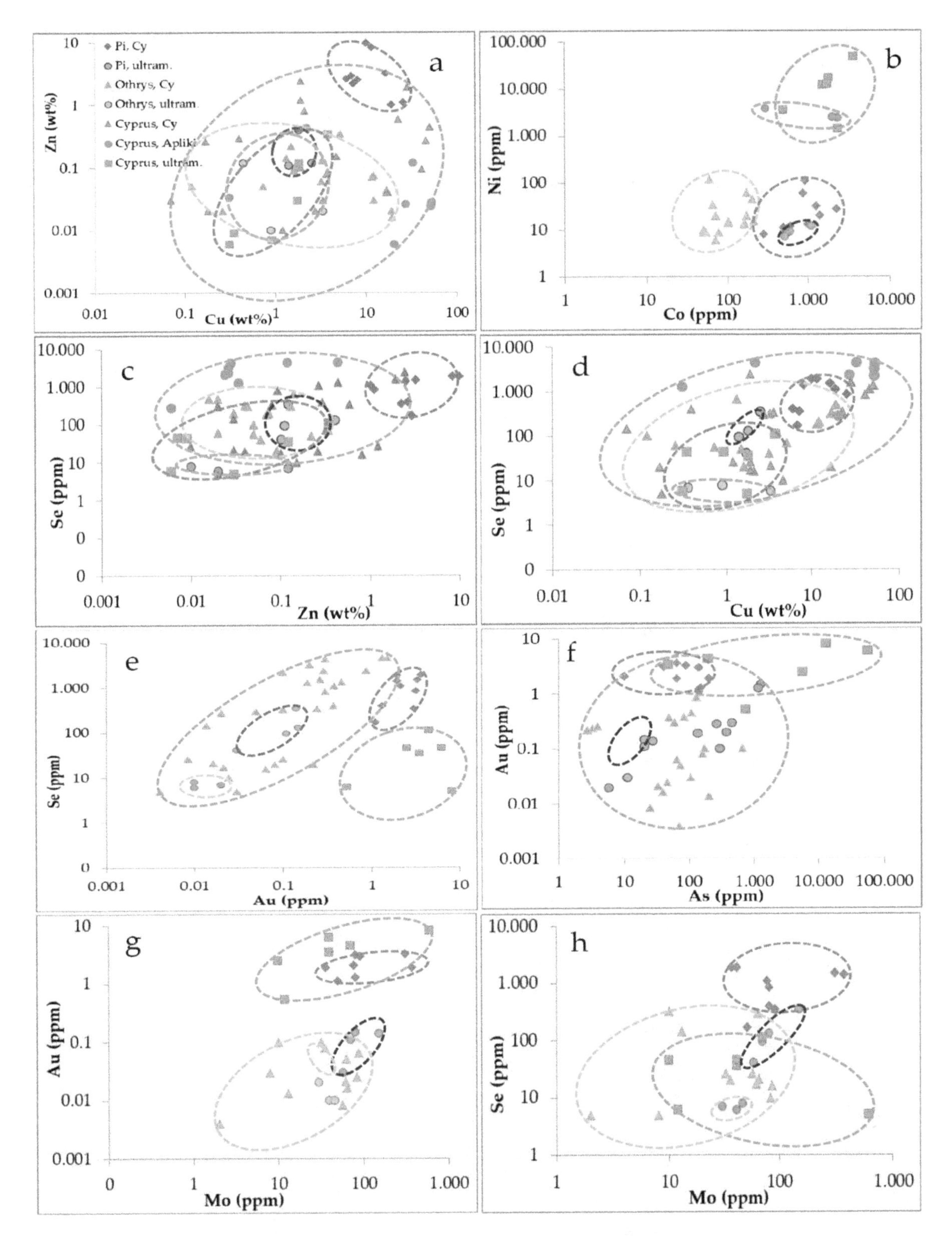

Figure 6. Plots of major and trace elements of Cyprus-type (labeled as Cy) sulphide ores and Fe-Cu ± Co ± Zn sulphides hosted in ultramafic rocks (labeled as ultram.) from the Pindos (Pi), Othrys, and Troodos ophiolite complexes. Although there are overlapping fields, the ores from the Pindos can be distinguished for the presence of ores with higher Zn (**a**,**c**) and Co (**b**) contents in Cyprus-type ores. The highest Se contents are recorded in the Pindos and the Apliki ores (Cyprus) accompanied by Cu and Au contents (**d**,**e**). A positive trend is also clear between Au and As (**f**). Data: Table 3; [3,7,16,18,20,29,30].

Table 3. Geochemical characteristics of sulphide ores hosted in the Pindos, Othrys, and Troodos ophiolites. Data: present study [20].

Location	wt %	wt %	wt %	ppm											
Description	**Fe**	**Cu**	**Zn**	**Co**	**Ni**	**Mo**	**Se**	**As**	**Au**	**Ag**	**Hg**	**Sb**	**Cu/(Cu + Ni)**	**Ni/Co**	**Pt/(Pt + Pd)**
Pindos															
Kondro Massive ore	26.9	6.9	2.9	1400	20	50	170	140	1.1	56	40	9.5	>0,99	0.014	__
	21.2	6.1	2.6	1250	32	80	400	150	1.3	34	70	7.1	>0,99	0.026	__
	26.5	16.4	3.2	600	12	310	1500	90	3.3	38	170	8.1	>0,99	0.02	__
	22.5	10.1	9.8	500	11	40	1900	64	3.6	32	280	10	>0,99	0.022	__
	11.9	25.4	1.1	2200	27	80	850	40	3.2	39	60	2.1	>0,99	0.012	__
	20.8	7.2	2.2	910	110	91	350	140	3	35	48	6.6	>0,99	0.12	__
	9.5	18.9	1	870	60	76	1100	10	2.1	40	128	1.3	>0,99	0.07	__
	26.4	11.6	8.3	280	8	36	1900	64	1.9	38	280	10	>0,99	0.028	__
	23.2	8.1	2.5	1000	13	370	1400	200	1.9	48	168	1.4	>0,99	0.013	__
Aspropotamos															
Disseminated Diabase breccia	7.1	0.006	0.007	27	<5	<5	<5	25	0.018	<1	<1	0.3	__	__	0.95
	6.9	0.007	0.009	15	<5	<5	<5	25	0.025	<1	<1	0.6	__	__	0.97
Tsoumes	49.8	1.8	0.4	600	9	80	130	20	0.15	6	10	1	>0,99	0.015	0.12
	33.4	2.5	0.12	1100	12	150	350	27	0.14	7	30	0.8	>0,99	0.01	0.87
	31.2	1.4	0.11	540	8	70	95	20	0.11	8	11	0.9	>0,99	0.015	0.22
	30.3	1.7	0.1	520	7	58	41	11	0.03	5	16	0.4	>0,99	0.013	0.87
Othrys															
Eretria	39.4	3.3	0.02	2300	2400	40	6	2	0.01	<1	1	0.2	0.93	1.04	0.99
	35.9	0.44	0.12	400	3700	30	7	<5	0.02	<1	0.3	0.2	0.49	12.33	0.61
	37.4	0.89	0.01	2000	2500	46	8	<5	0.01	<1	0.6	0.2	0.78	1.25	0.86
Cyprus															
Limassol	38.8	1.8	0.12	1500	12,000	40	35	47	3.5	2.1	2	18	0.6	8	__
	43.5	1.74	0.03	1800	17,100	610	5	12,800	8.2	8.2	<1	73	0.5	9.5	__
	43.8	3.8	0.34	2300	1400	70	110	190	4.5	4.5	<1	10	0.96	0.61	__
	38.7	0.91	0.007	3600	48,500	40	45	54,800	6.2	0.91	2	94	0.66	13.5	0.95
	54.4	0.35	0.009	1700	12,500	10	44	5600	2.5	0.35	1.5	13	0.21	7.35	0.5
	13.2	0.31	0.006	500	3500	12	6	730	0.53	0.31	1	6.6	0.53	7	0.61

Symbol __: Pt contents < 10 ppb.

6. Discussion

6.1. A Comparison between Magmatic Sulphides and Those Hosted in Mafic–Ultramafic Ophiolites

It has been well established that the formation of magmatic deposits is related to the segregation of sulphide melts by immiscibility from basaltic magmas, which are able to collect precious metals (PGE, Ag, Au) as well as other chalcophile elements (Se, Te, Bi, Pb, As, Sb) because of their high partition coefficient ($D_{sulphide\ melt/silicate\ melt}$) [34,35]. The compositions of the magmatic ores may be mainly controlled by (a) the metal abundances of the mantle source and the degree of partial melting [36,37], (b) the degree of fractional crystallization and potential crustal contamination during magma ascent [38], and (c) interactions with magmatic-hydrothermal or metamorphic-hydrothermal fluids [39–41]. Compositional data obtained from a large number of magmatic sulphide deposits [42] show clear differences from the Fe–Cu ± Co ± Zn ± Ni-sulphide mineralization hosted in ophiolite complexes.

The unusual Fe-Cu-Ni-Co type of sulphides hosted in mafic–ultramafic ophiolitic rocks is characterized by varying structural, mineralogical, and geochemical features, which are not of magmatic origin. Sulphides hosted in the magmatic sequence of ophiolite complexes [9–14] have been interpreted as reflecting the immiscible segregation of sulphide melts, because they exhibit magmatic texture features and steep positive chondrite-normalized PGE patterns [9–14]. However, the Pindos (Tsoumes) Fe–Cu–Co–Zn-sulphides, consisting mainly of pyrrhotite, chalcopyrite, bornite, and sphalerite and hosted within gabbro, close to its tectonic contact with serpentinized harzburgite, differ compared to the Othrys (Eretria) sulphides, which are located at the peripheries of podiform chromite bodies hosted in serpentinized harzburgite and consist of pyrrhotite, chalcopyrite, and minor Co-pentlandite. Furthermore, the Pindos massive sulphides differ from those in the serpentinized rocks of the Limassol Forest (Troodos), which are composed dominantly of troilite, maucherite, pentlandite, chalcopyrite, bornite, vallerite, magnetite, minor sphalerite, graphite, molybdenite, and gold. In addition, the sulphides in the Limassol Forest contain much higher Au and As, up to 8 and 62 ppm, respectively [15,18], compared to the Pindos sulphides (Table 3). Despite the above differences between the Pindos, Othrys, and Limassol sulphides, they are all characterized by higher Cu/(Cu + Ni) ratios (>0.99, 0.5–0.93, and 0.2–0.96, respectively) compared to most magmatic deposits, having Cu/(Cu + Ni) ratios ranging from 0.05 to 0.14 [42]. In addition, the Ni/Co ratio in magmatic deposits typically ranges from 15 to 50 [42], whereas the range of Ni/Co ratio is 1.0–12 for the Eretria, 0.6–14 for the Limassol, and 0.01–0.02 for the Pindos (Tsoumes) samples (Table 3).

Model calculations have shown that the relatively high Co tenor and low Ni/Co cannot be explained by an earlier phase of fractional crystallization or sulphide segregation [43]. Thus, the higher Ni and Co contents in sulphide occurrences associated with chromitite bodies in the Othrys (Eretria) peridotites may suggest re-mobilization of Fe, Co, and Ni during hydrothermal alteration of peridotite in the presence of aqueous H_2S, and precipitation of Ni–Co–Fe sulphides [44,45]. The presence of graphite-like material in chromitites associated with sulphides from the Othrys complex, along shear zones that served as fluid pathways through the chromitites [46,47], may support the mobilization and re-precipitation of Fe-Ni-Co sulphides. Furthermore, the occurrence of phosphides such as Ni–V–Co phosphide [48], $Mo_3Ni_2P_{1+x}$ [49], NiVP [50], and the associated sulphide V_7S_8 [51] in chromitite concentrates from the Othrys ophiolite are consistent with extremely low fO_2 (reducing environment) during serpentinization [51] and re-precipitation of sulphides.

Intergrowths between sulphides and magnetite often forming curved crystals, which reflect a simultaneous deposition, coupled with the occurrence of calcite and an unknown (Fe,Mg) silicate (Figure 4d–f) resembling Mg–hisingerite [52] may provide evidence for the conditions of the Pindos sulphide deposition. Such a (Fe,Mg) silicate is unusual compared to the common presence of Mg-enriched serpentine in other ophiolite complexes. It seems to be comparable with the secondary phyllosilicates described in altered ferroan metaperidotite ("Oxide Body"), from the Laramie Complex (Laramie city, WY, USA), and serpentinites containing the Si-free minerals, such as brucite and NiFe alloy (awaruite). Such secondary phyllosilicates contain approximately equal amounts of end-members of the serpentine [$(Mg,Fe^{2+})_3Si_2O_5(OH)_4$] and hisingerite [$Fe^{3+}{}_2Si_2O_5(OH)_4.nH_2O$] [28]. The substitution of Fe^{3+} ions into the serpentine structure is crystallographically favorable because of the smaller ionic radius of Fe^{3+} compared to that of Mg^{2+} [53]. Moreover, based on a thermodynamic model for hydrothermal alteration in the Fe-silicate system, it has been shown that the formation of serpentine–hisingerite solid solutions after primary olivine may occur at elevated $a_{SiO2(aq)}$ and low $a_{H2(aq)}$ at low temperatures (about 200 °C) [28]. In addition, it has been suggested that H_2 production is associated with Fe(III) incorporation into serpentine (or magnetite) [54–56].

The association of magnetite with sulphides from the Pindos (Tsoumes) resembles an unusual association of magnetite with sulphides of Cyprus-type ophiolite-hosted VMS deposit in Ortaklar, located in the Koçali Complex, Turkey [7]. They are similar in terms of the deposition order, as suggested by the observed textural relationships: Fe-sulphide (pyrrhotite or pyrite) → chalcopyrite → sphalerite and subsequently magnetite (Figure 4; [7]) and the Cu, Zn, Pb, and Au-Ag contents [7]. Such temporal relationships among the primary ore minerals have been attributed to the evolution of ore-forming fluids. Specifically, increasing oxygen fugacity (fO_2) and pH would deplete sulphide (H_2S) and facilitate the magnetite precipitation in the hydrothermal fluids [7]. In addition, the Pindos sulphides exhibit similarities with the Cogne magnetite deposit (Western Alps, Italy), which is the largest in a series of apatite and sulphide-free magnetite orebodies that are hosted in serpentinites belonging to western Alpine ophiolitic units [8]. The authors applying thermodynamic modelling of fluid–rock interactions concluded that fractionation processes such as phase separation were critical to generate hydrothermal fluids capable of precipitating large amounts of magnetite in various types of ultramafic host rocks [8]. Although variable textures described in the large Cogne magnetite deposit differ from those in the Pindos magnetite ore, the trace element content of magnetite from the Cogne deposit, characterized by high Mg and Mn and low Cr, Ti, and V [8], is comparable to those in the Pindos magnetite separates [57]. Additionally, Fe–Cu–Zn–Co–Ni mineralization has been reported in seafloor VMS deposits from modern oceans, as well as in their potential analogues on several ophiolite complexes, as exemplified in the Urals, supporting the origin of such deposits by hydrothermal processes [19]. Moreover, hydrothermal products including Cu–Zn–(Co)-rich massive sulphides, hosted in ultramafic rocks at the Rainbow (Mid-Atlantic Ridge), exhibit structure, mineralogy, and bulk rock chemistry similar to those found in mafic volcanic-hosted massive sulphide deposits [58].

In general, the main factors controlling metal associations in seafloor massive sulphide (SMS) deposits may be the temperature of deposition, seafloor spreading rate and r/w ratio, and zone refining [59]. The authors emphasized the significance of the final depositional conditions and evolution of mound and vent structures rather than the original geochemistry of the hydrothermal fluid; the composition of the substrate may become relevant in subseafloor mineralization, where sulphides are precipitated by the reaction of ascending hydrothermal fluids with substrate host rocks [59]. Individual deposits may show a mixture of geochemical signatures, which may be related to mafic and ultramafic rocks [19,58]. Assuming that the leaching of elements from substrate rocks is influenced by the structure of the oceanic lithosphere and by the nature of the hydrothermal convection (spreading rate), some specific geochemical features, such as Au enrichment, Au/Ag and Co/Ni ratios, may be related to the nature of the substrate, the presence of a magmatic influx of volatiles and metals, the morphology of vent structures, the ridge spreading rate, or a combination of these factors [19,58,59]. Although the observed textural and mineralogical features (Figures 3–5) are inconsistent with an origin of the sulphides at magmatic temperatures, the recorded differences, such as the higher Zn, Se, Mo, Au, Ag, Hg, and Sb and lower Ni contents in the Pindos compared to the Othrys sulphides (Table 3; Figure 6), may reflect inheritance of a primary magmatic signature.

6.2. *Genetic Significance of Trace Elements*

The massive Cyprus-type sulphides from the Kondro Hill are characterized by elevated Zn, Co, Se, Au, As, Ag, Mo, and Sb content (Table 1) compared to those of the Othrys and most of the Troodos sulphide ores (Figure 6). Apart from the major elements, namely Fe, Cu, and Zn, hosted in pyrite, chalcopyrite, bornite, and sphalerite, Au occurs as submicroscopic particles (<1 μm) in grains of As-bearing pyrite, chalcopyrite, and bornite [20,26]. Selenium and Te are found as individual fine minerals, such as selenides (clausthalite) and tellurides (melonite) in Cu minerals (Figure 2f; [26]). Although Se can be easily hosted as a solid solution in high-temperature chalcopyrite [58], the presence of clausthalite in a late generation of fine-grained chalcopyrite–sphalerite intergrowths, penetrating into an earlier stage ore, may indicate re-distribution of Se. A late growth of clausthalite is also supported by the occurrence of clausthalite and tellurides filling cracks in pyrite and Cu minerals [26]. Furthermore, on the basis of thermodynamic calculations, it has been demonstrated that the presence of selenides in the oxidation zones of sulphide ores of Uralian VMS deposits is related to their stability under oxidizing conditions [60]. Molybdenite or other visible Mo minerals in the Pindos and Othrys sulphides have not yet been reported. It has been established that Mo displays siderophile, chalcophile, and lithophile behavior, depending on the composition of the system (including fO_2 and fS_2), temperature, and pressure [61]. Further research is required to define the potential presence of invisible Mo minerals (less than 1 μ) in the Pindos sulphide-magnetite ores.

6.3. Stability of Sulphides

A salient feature of the sulphide minerals is a varying stability. The occurrence of euhedral pyrite crystals, in contrast to microcrystalline unhedral Cu and Zn sulphides (Figure 2), may indicate that pyrite was more stable during subsequent modification of the orebody. Although early large crystals of pyrite may be replaced by chalcopyrite or bornite and Fe-poor sphalerite intergrowths (Figure 2), the formation of cruciform aggregates of secondary minerals occurs only on bornite surfaces (Figure 2i–l), probably reflecting a difference in their stability. The preferential leaching of Cu and Zn sulphide phases and the neo-formation of high Cu sulphides on bornite in contrast to neighboring pyrite may be the result of a preferred dissolution of Cu sulphides over pyrite, due to differing surface potentials [62]. It has been suggested that bornite with sulfur in excess (x-bornite) is stable at high temperature [62]. The authors of this study show that if the so-called sulfur-rich bornites are annealed at lower temperature, chalcopyrite or chalcopyrite and digenite exsolve, depending on the annealing temperature and composition. In addition to this exsolution, a new phase forms below approximately 140 °C, which is referred to as x-bornite, and it is a metastable phase. Although x-bornite is a metastable phase, the presented data (Figure 2i–l; Table 1) may confirm that x-bornite can remain for a long time in natural environments, and epigenetic minerals, with a stoichiometry resembling geerite or spionkopite [63,64], can be formed under environmental conditions in a short time.

7. Conclusions

The compilation of the mineralogical, geochemical, and mineral chemistry data from the sulphide occurrences hosted in the Pindos ophiolite complex and those from other ophiolites lead us to the following conclusions:

- Elevated contents of Au as invisible submicroscopic Au in pyrite and Cu minerals in the Pindos sulphides may reflect main collectors of Au at the time of the sulphide mineralization.
- The occurrence of clausthalite (PbSe) and fine-grained gold in chalcopytite–bornite–sphalerite intergrowths of a subsequent stage mineralization in the Pindos sulphides indicates their re-mobilization/re-deposition.
- Sulphides (pyrrhotite, chalcopyrite, bornite, and sphalerite) associated with magnetite, at deeper parts of the Pindos (Tsoumes), exhibit Cu/(Cu + Ni), Ni/Co, and Pt/(Pt + Pd) ratios, suggesting either no magmatic origin or a complete transformation of a preexisting magmatic assemblages.
- Textural features and the presence of the (Fe/Mg) phyllosilicate resembling Mg–hisingerite, and calcite in the matrix of the Pindos sulphides, suggest precipitation of the sulphide-magnetite ore at the deeper levels from a Fe-rich and alkaline ore-forming system.

- The preferential leaching of Fe and S and neo-formed high Cu sulphides on bornite, in contrast to neighboring pyrite, may be the result of a preferred dissolution of Cu sulphides over pyrite, confirming literature data on differing surface potentials between those sulphides.
- Assuming that trace elements in epigenetic minerals are derived from the decomposition of primary minerals, and coupled with the higher Zn, Se, Mo, Au, Ag, Hg, and Sb and lower Ni contents in the Pindos compared to the Othrys sulphides, this may reflect inheritance of a primary magmatic signature.

Author Contributions: D.G.E., M.E.-E., and G.E. collected the samples, provided the field information, and contributed to the conceptualization of the manuscript. V.S. performed the SEM/EDS analyses. M.E.-E. discussed the mineralogical and chemical data with the co-authors and carried out the original draft of the manuscript. All authors have read and agreed to the published version of the manuscript.

Acknowledgments: We thank the reviewers for the constructive criticism and suggestions on an earlier draft of the manuscript. In particular, the review of this work by the Academic Editor Paolo Nimis is greatly appreciated. Many thanks are due to our colleague Costas Mparlas for the donation of certain sulphide samples from his collection and valuable discussions.

References

1. Hannington, M.; Herzig, P.; Scott, S.; Thompson, G.; Rona, P. Comparative mineralogy and geochemistry of gold-bearing sulfide deposits on the mid-ocean ridges. *Mar. Geol.* **1991**, *101*, 217–248. [CrossRef]
2. Franklin, J.M.; Sangster, D.M.; Lydon, J.W. Volcanic-associated massive sulfide deposits. *Econ. Geol.* **1991**, *75*, 485–627.
3. Hannington, M.D.; Galley, A.; Gerzig, P.; Petersen, S. Comparison of the TAG mound and stockwork complex with Cyprus-type massive sulfide deposits. *Proc. Ocean Drill. Program* **1998**, *158*, 389–415.
4. Barrie, C.T.; Hannington, M.D. Classification of volcanic-associated massive sulfide deposits based on host-rock composition. In *Volcanic-Associated Massive Sulfide Deposits: Processes and Examples in Modern and Ancient Settings*; Society of Economic Geologists: Littleton, CO, USA, 1999; pp. 1–11.
5. Galley, A.; Hannington, M.; Jonasson, I. Volcanogenic massive sulphide deposits. *mineral deposits of Canada Spec. Publ.* **2007**, *5*, 141–161.
6. Nimis, P.; Zaykov, V.V.; Omenetto, P.; Melekestseva, I.Y.; Tesalina, S.G.; Orgeval, J.J. Peculiarities of some mafic–ultramafic- and ultramafic-hosted massive sulfide deposits from the Main Uralian Fault Zone, southern Urals. *Ore Geol. Rev.* **2008**, *33*, 49–69. [CrossRef]
7. Yıldırım, N.; Dönmez, C.; Kang, J.; Lee, I.; Pirajno, F.; Yıldırım, E.; Günay, K.; Seo, J.H.; Farquhar, J.; Chang, S.W. A magnetite-rich Cyprus-type VMS deposit in Ortaklar: A unique VMS style in the 1373 Tethyan metallogenic belt, Gaziantep, Turkey. *Ore Geol. Rev.* **2016**, *79*, 425–442. [CrossRef]
8. Toffolo, L.; Nimis, P.; Martin, S.; Tumiati, S.; Bach, W. The Cogne magnetite deposit (Western Alps, Italy): A Late Jurassic seafloor ultramafic-hosted hydrothermal system? *Ore Geol. Rev.* **2017**, *83*, 103–126. [CrossRef]
9. Foose, M.P. *The Setting of a Magmatic Sulfide Occurrence in a Dismembered Ophiolite, Southwest Oregon*; Distribution Branch, USA Geological Survey: Reston, VA, USA, 1985; p. 1626.
10. Bacuta, G.C.; Kay, R.W.; Gibbs, A.K.; Lipin, B.R. Platinum-group element abundance and distribution in chromite deposits of the Acoje Block. Zambales ophiolite Complex, Philippines. *J. Geochem. Explor.* **1990**, *37*, 113–145. [CrossRef]
11. Lachize, M.; Lorand, J.P.; Juteau, T. Calc-alkaline differentiation trend in the plutonic sequence of the Wadi Haymiliyah section, Haylayn Massif, Semail Ophiolite, Oman. *Lithos* **1996**, *38*, 207–232. [CrossRef]
12. Karaj, N. Reportition des platinoides chromites et sulphures dans le massif de Bulqiza, Albania. In *Incidence sur le Processus Metallogeniques dans les Ophiolites (These)*; Universite de Orleans: Orleans, France, 1992; p. 379.
13. Prichard, H.M.; Lord, R.A. A model to explain the occurrence of platinum- and palladium-rich 3065 ophiolite complexes. *J. Geol. Soc.* **1996**, *153*, 323–328. [CrossRef]

14. Proenza, J.A.; Gervilla, F.; Melgarejo, J.; Vera, O.; Alfonso, P.; Fallick, A. Genesis of sulfide-rich chromite ores by the interaction between chromitite and pegmatitic olivine–norite dikes in the Potosí Mine (Moa-Baracoa ophiolitic massif, eastern Cuba). *Miner. Depos.* **2001**, *36*, 658–669. [CrossRef]
15. Panayiotou, A. *Cu-Ni-Co-Fe Sulphide Mineralization, Limassol Forest, Cyprus*; Panayiotou, A., Ed.; Intern. Ophiolite Symposium: Nicosia, Cyprus, 1980; pp. 102–116.
16. Economou, M.; Naldrett, A.J. Sulfides associated with podiform bodies of chromite at Tsangli, Eretria, Greece. *Miner. Depos.* **1984**, *19*, 289–297. [CrossRef]
17. Thalhammer, O.; Stumpfl, E.F.; Panayiotou, A. Postmagmatic, hydrothermal origin of sulfide and arsenide mineralizations at Limassol Forest, Cyprus. *Miner. Depos.* **1986**, *21*, 95–105. [CrossRef]
18. Foose, M.P.; Economou, M.; Panayotou, A. Compositional and mineralogic constraints in the Limassol Forest portion of the Troodos ophiolite complex, Cyprus. *Miner. Depos.* **1985**, *20*, 234–240. [CrossRef]
19. Melekestzeva, I.Y.; Zaykov, V.V.; Nimis, P.; Tret'yakov, G.A.; Tessalina, S.G. Cu–(Ni–Co–Au)- bearing massive sulfide deposits associated with mafic–ultramafic rocks of the Main Urals Fault, South Urals: Geological structures, ore textural and mineralogical features, comparison with modern analogs. *Ore Geol. Rev.* **2013**, *52*, 18–36. [CrossRef]
20. Economou-Eliopoulos, M.; Eliopoulos, D.; Chryssoulis, S. A comparison of high-Au massive sulfide ores hosted in ophiolite complexes of the Balkan Peninsula with modern analogues: Genetic significance. *Ore Geol. Rev.* **2008**, *33*, 81–100. [CrossRef]
21. Kostopoulos, D.K. Geochemistry, Petrogenesis and Tectonic Setting of the Pindos Ophiolite, NW Greece. Ph.D. Thesis, Univ. of Newcastle, Newcastle, UK, 1989.
22. Jones, G.; Robertson, A.H.F. Tectono-stratigraphy and evolution of the Mesozoic Pindos ophiolite and related units, northwestern Greece. *J. Geol. Soc. Lond.* **1991**, *148*, 267–288. [CrossRef]
23. Pe-Piper, G.; Tsikouras, B.; Hatzipanagiotou, K. Evolution of boninites and island-arc tholeiites in the Pindos ophiolite, Greece. *Geol. Mag.* **2004**, *141*, 455–469. [CrossRef]
24. Kapsiotis, A.; Grammatikopoulos, T.; Tsikouras, B.; Hatzipanagiotou, K.; Zaccarini, F.; Garuti, G. Chromian spinel composition and Platinum-group element mineralogy of chromitites from the Milia area, Pindos ophiolite complex, Greece. *Can. Miner.* **2009**, *47*, 1037–1056. [CrossRef]
25. Skounakis, S.; Economou, M.; Sideris, C. The ophiolite complex of Smolicas and the associated Cu-sulfide deposits. In *Proceedings, International Symposium on the Metallogeny of Mafic and Ultramafic Complexes, UNESCO, GCP Project 169*; Augoustidis, S.S., Ed.; Theophrastus Publications S.A.: Athens, Greece, 1980; Volume 2, pp. 361–374.
26. Barlas, C.; Economou-Eliopoulos, M.; Skounakis, S. Selenium-bearing minerals in massive sulfide ore from the Pindos ophiolite complex. In *Mineral. Deposits at the Beginning of the 21st Century*; Piestrzyiski, A., Ed.; CRC Press: Rotterdam, The Netherlands, 2001; pp. 565–568.
27. Sideris, C.; Skounakis, S.; Laskou, M.; Economou, M. Brecciated pipeform diabase from the Pindos ophiolite complex. *Chem. Der Erde* **1984**, *43*, 189–195.
28. Tutolo, B.M.; Evans, B.W.; Kuehner, S.M. Serpentine–hisingerite solid solution in altered ferroan peridotite and olivine gabbro. *Minerals* **2019**, *9*, 47. [CrossRef]
29. Metsios, C. Metsios, C. Mobility of selenium—Environmental Impact. Master's Thesis, National University of Athens, Athens, Greece, 1999; 132p. (In Greek)
30. Constantinou, G. Metalogenesis associated with Troodos ophiolite. In Proceedings of the International Ophiolite Synposium, Nicosia, Cyprus, 1–8 April 1979; pp. 663–674.
31. Constantinou, G.; Govett, G.J.S. Genesis of sulphide deposits, ochre and umber of Cyprus. *Trans. Inst. Min. Met.* **1972**, *81*, B34–B46.
32. Rassios, A. Geology and copper mineralization of the Vrinena area, east Othris ophiolite, Greece. *Ofioliti* **1990**, *15*, 287–304.
33. Robertson, A.H.F.; Varnavas, S.P. The origin of hydrothermal metalliferous sediments associated with the early Mesozoic Othris and Pindos ophiolites, mainland Greece. *Sediment. Geol.* **1993**, *83*, 87–113. [CrossRef]
34. Naldrett, A. *Magmatic Sulfide Deposits—Geology, Geochemistry and Exploration*; Springer: Heidelberg, NY, USA, 2004; pp. 1–727.
35. Barnes, S.J.; Mungall, J.E.; Le Vaillant, M.; Godel, B.; Lesher, C.M.; Holwell, D.; Lightfoot, P.C.; Krivolutskaya, N.; Wei, B. Sulfide-silicate textures in magmatic Ni-Cu-PGE sulfide ore deposits: Disseminated and net-textured ores. *Am. Miner.* **2017**, *102*, 473–506. [CrossRef]

36. Naldrett, A.J.; Barnes, S.-J. The behaviour of platinum group elements during fractional crystallization and partial melting with special reference to the composition of magmatic sulfide ores. *Fortschr. Miner.* **1986**, *63*, 113–133.
37. Maier, W.D.; Barnes, S.J.; Campbell, I.H.; Fiorentini, M.L.; Peltonen, P.; Barnes, S.J.; Smithies, R.H. Progressive mixing of meteoritic veneer into the early Earth's deep mantle. *Nature* **2009**, *460*, 620–623. [CrossRef]
38. Lesher, C.M.; Burnham, O.M.; Keays, R.R.; Barnes, S.J.; Hulbert, L. Trace-element geochemistry and petrogenesis of barren and ore-associated komatiites. *Can. Miner.* **2001**, *39*, 673–696. [CrossRef]
39. Barnes, S.-J.; Prichard, H.M.; Cox, R.A.; Fisher, P.C.; Godel, B. The location of the chalcophile and siderophile elements in platinum-group element ore deposits (atextural, microbeam and whole rock geochemical study): Implications for the formationof the deposits. *Chem. Geol.* **2008**, *248*, 295–317. [CrossRef]
40. Hinchey, J.G.; Hattori, K.H. Magmatic mineralization and hydrothermal enrichment of the High Grade Zone at the Lac des Iles palladium mine, northern Ontario. *Can. Miner.* **2005**, *40*, 13–23. [CrossRef]
41. Su, S.G.; Lesher, C.M. Genesis of PGE mineralization in the Wengeqi mafic-ultramafic complex, Guyang County, Inner Mongolia, China. *Miner. Depos.* **2012**, *47*, 197–207. [CrossRef]
42. Naldrett, A.J. Nickel sulfide deposits: Classification, composition and genesis. *Econ. Geol* **1981**, *75*, 628–655.
43. Konnunaho, J.P.; Hanski, E.J.; Karinen, T.T.; Lahaye, Y.; Makkonen, H.V. The petrology and genesis of the Paleoproterozoic mafic intrusion-hosted Co–Cu–Ni deposit at Hietakero, NW Finnish Lapland. *Bull. Geol. Soc. Finl.* **2018**, *90*, 104–131. [CrossRef]
44. Shiga, Y. Behavior of iron, nickel, cobalt and sulfur during serpentinization, with reference to the Hayachine ultramafic rocks of the Kamaishi mining distric, northeastern Japan. *Can. Miner.* **1987**, *25*, 611–624.
45. Alt, J.C.; Shanks, W.C. Serpentinization of abyssal peridotites from the MARK area, Mid-Atlantic Ridge: Sulfur geochemistry and reaction modeling. *Geochim. Cosmochim. Acta* **2003**, *67*, 641–653. [CrossRef]
46. Etiope, G.; Ifandi, E.; Nazzari, M.; Procesi, M.; Tsikouras, B.; Ventura, G.; Steele, A.; Tardini, R.; Szatmari, P. Widespread abiotic methane in chromitites. *Sci. Rep.* **2018**, *8*, 8728. [CrossRef]
47. Economou-Eliopoulos, M.; Tsoupas, G.; Skounakis, V. Occurrence of graphite-like carbon in podiform chromitites of Greece and its genetic significance. *Minerals* **2019**, *9*, 152. [CrossRef]
48. Ifandi, E.; Zaccarini, Z.; Tsikouras, B.; Grammatikopoulos, T.; Garuti, G.; Karipi, S. First occurrences of Ni–V–Co phosphides in chromitite from the Agios Stefanos Mine, Othrys Ophiolite, Greece. *Ofioliti* **2018**, *43*, 131–145.
49. Zaccarini, F.; Bindi, L.; Ifandi, E.; Grammatikopoulos, T.; Stanley, C.; Garuti, G.; Mauro, D. Tsikourasite, Mo3Ni2P1 + x (x < 0.25), a new phosphide from the chromitite of the Othrys Ophiolite, Greece. *Minerals* **2019**, *9*, 248.
50. Bindi, L.; Zaccarini, F.; Ifandi, E.; Tsikouras, B.; Stanley, C.; Garuti, G.; Mauro, D. Grammatikopoulosite, NiVP, a new phosphide from the chromitite of the Othrys Ophiolite, Greece. *Minerals* **2020**, *10*, 131. [CrossRef]
51. Bindi, L.; Zaccarini, F.; Bonazzi, P.; Grammatikopoulos, T.; Tsikouras, B.; Stanley, C.; Garuti, G. Eliopoulosite, V7S8, a new sulfide from the podiform chromitite of the othrys ophiolite, Greece. *Minerals* **2020**, *10*, 245. [CrossRef]
52. Eggleton, R.A.; Tilley, D.B. Hisingerite: A ferric kaolin mineral with curved morphology. *Clays Clay Miner.* **1998**, *46*, 400–413. [CrossRef]
53. Wicks, F.J.; O'Hanley, D.S. Volume 19, hydrous phyllosilicates: Serpentine minerals: Structures and petrology. In *Reviews in Mineralogy*; BookCrafters: Chelsea, MI, USA, 1988; pp. 91–167.
54. Andreani, M.; Munoz, M.; Marcaillou, C.; Delacour, A. Mu XANES study of iron redox state in serpentine during oceanic serpentinization. *Lithos* **2013**, *178*, 70–83. [CrossRef]
55. Klein, F.; Bach, W.; Humphris, S.E.; Kahl, W.-A.; Jons, N.; Moskowitz, B.; Berquo, T.S. Magnetite in seafloor serpentinite–Some like it hot. *Geology* **2014**, *42*, 135–138. [CrossRef]
56. Bonnemains, D.; Carlut, J.; Escartı'n, J.; Me'vel, C.; Andreani, M.; Debret, B. Magnetic signatures of serpentinization at ophiolite complexes. *Geochem. Geophys. Geosyst.* **2016**, *17*, 2969–2986. [CrossRef]
57. Eliopoulos, D.; Economou-Eliopoulos, M. Trace element distribution in magnetite separates of varying origin: Genetic and exploration significance. *Minerals* **2019**, *9*, 759. [CrossRef]
58. Marques, A.F.A.; Barriga, F.; Scott, S.D. Sulfide mineralization in an ultramafic-rock hosted seafloor hydrothermal system: From serpentinization to the formation of Cu–Zn–(Co)-rich massive sulfides. *Mar. Geol.* **2007**, *245*, 20–39. [CrossRef]

59. Toffolo, L.; Nimis, P.; Tret'yakov, G.A.; Melekestseva, I.Y.; Beltenev, V.E. Seafloor massive sulfides from mid-ocean ridges: Exploring the causes of their geochemical variability with multivariate analysis. *Earth-Sci. Rev.* **2020**, in press. [CrossRef]
60. Belogub, E.V.; Ayupovaa, N.R.; Krivovichevb, V.G.; Novoselov, K.A.; Blinov, I.A.; Charykova, M.V. Se minerals in the continental and submarine oxidation zones of the South Urals volcanogenic-hosted massive sulfide deposits: A review. *Ore Geol. Rev.* **2020**, *122*, 103500. [CrossRef]
61. Fitton, J.G. Coupled molybdenum and niobium depletion in continental basalts. *Earth Planet. Sci. Lett.* **1995**, *136*, 715–721. [CrossRef]
62. Kullerud, G. The Cu–Fe–S system. In *Washington Year Book*; Carnegie Institution of Washington: Washington, DC, USA, 1964; Volume 63, pp. 200–202.
63. Goble, J.; Robinson, G. Geerite, $Cu_{1.60}S$, a new copper sulfide from Dekalb Township, New York. *Can. Miner.* **1980**, *18*, 519–523.
64. Goble, J.R. Copper sulfides from Alberta: Yarrowite Cu_9S_8 and Spionkopite $Cu_{39}S_{28}$. *Can. Miner.* **1980**, *18*, 511–518.

Permissions

All chapters in this book were first published in MDPI; hereby published with permission under the Creative Commons Attribution License or equivalent. Every chapter published in this book has been scrutinized by our experts. Their significance has been extensively debated. The topics covered herein carry significant findings which will fuel the growth of the discipline. They may even be implemented as practical applications or may be referred to as a beginning point for another development.

The contributors of this book come from diverse backgrounds, making this book a truly international effort. This book will bring forth new frontiers with its revolutionizing research information and detailed analysis of the nascent developments around the world.

We would like to thank all the contributing authors for lending their expertise to make the book truly unique. They have played a crucial role in the development of this book. Without their invaluable contributions this book wouldn't have been possible. They have made vital efforts to compile up to date information on the varied aspects of this subject to make this book a valuable addition to the collection of many professionals and students.

This book was conceptualized with the vision of imparting up-to-date information and advanced data in this field. To ensure the same, a matchless editorial board was set up. Every individual on the board went through rigorous rounds of assessment to prove their worth. After which they invested a large part of their time researching and compiling the most relevant data for our readers.

The editorial board has been involved in producing this book since its inception. They have spent rigorous hours researching and exploring the diverse topics which have resulted in the successful publishing of this book. They have passed on their knowledge of decades through this book. To expedite this challenging task, the publisher supported the team at every step. A small team of assistant editors was also appointed to further simplify the editing procedure and attain best results for the readers.

Apart from the editorial board, the designing team has also invested a significant amount of their time in understanding the subject and creating the most relevant covers. They scrutinized every image to scout for the most suitable representation of the subject and create an appropriate cover for the book.

The publishing team has been an ardent support to the editorial, designing and production team. Their endless efforts to recruit the best for this project, has resulted in the accomplishment of this book. They are a veteran in the field of academics and their pool of knowledge is as vast as their experience in printing. Their expertise and guidance has proved useful at every step. Their uncompromising quality standards have made this book an exceptional effort. Their encouragement from time to time has been an inspiration for everyone.

The publisher and the editorial board hope that this book will prove to be a valuable piece of knowledge for researchers, students, practitioners and scholars across the globe.

List of Contributors

Qian Ge and Fengyou Chu
Key Laboratory of Submarine Geosciences, Ministry of Natural Resources, Hangzhou 310012, China
Second Institute of Oceanography, Ministry of Natural Resources, Hangzhou 310012, China

Z. George Xue
Department of Oceanography and Coastal Sciences, Louisiana State University, Baton Rouge, LA 70803, USA
Center for Computation and Technology, Louisiana State University, Baton Rouge, LA 70803, USA
Coastal Studies Institute, Louisiana State University, Baton Rouge, LA 70803, USA

Lei Lu
School of Resources and Environmental Engineering, Hefei University of Technology, Hefei 230009, China
MLR Key Laboratory of Metallogeny and Mineral Resource Assessment, Institute of Mineral Resources, Chinese Academy of Geological Sciences, Beijing 100037, China

Xiaochun Xu
School of Resources and Environmental Engineering, Hefei University of Technology, Hefei 230009, China

Zhi Zhao and Chenghui Wang
MLR Key Laboratory of Metallogeny and Mineral Resource Assessment, Institute of Mineral Resources, Chinese Academy of Geological Sciences, Beijing 100037, China

Yan Liu
Southern Marine Science and Engineering Guangdong Laboratory (Guangzhou), Guangzhou 511458, China
Key Laboratory of Deep-Earth Dynamics of Ministry of Natural Resources, Institute of Geology, Chinese Academy of Geological Science, Beijing 100037, China

Huichuan Liu
State Key Laboratory of Petroleum Resources and Prospecting, China University of Petroleum (Beijing), Beijing 102249, China

Federica Zaccarini and Giorgio Garuti
Department of Applied Geological Sciences and Geophysics, University of Leoben, A-8700 Leoben, Austria

Wencai Zhang and Aaron Noble
Department of Mining and Minerals Engineering, Virginia Polytechnic Institute and State University, Blacksburg, VA 24061, USA

Xinbo Yang and Rick Honaker
Department of Mining Engineering, University of Kentucky, Lexington, KY 40506, USA

Christina Stouraiti, Sofia Petushok and Konstantinos Soukis
Faculty of Geology and Geoenvironment, National and Kapodistrian University of Athens, 15784 Athens, Greece

Vassiliki Angelatou and Demetrios Eliopoulos
Department of Mineral Processing, Institute of Geology and Mineral Exploration, 13677 Acharnes, Greece

Circe Verba
National Energy Technology Laboratory, Albany, OR 97321, USA

Scott N. Montross
National Energy Technology Laboratory, Albany, OR 97321, USA
Leidos Research Support Team, Albany, OR 97321, USA

Jonathan Yang
National Energy Technology Laboratory, Albany, OR 97321, USA
Oak Ridge Institute for Science and Education, Oak Ridge, TN 37830, USA

James Britton
West Virginia Geological and Economic Survey, Morgantown, WV 26507, USA

Mark McKoy
National Energy Technology Laboratory, Morgantown, WV 26507, USA

Ioannis-Porfyrios D. Eliopoulos and George D. Eliopoulos
Department of Chemistry, University of Crete, Heraklion GR-70013, Crete, Greece

Luca Bindi and Paola Bonazzi
Dipartimento di Scienze della Terra, Università degli Studi di Firenze, I-50121 Florence, Italy

Tassos Grammatikopoulo
SGS Canada Inc., 185 Concession Street, Lakefield, ON K0L 2H0, Canada

Basilios Tsikouras
Faculty of Science, Physical and Geological Sciences, Universiti Brunei Darussalam, BE 1410 Gadong, Brunei Darussalam

Chris Stanley
Department of Earth Sciences, Natural History Museum, London SW7 5BD, UK

Prince Sarfo, Thomas Frasz, Avimanyu Das and Courtney Young
Metallurgical and Materials Engineering, Montana Tech, 1300 West Park Street, Butte, MT 59701, USA

Olga N. Kiseleva and Evgeniya V. Airiyants
Sobolev Institute of Geology and Mineralogy, Siberian Branch Russian Academy of Science, pr. Academika Koptyuga 3, Novosibirsk 630090, Russia

Dmitriy K. Belyanin and Sergey M. Zhmodik
Sobolev Institute of Geology and Mineralogy, Siberian Branch Russian Academy of Science, pr. Academika Koptyuga 3, Novosibirsk 630090, Russia
Faculty of Geology and Geophysics, Novosibirsk State University, Novosibirsk 630090, Russia

Lingkang Chen
College of Earth Sciences, Chengdu University of Technology, Chengdu 610059, China
College of Sciences, Guangdong University of Petrochemical Technology, Maoming 525000, China
School of Resource and Environmental Engineering, Jiangxi University of Science and Technology, Ganzhou 341000, China

Haixia Chen
College of Sciences, Guangdong University of Petrochemical Technology, Maoming 525000, China

Xiongwei Jin and Lanrong Qiu
School of Resource and Environmental Engineering, Jiangxi University of Science and Technology, Ganzhou 341000, China

Zhengwei He
College of Earth Sciences, Chengdu University of Technology, Chengdu 610059, China
State Key Laboratory of Geohazard Prevention and Geoenvironment Protection, Chengdu University of Technology, Chengdu 610059, China

Hurong Duan
College of Geomatics, Xi'an University of Science and Technology, Xi'an 710054, China

Demetrios G. Eliopoulos and George Economou
Institute of Geology and Mineral Exploration (IGME), Sp. Loui 1, Olympic Village, GR-13677 Acharnai, Greece

Maria Economou-Eliopoulos and Vassilis Skounakis
Department of Geology and Geoenvironment, University of Athens, 15784 Athens, Greece

Index

Printed in the USA
CPSIA information can be obtained
at www.ICGtesting.com
JSHW051510111223
53612JS00005B/74